Maasailand ecology: Pastoralist development and wildlife conservation in Ngorongoro, Tanzania

Cambridge Studies in Applied Ecology and Resource Management

The rationale underlying much recent ecological research has been the necessity to understand the dynamics of species and ecosystems in order to predict and minimise the possible consequences of human activities. As the social and economic pressures for development rise, such studies become increasingly relevant, and ecological considerations have come to play a more important role in the management of natural resources. The objective of this series is to demonstrate how ecological research should be applied in the formation of rational management programmes for natural resources, particularly where social, economic or conservation issues are involved. The subject matter will range from single species where conservation or commercial considerations are important to whole ecosystems where massive perturbations like hydro-electric schemes or changes in land-use are proposed. The prime criterion for inclusion will be the relevance of the ecological research to elucidate specific, clearly defined management problems, particularly where development programmes generate problems of incompatibility between conservation and commercial interests.

Also in the series

G. Caughley, N. Shepherd and J. Short (editors) Kangaroos: their ecology and management in the sheep rangelands of Australia

P. Howell, M. Lock and S. Cobb (editors) The Jonglei Canal: impact and opportunity

R. Hudson, K. Drew and L. Baskin (editors) Wildlife Production Systems: economic utilisation of ungulates

M. Stanley-Price Animal Re-introductions: The Arabian Oryx in Oman

M. Boyce The Jackson Elk Herd: intensive management in North America

R. Sukumar The Asian Elephant: ecology and management

MAASAILAND ECOLOGY

Pastoralist development and wildlife conservation in Ngorongoro, Tanzania

K.M. Homewood
Lecturer in Human Sciences
Anthropology Department
University College
London

and

W.A. Rodgers
Wildlife Institute of India

CAMBRIDGE UNIVERSITY PRESS
Cambridge
New York   Port Chester
Melbourne   Sydney

Published by the Press Syndicate of the University of Cambridge
The Pitt Building, Trumpington Street, Cambridge CB2 1RP
40 West 20th Street, New York, NY 10011–4211, USA
10 Stamford Road, Oakleigh, Melbourne 3166, Australia

First published 1991

Printed in Great Britain at the University Press, Cambridge

British Library Cataloguing in publication data

Homewood, K. M.
Maasailand ecology : pastoralist development and wildlife conservation in Ngorongoro, Tanzania.
1. Africa. Tropical regions. Ecology
I. Title II. Rodgers, W. A.
574.509678

Library of Congress cataloguing in publication data

Homewood, K. M.
Maasailand ecology : pastoralist development and wildlife conservation in Ngorongoro, Tanzania / K.M. Homewood and W.A. Rodgers.
p. cm. — (Cambridge series in applied ecology and resource management)
ISBN 0 521 40002 3
1. Range management – Tanzania – Ngorongoro Game Control Area Reserve. 2. Range ecology – Tanzania – Ngorongoro Game Control Area Reserve. 3. Livestock – Tanzania – Ngorongoro Game Control Area Reserve – Ecology. 4. Wildlife conservation – Tanzania – Ngorongoro Game Control Area Reserve. 5. Maasai (African people) 6. Human ecology – Tanzania – Ngorongoro Game Control Area Reserve.
7. Ngorongoro Game Control Area Reserve (Tanzania) I. Rodgers, W.A.
II. Title. III. Series.
SF85.4.T34H66 1991
333.74′09678—dc20 90-26097 CIP

ISBN 0 521 40002 3 hardback

SE

CONTENTS

PREFACE

Metolu lung' elukunya nabo engeno One head does not complete the wisdom
(Many heads are better than one – Maasai saying)

Applied research in the field of natural resource management must integrate science with the complex issues of economics, politics, and human rights. These are bound up with cultural values that lead to subjective viewpoints, dogma and prejudice, which can influence policy more than does ecological fact.

This book grew from our ecological research on pastoralism in a joint wildlife/human land use area. The management in Ngorongoro Conservation Area in northern Tanzania have for decades perceived a conflict between wildlife values and pastoralist activities. By 1980 the conflict was seen as severe enough to warrant expulsion of the pastoralists, but the Ngorongoro Conservation Area Authority needed objective documentation to back up action. UNESCO was to fund a management plan and we were commissioned to produce background information on the ecological facts. Our input was expected to be a standard environmental impact assessment: In what way do pastoralists affect the wildlife? Is this a major problem? If so, recommend pastoralist relocation.

The reality turned out to be more complex. The wildlife community is held to be one of the modern wonders of the world, and the Maasai are arguably the best known of Africa's pastoralist peoples. They have coexisted successfully for centuries. UNESCO acknowledged the importance of both, as well as of their interaction, in declaring Ngorongoro Conservation Area a World Heritage Site and a Man and Biosphere Reserve.

Little was known of the ecology of the Ngorongoro Maasai and their herds, the resources they need and the impacts they have on environment and wildlife. The study we carried out focused on these issues. The management plan that resulted was not accepted, and the past decade has seen growing uncertainty, resentment and conflict between pastoralist and wildlife manager. Now in 1990, a decision on Maasai occupance remains to be made. We make a strong case for their continued presence. Our studies show the Maasai add to the values of Ngorongoro, rather than detract from them.

Applied ecology is more than biological observation and experiment, and this book is more than the report of a field study that spanned a couple of years' observation and measurement. Short-term study findings take on a new significance when set in the context of the history and prehistory of land use conflict, of man/animal interactions, and of the state of knowledge on rangeland ecology, pastoralist development and wildlife management in East Africa. This integrated interdisciplinary approach is used to evaluate and interpret our findings and to explore possible objectives and strategies for future management in Ngorongoro Conservation Area. The pastoralists emerge as an integral part of a remarkable ecosystem, and we feel that its survival is bound up with the recognition of their place there. Ngorongoro is not an isolated case. It symbolises a growing pattern of land use conflict between pastoralist and conservationist all over the world. The same questions as to the nature of development, the tradeoff between productivity and sustainability, and the future of traditional ways of life, all arise in many other places. We discuss these issues in a wider African context in this book.

We acknowledge with gratitude the assistance of people who helped with our initial field studies and with the writing of this book. Especially we thank our Maasai hosts and friends, in particular the households of Andrea Lesian, Ole Ngodoo and Ole Senguyan who welcomed us in to their daily lives. The study owed its origins to the Ngorongoro Conservation Authority, and to the then Conservator Mr A. Mgina, who while wishing to remove a pastoralist problem, did right to initiate scientific study. Professor A. Mascarenhas of the Institute of Resource Assessment in the University of Dar es Salaam organised the planning process. We learnt a great deal from working with other planning team members, especially Kai Arhem, Joseph Ole Kuwai, Lazarus Parkipuny and Henry Fosbrooke. Henry was a mine of information and a continuing inspiration. We are grateful for the help we have had from past and present Conservators, particularly S. Ole Saibull and J. Kayera, as well as from Conservation Authority staff, particularly Joseph Ole Kuwai, P.J. Mshanga, Philip Ole Sayalel, Stephen Makacha, Lazarus Ole Mariki, Sebastian Chuwa, and Joseph Karomo. Saiguran Ole Senet was our field assistant and interpreter of language, custom and folklore: without him, the study would not have been possible.

Margaret and Per Kullander of Gibbs Farm gave us hospitality and help. Henry Kiwia, Jane Griffiths, Aadje Geertsema, Nicky Tortike and Pat Moehlman helped in the field. Part of the field work was financed by the University of London Central Research Fund. Robin Pellew suggested the book, and Richard Waller gave us encouragement and useful material at an early stage. Chris Bulstrode, Jeff Lewis and Steve Cobb took on the heroic task of reading a first draft. Dr Paul Howell went through it with a toothcomb and suggested useful finishing touches. Dr Paul Spencer helped us with his expert knowledge of Maasai society. Many friends and colleagues read sections and commented: any errors that remain are our own. Alan Crowden, Katherine Willis, Maria Murphy and Alison Litherland of Cambridge University Press nursed the book through to publication. Our families put up with us. We thank them all.

1

Management problems and applied ecology in Ngorongoro Conservation Area

The Authority is charged with the duty of conserving and developing the natural and human resources of the Ngorongoro Conservation Area

Ngorongoro Conservation Area Ordinance, Cap. 413.

This book discusses the ecology and management of Ngorongoro Conservation Area (NCA) in northern Tanzania. NCA is internationally renowned as a conservation area for its scenic beauty, its spectacular wildlife, and its important archaeological and paleontological remains. It is also outstanding for its pioneering joint land use policy, which is dominated by conservation aims but at the same time maintains a large population of Maasai pastoralists living from traditional cattle and small stock husbandry. NCA does, however, illustrate a number of problems, ranging from the biological to the political, that are common to other semi-arid rangelands. In particular, joint land use inevitably means conflict between different interest groups. Despite these problems NCA has worked for thirty years and is still seen as a pilot model for multiple land use management in the semi-arid rangelands of East African Maasailand, and to some extent for the savannas of sub-Saharan Africa.

The central management issue in NCA is the conflict between conservation and pastoralist interests that has surfaced in many ways throughout Maasailand and elsewhere in Africa. In biological terms the conflict centres on the relative demands and impacts of livestock, wildlife and people on the natural resources. Maasai and conservationists see livestock and wild ungulates as competing for grazing, with important consequences for population dynamics and productivity of wild and domestic herds. Conservationists fear overgrazing, trampling, and soil erosion as a result of livestock presence, and forest and woodland decline as a result of fire, timber and fuelwood use. Maasai fear the progressive erosion of traditional rights of access and resource use by an intruding Conservation Authority (NCAA). The spectacular variation, both spatial and temporal, of grazing, water, and mineral resources, and of disease risks, and the resulting mobility of both wild and domestic animals, have largely precluded simple solutions of internal zoning.

As well as this central management conflict, other groups have a strong interest in the rich resources of NCA. Successive colonial and national governments have managed parts of NCA for cattle ranching, wildlife hunting, tourism and large-scale agriculture. Illegal hunting for ivory and rhino horn have reached the status of organised crime in NCA.

Conservationists have several times sought to exclude the Maasai from the area, as was done in forming the Serengeti, Amboseli, Tarangire, Maasai Mara, Nairobi and other National Parks (Sindiga 1984). The case for expelling the Maasai from NCA has always been presented on ecological grounds of environmental degradation and competitive threat to wildlife species. Despite decades of studies on the vegetation and wildlife species of the NCA/Serengeti area, however, little was known until recently of the ecology of the Maasai or their livestock. Little or no ecological evidence has ever been presented to back up the argument of ecological damage. This gap in knowledge prompted our study.

In 1979 the Conservation Authority declared that the pastoralists would eventually have to leave the NCA for the better protection of the wildlife resource (NCAA Board of Directors, 1980). NCAA requested UNESCO to commission a planning study, which was largely carried out by staff of the University of Dar es Salaam (ourselves included). The resulting management plan was submitted to NCAA in 1982 (Institute of Resource Assessment 1982). This was the fourth joint land use management plan in the history of NCA, and (like its predecessors) was not officially accepted. Despite the founding Ordinance stressing the need to consider both people and natural resources (see chapter heading), the NCAA has for most of its history seen the split between Maasai and conservation interests as too great to allow compromise. It is ironic that when NCAA approached UNESCO to fund a management plan intended to lead to Maasai resettlement, UNESCO was in the process of designating NCA a World Heritage Site in recognition of the harmonious interaction of Maasai and biosphere within NCA. Also, despite periodic moves to expel the Maasai, the land availability, political and human rights problems of attempting to resettle some 25 000 people and their herds were not considered. The current NCAA administration is better aware of Maasai needs and rights (Kayera 1985) but there is still a very ambivalent attitude towards them and their status in NCA (Malpas and Perkin 1986). The International Union for the Conservation of Nature (IUCN) has recently declared NCA to be a Heritage in Danger, has solicited greater government support for the area, and has recently helped coordinate background research for a fifth management plan.

The Government of Tanzania recognises the current land use debate in NCA as one of national and international importance. A seven-person Government

Commission of Enquiry has now completed analysis of the IUCN research studies and of the question of long term pastoralist rights in NCA. The Commission report states that the Maasai are an integral part of the system, that they have not caused any significant reduction in conservation values, and that the Maasai have the right to stay. The report urges the NCAA to make major improvements in their inputs to pastoralist welfare and 'compatible' development. The report has now been submitted to the Minister for Lands, Natural Resources and Tourism: the political decision is still awaited. This is the stage at which all previous management plans have been turned down. Governments often see the pastoralist way of life as backward and incompatible with administrative goals such as tax collection, provision of health and education services, economic development and the promotion of national unity. In NCA conservation adds an extra dimension and many traditional wildlife conservationists still see problems in maintaining a pastoralist population. Officials associated with the Tanzanian Government and NCAA expressed this bluntly: 'They live like beasts and must be civilized', 'They harbour poachers', 'What will happen in the future when their numbers increase?' (direct personal communications to W.A. Rodgers in August 1989). Most social scientists and development agencies see no major conflict. We believe the worriers overreact to minor and isolated problems and overlook the past 30 years of largely successful compromise. Our purpose in writing this book is to argue a highly charged issue, involving one of the world's most important heritages, in terms of ecological fact and theory rather than strong feelings.

This book describes our study of the ecology of NCA Maasai pastoralism and sets it in the wider context of the ecology of NCA and of Maasailand in general. The book explores the conservation values of NCA and those ecological issues that have given rise to conservationist concern. It documents the nature and extent of pastoralist and other impacts on environment and wildlife. It looks at past management inputs affecting the Maasai and plots the course of pastoralist development in the NCA. The book seeks to establish which factors threaten the continued existence of conservation and of pastoralism in NCA, and conversely those factors that are either compatible with, or positively reinforce, the aims of both. The book leads up to a synthesis of the various facets of NCA ecology – range, wildlife, livestock and human – and an integrated view of land use prospects in NCA. We marshall evidence from historical, political, anthropological, development and archaeological as well as biological studies to explore the ecology of NCA and the future of joint pastoralist/conservation land use here and elsewhere.

The book is a mixture of applied biology and the management issues of policies, politics and economics. Biological knowledge will mean little if the major policy and management crisis the NCA faces is not resolved. We attempt to use basic biological facts to suggest preferred policies and management strategies.

Following this introduction, chapter 2 gives an outline of the physical environment, plant and animal communities, and the archaeological and palaeontological resources that contribute to the internationally recognised conservation and heritage value of NCA. Chapter 3 describes the Maasai population of NCA and their way of life. It begins with a discussion of the Maasai social system and the implications of this system for land use and livestock management. Chapter 3 also describes the successive pastoralist groups that have used the resources of NCA over the last few thousand years, and sketches the origins of the Maasai who now inhabit the area. Chapter 4 documents the history and politics of land use controversy in NCA. The sequence of colonial and national administrations that have manoeuvred to control the area since the turn of the century is discussed. Chapter 4 goes on to outline the present-day perceptions of different interest groups as to the value of NCA, the proper use to which its resources should be put and the relative importance of its various management problems. This material acts as the background for chapter 5, which summarises the main questions underlying current management debates in NCA. Chapter 5 then presents the rationale for our research and sets out the scope, methods, study sites and schedule of our field work.

Chapter 6 describes the unusual productivity of the NCA rangelands and reviews the state of knowledge on the dynamics of the grasslands, woodlands and forest of NCA. The management problems and debates specific to each vegetation type are discussed and underlying biological facts established. Particular emphasis is given to the problems of evaluating different types of rangeland and woodland degradation. Chapter 7 summarises the wealth of knowledge on the wild mammals of the NCA/Serengeti area. Their population and community interactions are reviewed, particularly the dramatic population eruption of the migratory wildebeest herds, and the mass antelope migrations that make NCA a crucial part of the Serengeti Ecological Unit and such a spectacular wildlife showcase. Specific wildlife management policies and problems are analysed.

Chapter 8 deals with livestock. It seeks to explain the special character of pastoralism (as opposed to other more intensive methods of husbandry) and to establish its importance in sub-Saharan Africa. The ecology of NCA livestock is set out in detail: their land use patterns, their population dynamics, their performance. Chapter 9 explores the conflict and complementarity between wildlife and livestock, assesses the impacts of wild and domestic herds on the environment and on one another's performance in NCA, and evaluates the productivity of Maasai cattle in NCA. Finally chapter 9 looks at the status of wildlife versus livestock populations in NCA and compares the situation there to that in other joint land use areas in East Africa.

Chapter 10 looks at the ecology of the Maasai in NCA. It describes past Maasai livestock development projects and their impacts on the ecology of Tanzanian and Kenyan Maasailand. A detailed discussion of the demography of the NCA Maasai clarifies past and probable future responses to ecological conditions and development interventions. Chapter 10 concludes with a study of the Maasai food system in NCA and summarises the trends and special problems of pastoralist subsistence in the Conservation Area.

Chapter 11 draws together the threads from all the previous chapters and weaves them into a final synthesis looking at three main issues. These are the balance of conflict and complementarity between conservation and pastoralist development; the past and future roles and forms of tourism in NCA; and the potential for different forms of wildlife utilisation. The chapter integrates this material to give an overview of land use prospects. Chapter 12 considers recommendations for conservation-compatible interventions in pastoralist development, and chapter 13 ends the book with our conclusions on the positive need for integrated Maasai land use in the ecology of NCA.

This book analyses the particular case of Maasai ecology in NCA. However, its implications go beyond NCA. The book takes the ecology of human land use as a central approach to understanding processes and linkages that are of management importance. In doing so it attempts to bridge the gap between anthropological and development-oriented studies that fail to deal with the biological aspects of conservation arguments, and the biological studies that commonly ignore political and sociological realities which motivate (or hamper) management action.

Several topical areas of concern in African ecology are discussed. The Sahelian droughts and famines of the 1970s and 1980s made clear the need for a better understanding of the ecology of pastoralism in arid and semi-arid rangelands of sub-Saharan Africa. In particular, the role of pastoralist impacts in environmental degradation has been the subject of lively debate. This book provides a case study of an area where pastoralist-induced degradation has often been assumed but not proven. It shows how general ecological principles superficially applied can be misleading, and how ecological theories may be misused to justify politically expedient rather than biologically sound management measures.

A second, and related, issue is the extent to which western livestock and range management and development are relevant in an area of unpredictable climate and fluctuating primary productivity. The concept of ecological carrying capacity if anything 'cramps our understanding' of such systems through its emphasis on their long-term means rather than on their variability (Shepherd and Caughley 1987). It is the variability of semi-arid systems, and largely but not solely the variance around annual rainfall means, that precludes the attainment of equilibrium densities, of sedentary populations and societies, and that forces flexible management of such systems. Western techniques of range management may not be sufficiently flexible to work in many arid and semi-arid rangelands without enormous capital investment. Western standards of productivity may be inappropriate if they fail to take account of variables of central importance to pastoralist subsistence. Indigenous breeds, techniques and strategies are increasingly acknowledged for their true worth (Niamir 1990). Introduced technologies have a poor record in sub-Saharan rangelands, both on performance and on their unexpected environmental impacts. This begs a series of questions as to the course that pastoralist development can and should take in this and other parts of Maasailand.

The emphasis on variability applies not only to natural factors such as rainfall, but also to political and economic forces. Over the period we have known Ngorongoro we have seen the tourism component turn full circle. During the first five years of the 1980s tourist numbers were at an all time low. Lodges and infrastructure were deteriorating, political and sociological wisdom decried foreign tourism, and NCA ran at a considerable monetary loss. Now in 1989 tourism is seen as the 'only way for Tanzania to generate foreign exchange and improve her economic status' (Minister for Lands, Natural Resources and Tourism, pers. commun. August 1989). NCA tourist numbers have grown rapidly, new lodges are planned, NCA tourist revenues are the highest ever. Tourism does not directly affect pastoralism. However, by changing economic perceptions of the value of NCA to the nation, tourism can change policies of land use which themselves radically affect pastoralist futures. Tourism has shown major fluctuations in other East African countries, and an upturn in tourist figures and revenue is not necessarily a lasting and problem-free trend. Just as with the biological factors, it is perhaps the variability, rather than the short-term mean, which should guide understanding of the dynamics of the system.

The book also covers a number of topical debates in the field of wildlife conservation. Orthodox conservation management in East Africa has excluded local populations from access to and resource use within conservation areas. In Maasailand such action was justified firstly on the grounds of pastoralist misuse of rangelands and consequent environmental degradation, and secondly on the basis of compensation derived from the revenue generated by foreign tourism. The environmental degradation issue is under question, as are the social and ecological impacts and economic justification of foreign tourism. At the same time, there is a growing controversy as to whether traditional forms of human land use should not be maintained in conservation areas. In NCA, the present landscape and wildlife populations have evolved through millennia of hunter-gatherer and pastoralist impacts. Human presence does not necessarily detract from the quality of naturalness or wildness, nor from the natural ecological processes that conservationists value. Traditional forms of land use may even contribute to the range of states that conservationists wish to maintain. Conservation organisations are also beginning to realise that conservation areas must enlist local support rather than antagonism if they are to survive in the long run. In particular, the wildlife of East African savanna parks depends to a great extent on the pastoralist rangelands that surround them. Joint land use systems such as the one pioneered by NCA have their problems, but may come to be seen as a vigorous and viable solution, and perhaps the only chance for long-term conservation of savanna areas.

2

Ngorongoro Conservation Area

Ol tau l'enkop-ang The heart of our land

(Maasai expression: Waller 1979)

This chapter describes the environment of the NCA and its plant, animal and human resources. It begins with a basic catalogue of the main geomorphological land units. Climate is discussed in terms of the concepts of seasonality, year to year variability and drought. The special nature of the NCA soils, the main vegetation and habitat types, and their wildlife communities are described. The palaeontological and archaeological remains which contribute to NCA's special character and its importance to the world community are outlined. This review of NCA natural and human resources lays the foundation for more detailed examination, in later chapters, of the dynamics of the system, of interactions between species and of conflicts between differing management objectives. At the same time it illustrates both the uniqueness of NCA and the extent to which the area is representative of other African rangelands.

Natural resources

The natural resources of NCA are important to the management of conservation and development on two levels. Firstly, diversity of soils, topography and landform underlie the ecological diversity that supports the rich and abundant wildlife and pastoralist communities. Secondly, the same physical diversity creates a spectacular landscape which of itself exerts a powerful hold on international conservation interest.

(a) NCA land units

Ngorongoro Conservation Area is ecologically continuous with Serengeti National Park (SNP) (Fig. 2.1). These, together with adjacent rangelands, form the 25 000 km^2 Serengeti Ecological Unit, defined by the annual migratory movements of some three million ungulates. Ngorongoro Crater itself accounts for some 250 km^2 out of the total 8292 km^2 NCA.

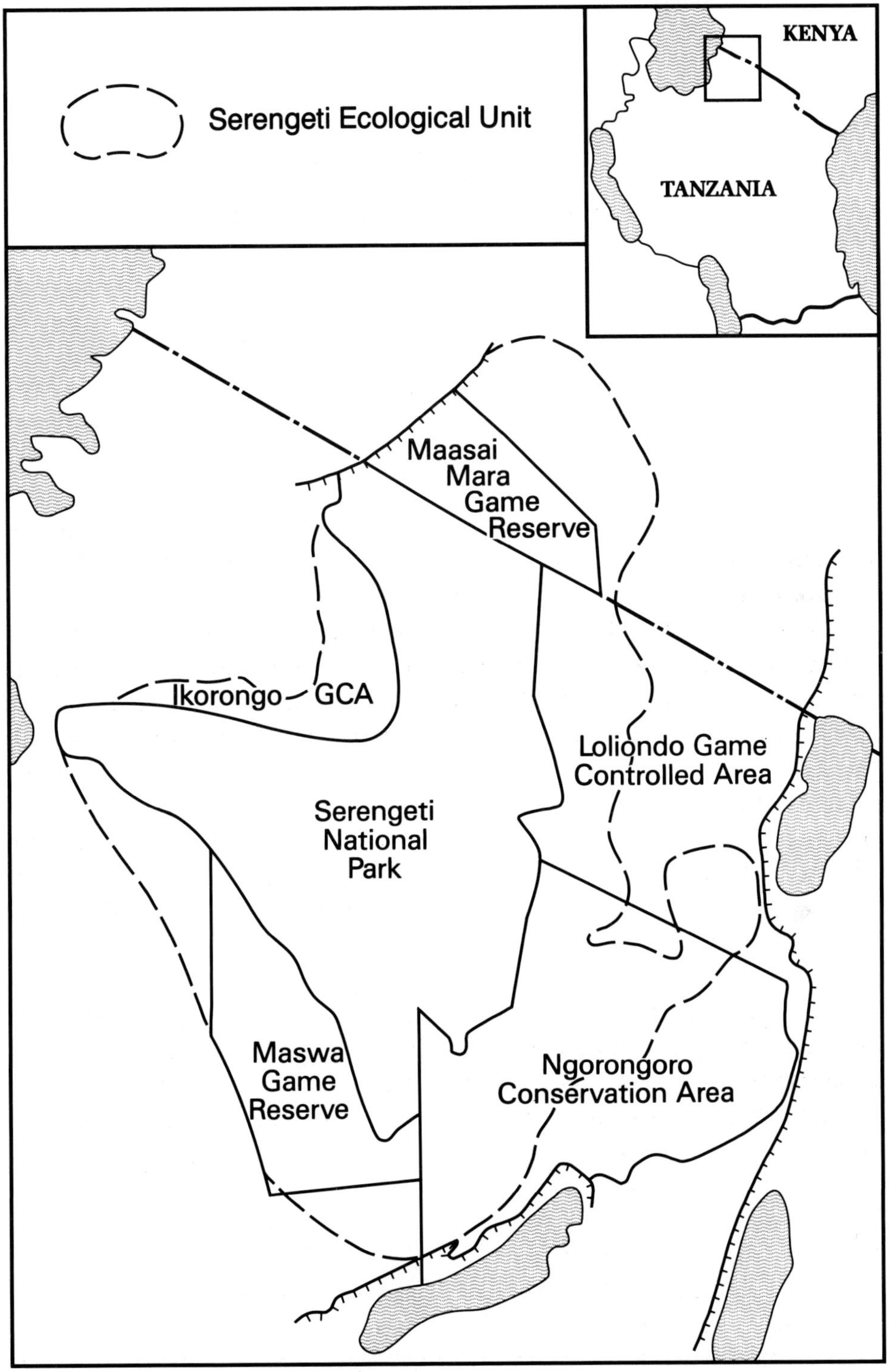

Fig. 2.1. Position of Ngorongoro Conservation Area and the Serengeti Ecological Unit (after Herlocker 1972).

Habitats range from lowlying arid grassland plains, swamps and saline lakes, to volcanic peaks with Afro alpine and montane communities. The geology of NCA is described by Pickering (1960, 1968) and Hay (1976). Figure 2.2 shows the evolution of present-day features (redrawn from Fosbrooke 1972).

NCA can be conveniently divided into five broad geomorphological units (Table 2.1, Fig. 2.3, see back endpapers for place names).

1. The Crater Highlands comprise a gently rolling volcanic plateau between 2100 m and 2800 m which drops steeply on all sides. Rising from the plateau are a number of extinct volcanoes, of which the southernmost, Oldeani and Lemagrut, rise from Lake Eyasi at 1000 m to over 3000 m in height. Their upper slopes are forested and give rise to important springs. Ngorongoro Crater, to the north of these two mountains, is really not a crater but a caldera caused by the collapse of a once massive volcano. Its floor is some 18 km in diameter, forming a circular enclosed plain of 250 km^2 at 1700 m with a soda lake and several permanent springs and swamps. North of Ngorongoro Crater lies the smaller caldera of Olmoti (again with permanent springs and a small swamp). The country then rises gently to the northern volcanic peaks of Olosirwa (the highest peak at 3680 m), Loolmalasin and Nairobi. The last of these has its own small caldera of 30 km^2 in which lies Lake Empakaai. On the northeastern corner of the highlands lies a younger volcano, Kerimasi; Oldonyo Lengai (Maasai: the Mountain of God), a still active and strikingly steep-sided carbonatite volcano, is just outside the NCA boundary. The steep edges of the plateau are dissected by deep ravines, which suggest that heavier rainfall in the past has eroded deeply into the volcanic ash deposits.

2. The Angata Salei Plain forms a flat 10 km-wide corridor separating the Gol Mountains from the Crater Highlands. It runs northeast–southwest, broadening to the north, and acts as a wind tunnel carrying fine powder ash from the Oldoinyo Lengai ash cone. Soils are thus fine, unconsolidated and susceptible to erosion by wind and trampling. Mobile sand dunes or barchans can be seen at the western end of the plain near Olduvai Gorge.

3. The Oldoinyo Ogol Hills or Gol Mountains lie across the northwestern boundary of NCA. The hill ranges are of ancient Precambrian rocks and their steep eastern scarp faces are probably relics of earlier rift systems. They rise abruptly to 2200 m from the surrounding recently deposited ash soils of the plains at 1750 m. The ranges form an irregular east–west pattern with intervening grassy plains and minor valleys.

4. The Serengeti Plain occupies some 3000 km^2 in the west and southwest of NCA and extends westwards into the Serengeti National Park. The plain funnels gently from high points to the north and south down to 1580 m at Olduvai Gorge. The gorge is a deeply incised seasonal streambed draining Lake

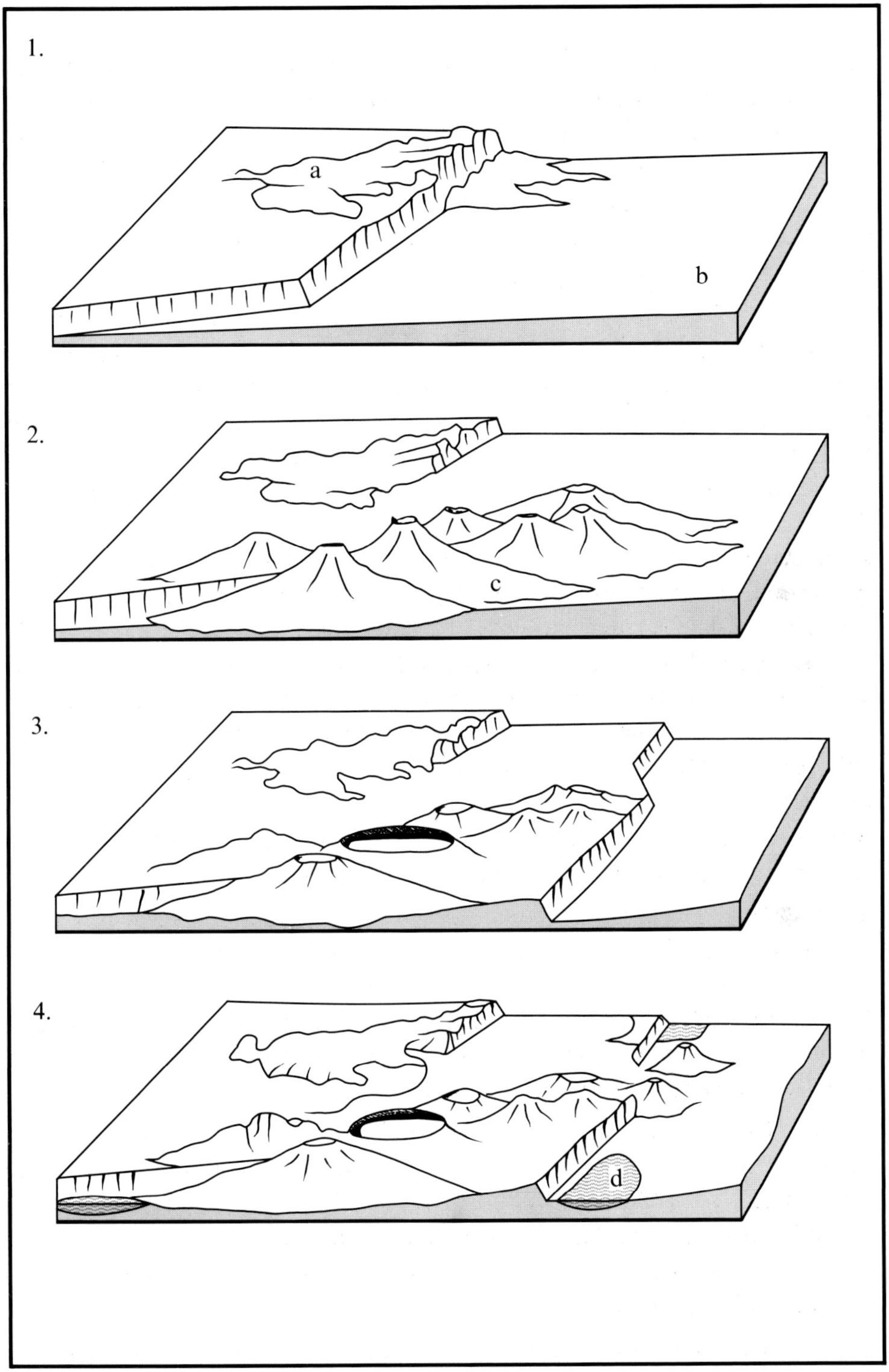

Fig. 2.2. Evolution of Ngorongoro by rifting and volcanic activity (redrawn from Fosbrooke 1972). 1. First faulting period. 2. Volcanic activity. 3. Second faulting period; caldera formation. 4. Erosion, continued volcanic activity, formation of present day lakes in Rift Valley floor. a. Gol mountains b. Rift Valley floor c. Volcanoes d. Rift Valley Lakes.

Table 2.1 *Main land units of NCA*

	Area (km^2)	Area (%)	Altitude (m)	Main land form geology	Main vegetation type
1. Crater Highlands	2690	33	2000–3000	Volcanic plateau, extinct volanic peaks, calderas	Forest, highland woodland, derived tussock grasslands
2. Angata Salei Plain	730	9	1750	Corridor between Gol and Crater Highlands; flat, fine wind-blown ash soils	Short and medium grass associations
3. Oldonyo Ogol hills (Gol Mountains)	700	9	1750–2200	Old rifted and eroded hills of Precambrian rock	Short grass on hilltops *Acacia*/*Commiphora* bush slopes
4. Serengeti Plain	2750	34	1500–1750	Extensive plain overlaid with volcanic ash soils	Short grass association
5. Eyasi Scarp/Kakesio	1350	16	1000	Steep scarp slope dropping to Rift floor; rolling plains and low ridges	Bush, woodland, wooded grassland

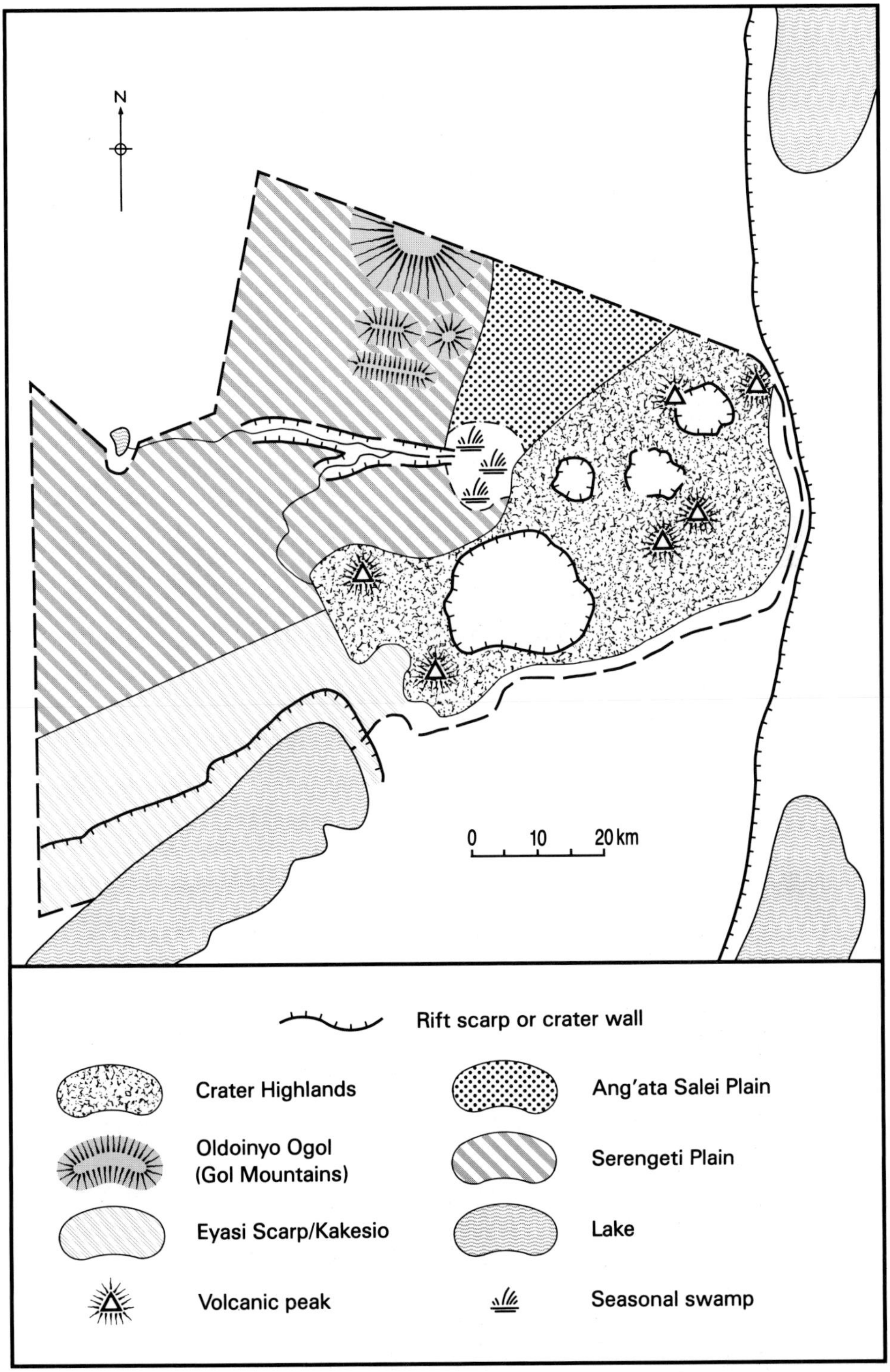

Fig. 2.3. Main land units of NCA. ——— = boundary of NCA.

Lagarja (Ndutu) on the Serengeti boundary into the Olbalbal depression and swamp at the western foot of the Crater highlands. In sharp physiographic contrast to the Crater Highlands, the Serengeti Plain is a wide featureless expanse only broken in places by outcrops of small inselbergs or 'kopjes', erosion terraces and shallow faults.

5. Lake Eyasi Scarp forms the border of the NCA in the southwest where the Serengeti plains gradually rise and break up into increasingly rolling and rocky country by Endulen and Kakesio. This land unit includes the scarp ridge above a broken, rocky and in places sheer escarpment falling to Lake Eyasi, an internal drainage soda lake just outside the NCA boundary.

(b) Climate

The most important aspect of climate in this region, as in all other largely semi-arid areas, is rainfall. Rainfall governs vegetation production and the availability of water. The quantity and quality of plant food and water are major factors in both wild and domestic stock's ability to use the area. Rainfall is highly seasonal and extremely variable within and between seasons and years. It is determined by large-scale tropical weather patterns, and modified by local topography. As local topography is so pronounced, there are great variations of rainfall pattern within NCA.

Tropical rainfall is associated with the rising moist air and low pressure belt of the Inter Tropical Convergence Zone (ITCZ), where trade winds of the northern and southern hemispheres meet. The ITCZ migrates north and south of the Equator following the seasonal movement of the sun with a lag of one to two months. Rainfall is thus associated with the season of hot overhead sun. On the Equator there are two such rainfall peaks associated with solar passage. Further from the Equator these two peaks merge into a single and progressively shorter rainy season associated with the correspondingly shorter period of overhead sun. The further from the Equator, the shorter the rainy season, the lower the total rainfall and the less predictable the year to year variation.

This general pattern is modified by continental physiography. In East Africa the southward movement of the ITCZ produces rather dry northeasterly winds that have crossed extensive arid areas; as it moves back north the relatively stronger and wetter southeasterlies develop. However, during the Northern Hemisphere summer the strong heating effect over the Himalayan land mass causes moisture-laden winds blowing from the southeast across the Indian Ocean towards East Africa to be diverted towards the Indian subcontinent. East Africa thus has a drier climate than might be expected by virtue of its position.

Other than the eastern seaboard the main high rainfall areas of East Africa are those with sharp relief facing prevailing winds, those associated with Lake Victoria and to a lesser extent those with other major lakes. In NCA prevailing winds pass first over or close to the Crater Highlands, whose steep eastern windward side thus receives high rainfall. Most of NCA lies in their rainshadow and receives very much less. Further west in the Serengeti National Park the annual rainfall increases again with the proximity of Lake Victoria.

The climate of the whole Serengeti and western Ngorongoro area has been discussed by Norton-Griffiths, Herlocker and Pennycuick (1975) and Pennycuick and Norton-Griffiths (1976) and is summarised in Sinclair and Norton-Griffiths (1979).

The eastern slopes of the Crater Highlands receive over 1000 mm rain p.a. and Frame (1976) recorded some 1500 mm p.a. at Empakaai. This decreases to 800–1000 mm on the plateau and to below 600 mm p.a. in the rain shadow area (Olduvai Gorge for example). In any one year the average number of rainy days is low (from *c.* 45 in wetter areas to *c.* 30 at Olduvai). In theory, NCA gets bimodal rainfall with the short November northeasterly rains and the longer March–April southeasterly wet season. However, most authors generally place it in the zone of transition to a single rainy season (Griffiths 1962, Pratt and Gwynne 1977). The great variation in timing of the rainy season, typical for semi-arid areas, means that many months of the year from November to May show an appreciable average rainfall. These average monthly data mask the annual pattern of two potential rainy seasons separated by a hot dry period in December–February (Fig. 2.4). NCA thus comprises a series of different rainfall regimes with correspondingly different plant production patterns.

Waller's analysis (1976: 30) shows partial or major rain failure in some part of Kenya Maasailand once every two or three years during 1912–1930. The concept of drought has received considerable attention in the last decade (see for example Glantz 1987, Rasmusson 1987). It cannot be defined in purely physical terms, as its severity depends not only on rainfall totals and timing but also on land use, and on the social and ecological options and experience available. Many authors have looked for periodicity in rainfall patterns which bring recurring droughts of significance to pastoralists, but cyclical patterns were not evident in Pennycuick and Norton-Griffiths' analysis, nor in data from nearby ecosystems (e.g. Tsavo-Mkomazi: Cobb 1976). However, other characteristics are revealed by long-term data for stations throughout sub-Saharan Africa (Nicholson and Entekhabi 1986). These distinguish East (and Southern) African drought patterns from those of West Africa. West Africa shows long runs of wet or dry conditions, often with dry periods of 10–18 years, which include short-lived acute severe drought episodes. By contrast, East and Southern Africa are characterised by short-term fluctuations and tend to show

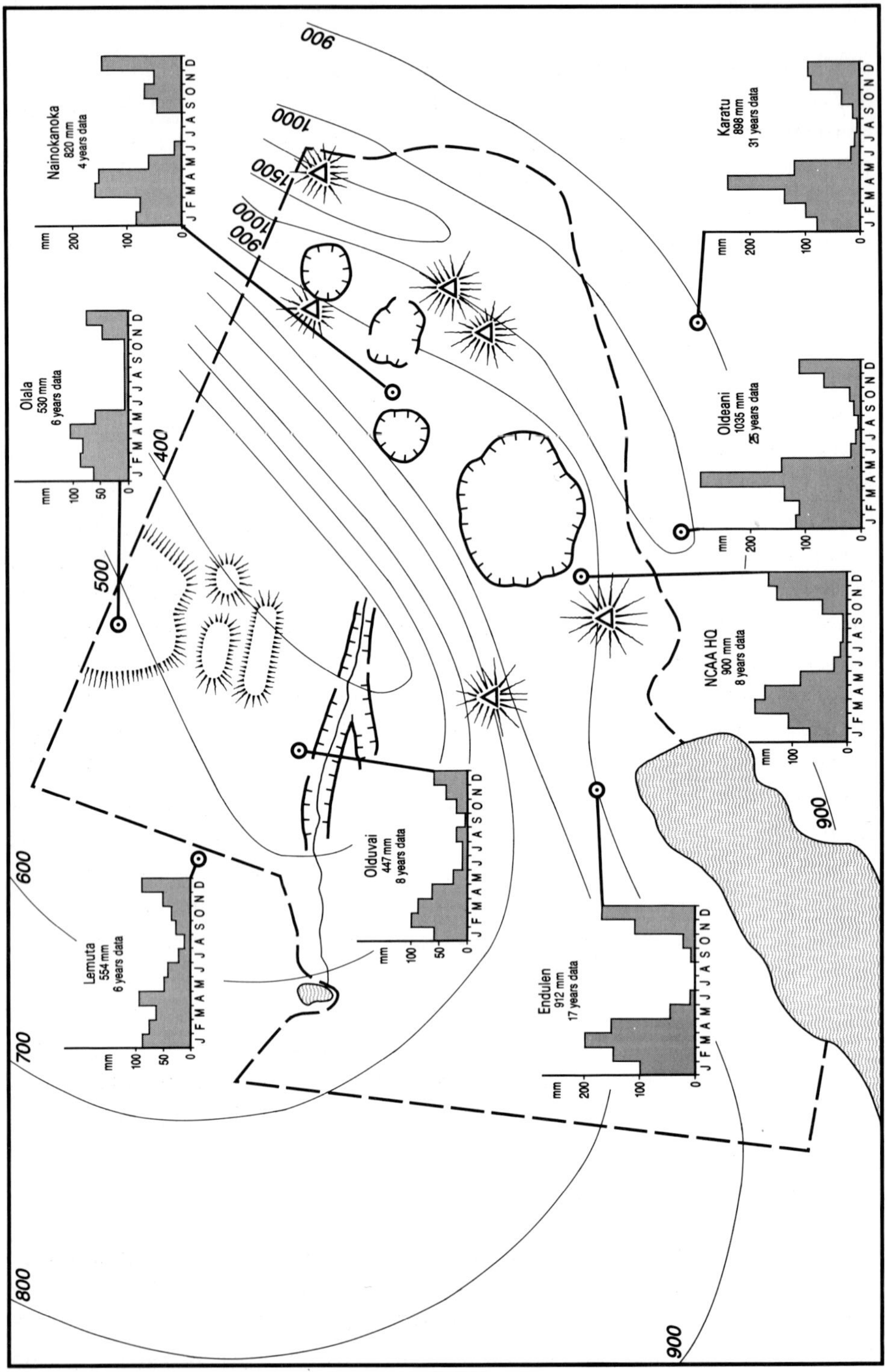

Fig. 2.4. Isohyets of mean annual rainfall (mm) in NCA. Inset histograms show seasonal distribution of rainfall at individual stations.

short-term severe droughts of one to three years duration, with overall wet or dry spells of two to six years duration. This is borne out for Kenya Maasailand in 1912–1930 by Waller (1976); and for East Africa generally in 1933–1984 by Rasmusson (1987) quoting Ogallo and Nassib (1984). Much of NCA has a 20% probability of below 500 mm rainfall, meaning that over the long term, annual rainfall will be below 500 mm one year in five, (though this is unlikely to hold for any given five-year run).

Both general East African patterns and local data show that periods of four years' consecutive below-average rainfall are not unusual (Pennycuick and Norton-Griffiths 1976). The years 1981–84, 1973–76 and 1952–56 seem to have been periods of low rain throughout most of NCA while there is great variation between sites for other periods. The years preceding and during our study in Ngorongoro in 1981–1983 were thus drier than average across the whole area. All study sites monitored received less than two-thirds of the 'average' rainfall.

On average, temperatures decrease by one degree Centigrade for every 200 m increase in altitude. With the 2500 metre range in altitude and the complex topography there is considerable microclimatic variation in temperatures to be exploited by human and animal communities, in contrast to conditions in the comparatively flat, low Serengeti. There is, however, a dearth of temperature data. The lowlying dry plains are hot, and classified as arid (Norton-Griffiths *et al.* 1975). Olduvai shade temperatures in January may reach 38 °C. Hot dry winds increase evaporation, reduce precipitation effectiveness and create an edaphic desert effect over much of the plains. Tropical mountain regions have extremes of variation between day and night temperatures (e.g. Coe 1967) and frost is commonplace at night in the higher areas of the highlands, especially during May–July. The effects of temperature on cattle production are discussed in chapter 8.

The last few years have seen an unfortunate reduction in the quality of climatic data collection in the NCA and Serengeti. Many records for the 1980s should be treated with caution, but the situation has improved recently with the institution of the Ngorongoro Ecological Monitoring Program (NEMP) in 1987.

(c) Soils

East African rangelands mostly overlie old acid Precambrian basement rocks which generate infertile soils (Pratt and Gwynne 1977). Around 95% of the land area of Africa has infertile soils of this type (D'Hoore 1964; Allan 1968). East Africa however has sizeable volcanic areas associated with the Rift Valley. These are high in mineral nutrients and can give rise to very fertile eutrophic brown soils, which have good structure, high mineral and organic content. Such soils make up less than one per cent of the African soil mantle. Many of the rangelands overlying these East African volcanic areas are traditionally pastoralist grazing lands and some of them, like NCA, remain so. NCA soils are thus predominantly of mineral and (in the highlands) organic fertility considerably higher than those of most other African rangeland areas. At the same time, like many East African areas, NCA has a close spatial association of many different soil types thanks to its geological and topographic diversity. The special nature of these soils, their mineral content and their striking patterns of production, as well as their vulnerability or resistance to environmental degradation in NCA, are explored in chapters 6, 7 and 8. Here, broad features are outlined.

In the plains and Ngorongoro Crater, soil parent materials are of volcanic origin with a high calcium, potassium and sodium content but low magnesium (Anderson and Talbot 1965, Anderson and Herlocker 1973, de Wit 1978). Spatial variations in the mineral content of the plains soils and associated vegetation are major factors affecting the migratory patterns of grazing ungulates (McNaughton 1988, 1990). The highly porous and free-draining nature of the soils allows even the existing low rainfall levels to leach out the more soluble salts from the surface and redeposit them as a calcareous hardpan at a shallow 0.5–1.0 m depth. The resulting volcanic tuff soils have a shallow highly alkaline and saline soil above an impermeable hardpan. This makes growing conditions difficult for much plant life but highly productive for grass growth when water is available. These soils have little texture or organic matter content and are susceptible to erosion. Windblown soil is rapidly redeposited elsewhere on the plains.

The soils of the basement Gol Mountains are highly leached infertile loams. Older tuffs still remain on flatter hilltops and the more gentle lower inclines, but erosion is severe elsewhere on Gol slopes.

The soils of the highlands have been poorly studied, though data on other volcanic mountains nearby may be relevant (e.g. Mt. Meru: Lundgren and Lundgren 1972). These are fertile red-brown eutrophic soils of volcanic origin and are acid with a high organic content. They may or may not have localised hardpans at depth. Their great depth and high humus content make them highly productive and less prone to erosion than for example the Gol soils.

(d) Water

The distribution of both natural and artificial water has a major bearing on distribution patterns of both wildlife and pastoralists and their stock. Ngorongoro water has been described in detail by Kametz (1962) and reviewed most recently by Cobb (1989). The crater highlands catchment acts as a major source of water for the wildlife, livestock and people of the adjacent plains to the west as well as for the agricultural districts in the south and east. At least 23 separate permanent streams supply some 500 km^2 of productive farming land (Gilchrist 1962). The importance of the catchment properties of the Northern Highlands Forest Reserve, 765 km^2 along the eastern slopes of NCA, was recognised by the German colonists who reserved the forests in 1914.

Much of the drainage pattern is of small internal streams, either into crater lakes, such as the Munge River flowing from Olmoti to Ngorongoro Crater, or into depressions such as Olbalbal which may hold water for up to 10 months of a wet year. Most surface waters are alkaline, leading to soda lakes, which periodically dry (Fosbrooke 1972). Further west the Serengeti climate is more humid and the soils less porous so surface water becomes more frequent. These waters are now within the National Park. While they are available to and used by the wildlife, domestic stock no longer have legal access to them.

The highly porous soils and low rainfall mean underground waters are of great importance, whether as natural springs or artificial boreholes. Borehole water is frequently brackish or even highly saline. The deep groundwaters and permanent spring waters of the plains west of the Crater Highlands have significantly higher salinity and fluoride levels than shallow groundwaters and other seasonal sources in the same area, while the water sources of the Crater Highlands are of comparatively high quality (Aikman and Cobb 1989, Cobb 1989). Fluoride is a particular problem. Only 25% of the samples taken throughout NCA were within the upper limit of fluoride levels normally accepted for human drinking water (Cobb 1989), and fluoridosis in livestock is common, causing brittle bones and often diagnosed through broken foot bones. Borehole function is of course dependent on pumps, spare parts, maintenance and diesel, and as a result these systems are frequently out of action (chapter 12). The development of and political bargaining for water resources has been a major feature of development plans both for Ngorongoro and more generally for all of Maasailand in both Kenya and Tanzania. Actual and potential water developments and their possible ecological and socioeconomic consequences are described in chapter 12.

(e) Vegetation and habitat types

This section presents a brief description of the main vegetation and habitat types in NCA, their appearance, composition and distribution. The importance of the vegetation to wildlife, pastoralist and outside observer is discussed in chapter 4. Chapter 6 looks at specific habitat types and the special nature of their plant production and dynamics in detail, exploring ideas of succession, degradation, resilience and the possibilities of management intervention.

The vegetation of NCA has been described by Herlocker and Dirschl (1972) and there are detailed analyses of the plains vegetation in Kreulen (1975), Schmidt (1975) and Banyikwa (1976). A full checklist of NCA plants has recently been compiled (Chuwa, Mwasumbi and Rodgers 1985) and further extended by more recent collections (NEMP 1989). Frame (1976) gives a generalised account of the ecology of the northern Crater Highlands at Empakaai. The vegetation of the Ngorongoro Crater is further discussed in Anderson and Herlocker (1973) in relation to soil types. The forested areas are still poorly described, as is much of the Crater Highlands.

The NCA flora is not particularly large (around 1000 spp. collected in 8000 km^2) but the habitats are strikingly diverse. Communities are not rich in species and there are few endemics. The grass *Holcolemma transiens* has all or most of its range within NCA. *Ethulia ngorongoroensis* M.G. Gibert *sp. nov.* is a small herb so far known only from NCA short grass and bushland, and the climber *Neonotonia verdcourtii* Isely = *Glycine sp. A* has as yet been found only in the crater and rim grasslands. Other than this the herb *Achyranthes fasciculata* is known only from Mbulu and NCA and there are two rare grasses (*Odontelytrum* known only from Yemen and NCA, *Odyssea* from Congo and NCA; Flora of Tropical East Africa 1974).

The forest in particular is species-poor, especially compared with the forests of the Usambara Mountains (Rodgers and Homewood 1982) and the Uzungwas and Ulugurus (Rodgers, Owen and Homewood 1983). In these other areas the formation of the mountain ranges by rifting rather than volcanic activity

Table 2.2 *NCA vegetation types and their extent (based on Chausi 1985 and Herlocker and Dirschl 1972)*

Vegetation type	Area (km^2)	Total (%)	Forest reserve (km^2)
Heath	142	1.7	83
Bamboo	39	0.5	39
Evergreen forest	622	7.6	622
Highland woodland	855	10.4	70
Highland grassland	1036	12.6	86
Lowland woodland	725	8.8	
Medium grassland	1690	20.6	
Short grassland	1061	12.9	
Sand dune grassland	2054	25	
Total (uncorrected)	8224		

may have allowed the initial retention of a far richer biotic community. Their long history, (*c.* 25 million years, cf. one million years for the Crater Highlands) has fostered the evolution of a much higher proportion of endemics in all classes as well as giving time for further colonisation by new species.

Herlocker and Dirschl (1972) provide a comprehensive description of NCA vegetation types and map their distribution. Table 2.2 is based on the areas calculated by Chausi (1985) from their map.

These vegetation types can be best described pooled under the headings of

1. Forest
2. Highland shrub and grassland
3. Bushland and woodland
4. Plains grasslands

The main ecological features of these categories are briefly described here. Their distribution is mapped in Fig. 2.5 and principal species are listed in Table 2.3.

1. Forest. Forest communities are restricted to the wetter eastern slopes and probably give way to shrublands where annual rainfall totals less than 900 mm. The Northern Highlands Forest Reserve runs from the northeast corner below Kerimasi to Oldeani Mountain. In terms of general African vegetation types (White 1983), the higher altitude stands (> 2000 m) are classified as 'undifferentiated montane forest'. The lower are 'dry transitional montane forest' and probably represent one of the largest remaining areas of this forest type in East Africa.

Tree cover is patchy and there are large areas of dense shrub layer with scattered or no emergents. Oldeani is largely covered with bamboo. Within the forest are several glades of open grassland (NEMP 1989), possibly of edaphic origin or linked with past settlement sites (cf. Wood 1974, Lundgren and Lundgren 1972 and Odner 1972). The steep slopes of the Ngorongoro massif are dissected by several deep ravines with thicket ('Afromontane evergreen thicket': Table 2.3).

The upper margins of the forest (near NCAA HQ at Ngorongoro; on the opposite side of the Crater rim; also the slopes of Empakaai) show an interesting mix of forest, *Crotalaria–Vernonia* shrubland and grassland. These distinctive community patterns probably reflect past cultivation (Frame 1982, see also chapter 6) as well as successional stages following continuing grazing and burning by pastoralists (Struhsaker *et al.* 1989). Around NCAA HQ there are patches of *Eucalyptus* plantation, which were developed for fuelwood supplies for administrative and tourist villages, but are no longer maintained.

The lower margins of the forest facing Karatu and Oldeani settlements are

Table 2.3 *Species composition of the principal vegetation types in NCA*
(a) Forest

	Canopy	Undergrowth	Variants
1. Undifferentiated montane forest			
	Bersama abyssinica	*Abutilon longicuspa*	*Juniperus procera* (on Lemagrut and Loolmalasin) *Arundinaria alpina* bamboo on Oldeani
	Cassipourea malosana	*Crotalaria arborea*	
	Ekebergia capensis	*Discopodium panninervium*	
	Hagenia abyssinica	*Urtica massaiense*	
	Nuxia congesta	*Vernonia auriculifera*	
	Olea capensis		
	Olea europea		
	Podocarpus		
2. Dry transitional montane forest			
	Albizia gummifera	Varied	Open glades
	Croton macrostachys	(many Rubiaceae)	Patches of shrubland
	Croton megalocarpus		
	Fagaropsis africana		
	Ficus thonningii		
	Teclea nobilis		
3. Afromontane evergreen thicket (Ravine forest, western slopes, very degraded)			
		Buddleia polystachya	*Juniperus procera* at >2500 m
		Clutia abyssinica	
		Heteromorpha trifoliata	
		Osyris lanceolata	
		Rhus natalensis	
		Rumex usambarensis	
		Scutia myrtina	

(b) Montane grassland and shrubland

	Tussock species	Turf/Mat species	Shrub/herb species
1. Grassland			
	Eleusine jaegeri	*Andropogon greenwayi*	*Artemisia affra*
	Pennisetum sphacelatum	*Cynodon dactylon*	*Helichrysum schimperi*
	(=P.schimperi)	*Digitaria abyssinica*	*Lupinus princei*
	Streblochaete	*(=D.scalarum)*	
	longarista		*Salvia merjamie*
	Themeda triandra	*Pennisetum clandestinum*	*Satureja punctata*
			Senecio spp.
			Trifolium massaiense
2. Shrubland			
	Pennisetum sphacelatum	As above	*Erica arborea*
	Themeda triandra		*Lantana triflora*
			Lippia javanica
			Stoebe kilimandscharica

Table 2.3 (*cont.*)
(c) Bushland and woodland

	Tree	Herb	Grass
1. Acacia lahai			
	Acacia lahai	*Erlangea tomentosa*	*Cynodon dactylon*
	Gnidia glauca	*Hypericum revolutum*	*Pennisetum clandestinum*
		Leonotis leonotis	*Digitaria abyssinica*
2. Acacia drepanolobium			
	Acacia drepanolobium	*Crotalaria* spp.	*Themeda triandra*
	Acacia hockii	*Indigofera bogdani*	
3. Gol slopes			
	Acacia tortilis	*Rhynchosia minima*	*Digitaria macroblephara*
	Acacia nilotica	*Aspilia mossambicensis*	*Themeda triandra*
	Erythrina abyssinica	*Malvaceae* spp.	*Aristida* spp.
	Commiphora trothae		
	Commiphora subsessifolia		
	Euphorbia nyikae		
4. Ilmesigio lower slopes			
	Acacia drepanolobium	Varied	*Hyparrhenia* spp.
	Albizia anthelmintica		*Themeda triandra*
	Grewia bicolor		
	Rhus vulgaris		

(d) Plains grassland

	Grasses/sedges	Herbs	Variants
	Aristida keniensis	*Crotalaria keniensis*	*Hypoestes forskalii* along drainage lines
	Cynodon dactylon	*Euphorbia inaequilaterale*	
	Chloris virgata	*Indigofera bogdani*	
	Digitaria abyssinica	*Hirpicum bequinotii*	
	Harpachne schimperi	*Solanum incanum*	
	Sporobolus ioclados		
	Kyllinga spp.		

Note:
Names are taken from Flora of Tropical East Africa (1974) or Herlocker and Dirschl (1972).

exploited illegally for fuelwood, poles and small timber, probably by the villages and estates outside NCA (Chamshama, Kerkhof and Singunda 1989, Struhsaker *et al.* 1989). Aerial photographs from 1958 to 1972 showed severe canopy loss (> 75%) in several places (NEMP 1989). Considerable bare soil and erosion is evident.

2. *Highland shrub and grassland.* The higher levels of Ngorongoro just reach the typical Afro-alpine communities described for taller mountains (Hedberg 1951, Elliott 1948). In general the highland areas are dominated by a tussock grassland with variable shrub–herb content typified by our Sendui study site (see below). The higher altitude tussock grassland (above 2200 m) is dominated by *Eleusine jaegeri* with tussocks to 1.5 m tall and 1 m in diameter. Between tussocks there is a sparse mat of short palatable grasses. *Pennisetum sphacelatum (= P. schimperi)* is commoner below 2200 m and found in patches above. A wide range of herbs, especially legumes, composites and labiates are found in the upland grasslands (Table 2.3(b)). Herbs may dominate completely in areas

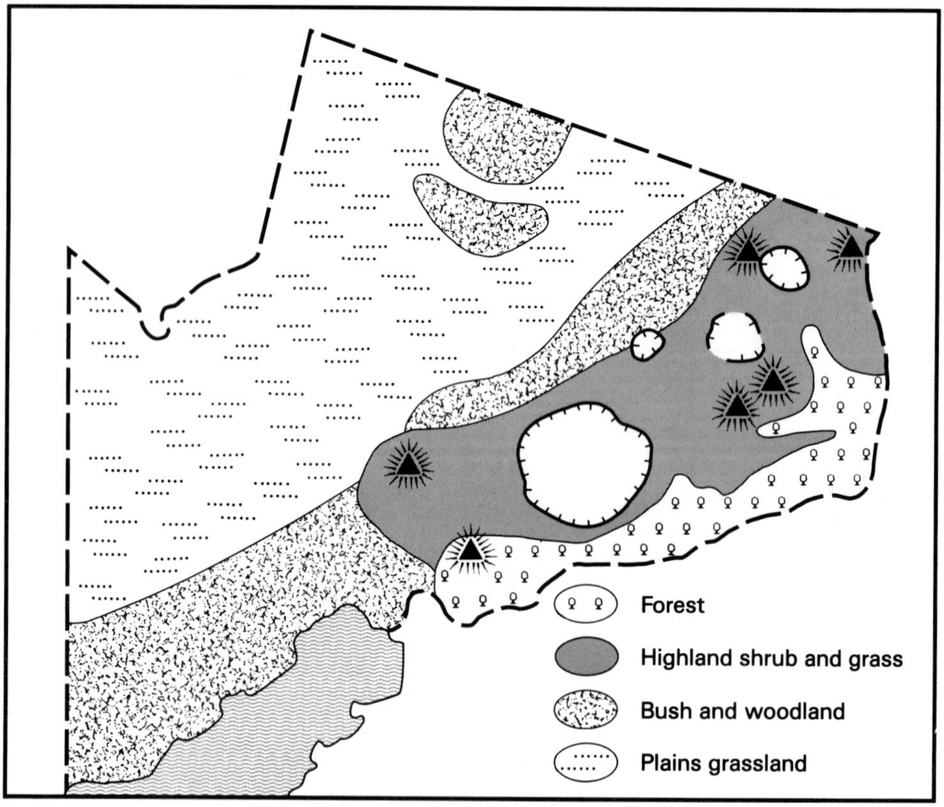

Fig. 2.5. Main vegetation types in NCA. Crater floors have a mixture of plains and highland grassland, with forest and woodland patches.

subject to erosion or intensive trampling. Although there is little heathland, a giant heather (*Erica arborea*) is found at higher levels.

A short grass plains flora similar to that of Gol (see below) dominates the slope and floor of Embulbul Depression. Some green biomass is present throughout the year. Mole rat (*Tachyortes daemon* Thomas) earth heaps are a conspicuous feature of this area.

3. *Bushland and woodland.* All the drier areas of NCA would, in the absence of fire and browsing, carry a woody climax vegetation cover (with the exception of the tuff soils of the plains whose hardpans present rooting problems). Where this climax woodland and bushland persists in NCA it is made up of microphyllous *Acacia–Commiphora* communities similar to those found throughout the semi-arid areas of eastern Africa. There are some 16 species of *Acacia* and five of *Commiphora* present in NCA so this vegetation type is complex. Two communities are quite distinctive. The *Acacia lahai* woodlands of moister slopes at an altitude of 2000 m are monospecific closed stands of up to 15 m in height (Table 2.3(c)). Their shade leads to a much greener and moister ground layer than elsewhere in NCA. Stands of *Acacia xanthophloea*, the yellow-barked fever tree, indicate abundant groundwater, as in Lerai Forest in Ngorongoro Crater. Kaihula (1983) gives details of their structure and dynamics.

The medium level slopes of the highlands have a low scattered tree cover of whistling thorn (the ant-gall species *Acacia drepanolobium*). Gol slopes have a mixed *Acacia* woodland. Table 2.3(c) lists common species for these two communities. The undulating country by Lake Lagarja and Endulen has a mixed *Acacia* savanna (principally *A. tortilis*), while the drier areas of Olduvai have *Acacia mellifera–Commiphora* bushlands with much *Sanseviera*. The scarp slopes of Eyasi have a mixed species bushland.

The slopes of the Gol Mountains have a woodland and wooded grassland cover with trees to 8 m and emergents (e.g. *Erythrina abyssinica*) to 10 m. The shrub layer has smaller individuals of most tree species. The ground layer is very variable, and large areas on steeper slopes may be all eroded rock. The herb layer rarely exceeds one metre in height, and climbers are rare.

4. *Plains grasslands.* The flat topped hills and the grassy plains between hill masses (such as Angata Kiti and Angata Salei) have a shallow volcanic tuff and dust soils, with a characteristic short grass association (SGA: <15 cm) interspersed with erosion steps and clumps of taller herbs. The Ngorongoro portion of the Serengeti Plains has only these short grass communities (Table 2.3(d)), but as rainfall increases westward and soil hardpans become less limiting, the grass communities become taller, reaching the red oat-grass *Themeda triandra* savanna with a height of up to 80 cm typical of fire-climax

East African rangeland by mid Serengeti. The flat hill tops often have an open shrub cover of *A. drepanolobium* over short grass. The transition zone between plains and hill slopes carries a denser grass cover of up to 50 cm high, with species from both communities.

(f) Wildlife resources

To the layman and the conservationist it is the wildlife resource that is the most well known and probably the most valuable component of NCA. This section briefly describes the resource; further aspects of wildlife ecology, conservation and management are discussed in chapter 7.

It is the Crater populations that have captured public imagination but there are many other wildlife values in NCA. The Serengeti migratory wildebeest population (currently some one million animals) spends much of the rainy season on the plains of west NCA, outside Serengeti National Park. The wildlife resources of NCA are best described in terms of five distinct communities:

1. The forest wildlife of Oldeani and Northern Highlands Forest Reserve.
2. The arid-land populations of south Natron and Lake Eyasi.
3. The Ngorongoro Crater populations, which may intermingle with:
4. The highland grassland populations, which may join into:
5. The migratory plains game populations, which use the NCA in the rainy season and move westwards and northwards in the dry season.

1. The NCA forest fauna, like the flora, is poor compared with those of older block mountain forests such as the Usambaras or Ulugurus (Rodgers, Owen and Homewood 1983), particularly in terms of forest primates, small mammals and birds. The fauna of the montane forests of the wetter eastern slopes of Ngorongoro and Oldeani is nevertheless of conservation value for its conspicuous large herbivores: buffalo, rhino, elephant and bushbuck. No detailed studies have been undertaken on wildlife in this or any other similar forest in East Africa, though preliminary accounts of the mammals and avifauna of Kilimanjaro forests appeared in a special edition of Tanzania Notes and Records (1965, reprinted 1974).

2. The arid lowlands surrounding Lake Natron have small populations of wildebeest and zebra, but also the more typically dry country Grant's gazelle, oryx and lesser kudu. Rhino and elephant populations were hunted out years ago. Lake Eyasi escarpment and the bushlands of Maswa and Endulen still have a greater kudu population, although all animal species are subject to illegal hunting (chapters 7, 11; Makacha, Msingwa and Frame 1982).

3. Populations of herbivores in Ngorongoro Crater have been briefly described by Estes and Small (1981) and more recently by Hanby and Bygott (1989) and Boshe (1988). Large mammal species are listed in chapter 7 together with estimates of population size. Other than for Grant's gazelle, giraffe, eland and impala, the Crater holds the majority of NCA dry season wild ungulate populations (Table 2.4). All herbivore populations are subject to some fluctuation, whether with seasonal migrations (wildebeest, zebra – Estes 1966, 1969, Boshe 1988) or longer-term factors (buffalo – Rose 1975). Rhinoceros populations were studied intensively in the mid-1960s (Goddard 1967). Poaching has eliminated rhino in many areas of northern Tanzania since 1976 and has severely reduced the Crater population (see chapter 7). The predator populations of the Crater have been studied mostly as part of intensive programmes on the predators of the entire Serengeti (from Schaller 1972 through to Pusey and Packer 1987 on lions; Kruuk 1972 – hyaena; Estes and Goddard 1967 – wild dog). The Crater has a very high density lion population, reflecting the highest resident biomass in Africa of their preferred prey (van Orsdol 1981). The lion population is currently very stable. Following their recovery from a disease-related crash in the 1960s (*Stomoxys* fly outbreak – Fosbrooke 1962) the lions reached their present population size in 1975. Since then a high proportion of subadults emigrate permanently each year; no immigration has been recorded in the last decade (Pusey and Packer 1987). Other species may have permanent resident populations (e.g. serval – Geertsema 1985), but some predator species (wild dog, cheetah) are present only intermittently as the result of periodic recolonisation of the Crater.

4. The highland grasslands populations include animals permanently resident in Empakaai and Olmoti craters and surrounds, and seasonal emergents from the forest and Ngorongoro Crater. Livestock are common in this zone; wildlife by contrast are not numerous (Frame 1976, 1982). Zebra are the commonest wild ungulates and are often seen near domestic herds. Eland, reedbuck and steinbok are frequently encountered. Wildebeest and gazelle are more common on the western slopes and Malanja depression above the Serengeti Plains.

5. The migrant plains game populations of the Serengeti invade the western Ngorongoro plains (Serengeti, Salei, Gol) during the rainy season, December to March/April, when over two million migrants (mainly wildebeest, but some zebra and gazelle) use this area. One survey estimated that over 75% of the wet season grazing by the Serengeti migratory herds took place outside the SNP boundary, the great majority being in NCA (Watson and Kerfoot 1964). Table 2.4 shows the main species. NCA, which comprises 8% of Arusha Region, accounted for 86% of the plains wildlife and 67% of all wildlife for the Region.

Table 2.4 *Main large ungulates of NCA*

Species	NCA count[a] 1980	Arusha region[a] 1980	95% limits[a] for regional estimate	1987 census of NCA[b]		
				Wet season	Dry season	Dry season Crater only
Wildebeest	830 800	1 067 575	28%	1 109 011	8 318	7415
Thomson's gazelle	373 800	144 504	20%	149 715	6 161	4 677
Zebra	69 700	389 543	7%	62 959	7 187	4 332
Buffalo	10 200	21 800	13%	3 102	3 484	2 855
Grant's gazelle	10 000	30 782	11%	10 303	7 587	1 135
Eland	2 900	30 679	12%	5 436	168	7
Giraffe	2 719	24 145	14%	1 666	1 226	
Impala	1 800	90 083	13%	3 301	452	
Coke's hartebeest	1 000	15 405	10%	345	160	112

Notes:

[a] Ecosystems Ltd. 1980: pp. 57, 62;

[b] Boshe 1989: p. 5, combining ground survey estimates for the Crater with aerial estimates for the rest of NCA.

A large proportion of the total is made up of migrants. Changing patterns of the duration and intensity of use of the NCA by migratory ungulates are explored later (see chapter 7).

Archaeological and palaeontological resources

The NCA is particularly rich in archaeological remains which are not only of scientific value, but of considerable tourist interest. Olduvai Gorge cuts through some two million years of deposits (Reck 1933, Leakey, L 1965, Leakey, M. 1971; see Poirier 1987:136–9 for a brief up-to-date summary of the importance of Olduvai). Louis and Mary Leakey carried out their famous studies at Olduvai from 1931 until late 1983. There are many archaeological sites of significance in and near NCA which together span a period of human evolution of some 3.5 million years (Mturi 1981). Laetoli, above Lake Lagarja (Ndutu) at the western foot of Lemagrut, is the site of well-investigated beds spanning 300 000 years to 3.5–3.8 million years BP (Leakey and Harris 1987).

As well as providing hominid bones these beds have yielded a vast array of other vertebrate remains, many of which still await full description (see Leakey, L. 1965, Leakey, M. 1971, Leakey and Harris 1987). These, together with studies of pollen grains and of the nature of sediments at Olduvai and Laetoli, have allowed the description of past climates and vegetation types. Palaeontologists present a picture of a dynamic mosaic of savanna grassland and woodland types (Andrews 1989). The distribution and relative extent of different vegetation types has changed throughout the last few million years with changes in climate and volcanic activity (particularly of Ol Doinyo Lengai, whose windblown ash conditions vegetation growth on the short grass plains). The Laetoli Pleistocene fauna resembles that of modern Serengeti woodland habitats in distribution of body sizes, locomotor types and dietary adaptation (Andrews 1989). Harris (1985) presents a picture of open grassland with scattered trees, with evidence of migration into or through Laetoli of a great diversity of species at the onset of the Pleistocene rainy season. Different ash layers have preserved what he tentatively interprets as traces of a resident dry season fauna (lagomorphs, guinea fowl and rhinos) with other layers perhaps corresponding to a wet season influx (larger bovids, equids, and elephants). The Olduvai Pleistocene faunas differ more markedly from those of present day Serengeti grassland and woodland habitats, especially in the dearth of species in the 10–45 kg range, in the relatively high proportion of small mammals with locomotor adaptations for low vegetation and ground-dwelling niches, and in the high proportions with insectivorous and grazing dietary adaptations (Andrews 1989). The distinctive Olduvai faunas may indicate a different and as yet poorly understood habitat, or represent some bias in preservation and recovery of fossil material.

Table 2.5 *Summary of Olduvai Gorge palaeontological record (after Poirier 1987)*

Bed	Age (years BP)	Record	Site types
Upper Beds	100 000–400 000	*Homo sapiens* Tool remains	
Bed IV	400 000–700 000	*Homo erectus* Acheulian hand axes and cleavers	
Bed III	700 000–1.2 million	No fossils Few artefacts	
Bed II	1.2–1.6 million	*Homo erectus* and *Homo habilis*	1. Thin layer of debris on old land surface indicating living or occupation site 2. Artefacts associated with large mammal skeleton or groups of smaller animals indicating butchering or kill sites 3. Artefacts and faunal remains dispersed through thick layer of clay/fine grained tuff 4. Occupation debris incorporated in filling of former river or stream channel
Bed I	1.65–1.8 million	*Australopithecus boisei* and *Homo habilis* tools and living site	
Volcanic	1.9 million	Sterile	

The first major find from Olduvai Gorge was in 1959 when Mary Leakey found the skull of *Zinjanthropus* (now renamed *Australopithecus boisei*). The discovery and naming of two species of *Homo*, *Homo habilis* and *H. erectus*, from the period of half to two million years BP, gave further evidence of the wealth of early hominids. Thirty-three fossil hominid specimens have been recorded from Laetoli, and the discovery of hominid footprints preserved in volcanic ash from some 3.7–3.5 million years BP has helped establish the origin of the human bipedal gait. Descriptions of the Laetoli hominids and discussions of their importance to our understanding of human evolution have yet to be published. Olduvai has a series of four primary lower beds (I–IV) and three upper beds (Masek, Ndutu and Naisiusiu beds) ranging from 1.9 million years BP to 14 000 years BP in Naisiusiu (Table 2.5).

Olduvai lower bed I (1.9 my BP) had a much wetter climate than the area experiences today, with a 10 km soda lake along much of the gorge. This bed provides evidence for the earliest known formalised stone tool industry – the 'Oldowan' culture. Potts (1988) has investigated the established view that Olduvai sites represent 'home bases' for early hominids, and after an analysis of geological, ecological and archaeological evidence puts forward the possibility that the Olduvai sites represent stone caches where hominids processed carcasses for food. The 'Oldowan' culture developed through increasingly sophisticated forms until Bed IV, when it was progressively replaced by Acheulean culture. The Ndutu beds have a middle Stone Age industry using advanced microliths and the Naisiusiu beds a Later Stone Age industry using advanced microliths. Lake Lagarja has several sites of Acheulean complexes dating back half a million years (Mturi 1976).

Nasera Rock (Apis Rock) on the edge of the Gol Mountains has a middle Stone Age shelter with artefacts up to 20 000 years BP (Mehlmann 1977), but also has evidence of pastoralist occupation some 2000 years ago. Lake Eyasi to the south of NCA has several archaeological sites. The Mumba shelter redescribed by Mehlmann (1979) has a Pleistocene stone age culture of some 30 000 years ago plus a burial site and associated pottery from 2–5000 years ago.

Other NCA sites dated within the last few thousand years are discussed in more detail in the next chapter, which outlines what is known of the earliest cultivators and pastoralists to have used NCA. Quite apart from their inherent scientific interest, all of these sites add to the perceived conservation values and viewing attractions of NCA. They also have a role in current management debate. It is important to realise that Ngorongoro has had a human impact for thousands of years, and that this has been essentially a pastoralist impact for up to 2000 years.

Conclusion

The dramatic geology and topography of NCA, the high productivity of its rangelands, and the resulting spectacular density and diversity of large herbivores, are in sharp contrast to the great majority of sub-Saharan rangelands. NCA covers all ecoclimatic zones ranging from the Afro-alpine through humid to semi-arid and arid. The composition of the individual plant communities is representative of wider East African rangeland types: their close juxtaposition, and their potential for flexible exploitation by mobile opportunistic grazers, are unique.

Archaeological and palaeontological remains are of international importance. They give evidence of human use of the area over thousands of years, and prehuman hominid use of the area dating back 3.5 my or more. The present day landscape and wildlife populations have evolved alongside and under the influence of human groups. For the last 2000 years the main inhabitants have been pastoralists.

3

Maasai of Ngorongoro

Iltung'ana loo ngishu The people of cattle.
(Maasai describing Maasai: Galaty 1982)

The Maasai, their herds and the ecological implications of their land use are central to the theme of this book. While much of our research concentrated on plant and animal communities, it is the interactions of these plant and animal systems with people that are at issue. The key to understanding Ngorongoro is the fact that pastoralists and their herds have been part of the ecosystem for millennia. The ecology of the area is bound up with the Maasai and their land use.

This chapter sets out to give some idea of what it is like to be a Maasai pastoralist in Ngorongoro. Starting with a description of Maasai settlements in NCA, and of the pattern of daily life there, we follow the Maasai herds through the changing seasons in a year's cycle. This material is mainly drawn from our own observations gathered while living in or near Maasai settlements in NCA. We explain the structure and function of the settlements and set out the social systems of section, clan, and age-set. These are fundamental to the way Maasai society deals with the opportunities and constraints presented by their environment, and as such have inescapable implications for the issues of population, subsistence and development covered in later chapters.

This chapter goes on to explore the prehistoric origins of pastoralists in East Africa and of the Maasai in Ngorongoro. This helps to explain the interrelations between the Maasai and neighbouring peoples. It also provides the background to the political, legal and administrative issues within NCA that are covered in chapter 4.

Maasai life in Ngorongoro

(a) The Maasai homestead

Maasai homesteads (singular *nkang*, plural *nkangitie*; Fig. 3.1) consist of a number of houses, built around central holding corrals for stock,

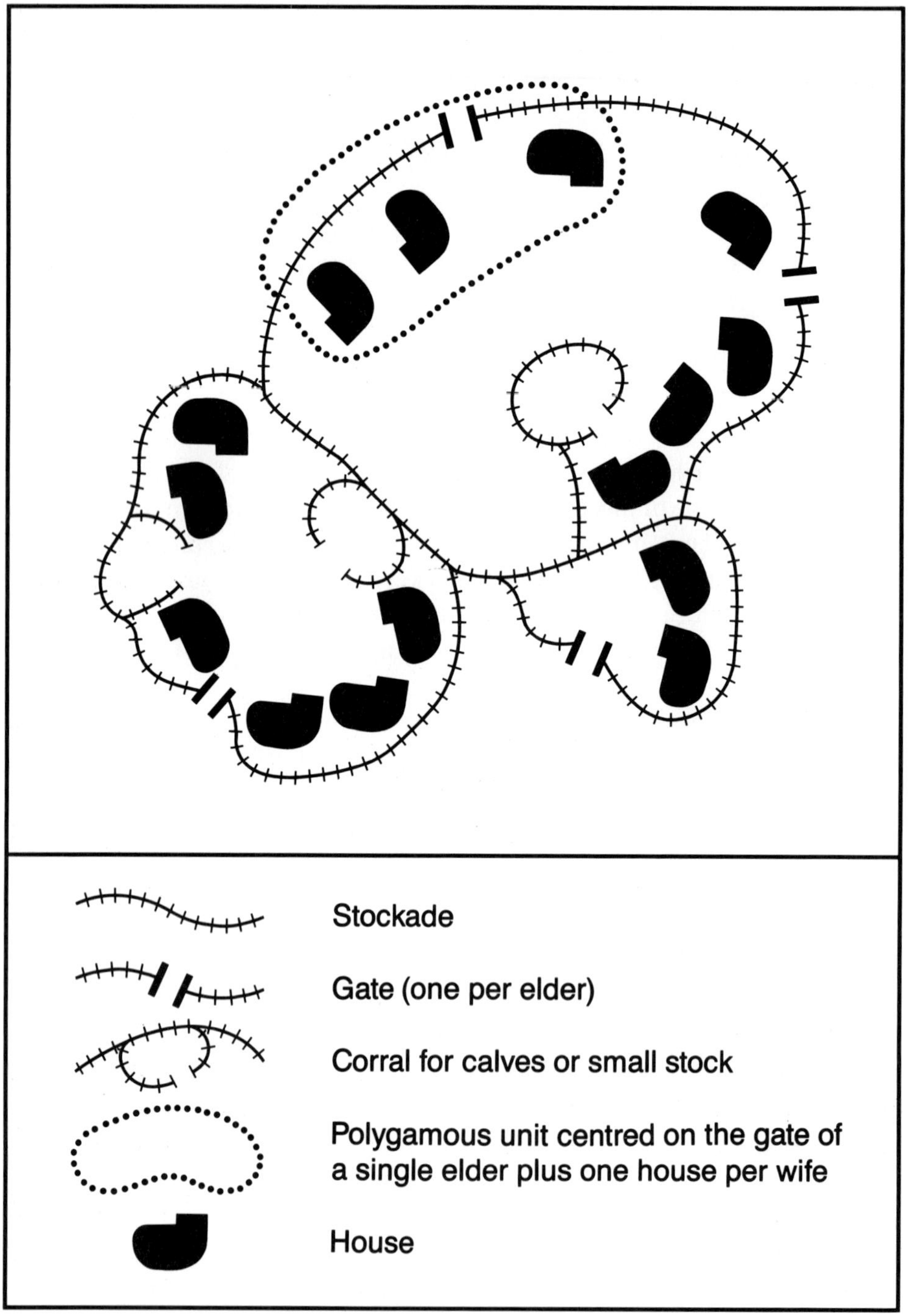

Fig. 3.1. Maasai boma.

and encircled by a continuous stockade made of thornbush or logs depending on the local vegetation. In this book the Swahili word boma is used to describe these homesteads, as this is a term commonly applied to them throughout East Africa. The boma may have one or more openings through which stock enter and leave. There is usually one such gate for each herdowner in a permanent cattle camp, so the number of gates generally indicates the number of independent families associated with the boma. Each gate thus serves one household, that is to say a domestic group centred on the husband/father as stockowner and head of the family, and including his wife or wives and their dependants. At the time of our study two-thirds of married elders in NCA had more than one wife. Individual households usually had two or more component houses, with each wife building her own house for herself, her children, dependants and the intermittent presence of her husband. Such polygynous families rarely live alone: the Maasai ideal is to belong to a boma comprising several households which cooperate over grazing decisions and herding labour, and help each other in time of need. Bomas are flexible in structure and composition. They are looseknit and adaptable groupings of broadly cooperative but largely independent and self-sufficient families:

> At night [the boma] is a stockade against intruders. During the day, with the herds dispersed and visitors coming and going, it tends to merge with the wider local community. Over the months, with families migrating typically perhaps twice a year, its composition changes. Newcomers move into empty huts and establish homesteads, and then quite often move on while others stay indefinitely. While they are together, [boma colleagues] share meat . . ., search together for any lost animal before it gets too dark, and lend children temporarily for herding . . . Spencer 1988:15

In the long run the family and locality are of greater importance as a social network than the boma, but bomas traditionally do not consist solely or even predominantly of members of one kinship group. Fosbrooke (1948) saw it as exceptional for a Maasai approaching elderhood to continue to live with his father; commonly he would claim his patrimony and move (possibly with his mother, if his father so wishes) to form a separate gate in some other boma. Nestel (1986) describing the Kajiado Maasai states that eldest sons settle in the father's boma; she found that younger sons preferred to set up independent bomas with agemates, but Spencer (pers. commun.) queries this. The individual bomas in our study illustrate a range of possible cases (chapter 5). They were made up of the families of a man and his married sons (Ole Senguyan's boma in Sendui), of several married brothers (Ole Lekando's boma in Ilmesigio), or in one case of otherwise unrelated agemates (Andrea Lesian's boma in Nasera, Gol). In each of these areas closely adjacent bomas (and therefore the associated grazing) were occupied by different descent groups, and elders from neighbouring bomas cooperated in herd management, grazing decisions, and allocation and supervision of herdboys. Day to day collaboration over livestock management thus tends to operate across, as much as within, kin groups in the short term, though in the longer term it is kin groups, especially close family, which form the lasting basis for cooperation (Spencer pers. commun.).

In addition to more permanent bomas, other sites are occupied seasonally during movements to regular wet or dry season ranges. Temporary bomas are also built to provide shelter for a few days or weeks in more arid areas where herd movements are less predictable.

(b) The Maasai day

Before dawn the woman of the house gets up and makes up the fire which has been smouldering through the night. As day breaks she goes out together with any younger girls of the house to begin the milking. Calves are brought out one at a time from the calf corral or, in the case of very young calves, from the house, and led to their mothers. They are usually allowed to suckle from two teats and the milker relies on the letdown stimulated by the calf to milk the other two teats into a gourd. It is considered greedy and wasteful to strip too much milk from the cow: the need to allow the calf enough milk for healthy growth outweighs short-term desire for more milk to feed people. After the milker has taken enough from any one cow she allows the calf to continue suckling from its mother and fetches the next calf. Quite commonly some of the older calves break out of their separate corral during the night and suckle their mothers dry: these cows are not milked. The milking may go on for an hour or an hour and a half. During this time the men get up and join the women and cattle in the main corral, looking over the animals, grooming them for ticks, inspecting and treating any wounds, finalising the day's grazing plan, and giving the herdboys their instructions. The woman of the house leaves the girls to finish the milking and prepares breakfast in the house, often cooking up a thin porridge of milk and maize flour for the herdboys and other children. The wife prepares milk, milky porridge or tea for her husband and any of his guests; warriors are given pure milk as is considered correct for their status (Talle 1990). The younger herders of eight or ten who may be away until evening with small stock or calves are given a calabash of milk or porridge to take with them, as are the few children walking long distances to attend dayschool at the 'village' centre. Once they have taken their breakfast, older herders commonly go without eating until evening, or may sometimes milk an animal during the day.

By now it is broad daylight, and the thornbush or log barrier sealing the gate is dragged aside. The cattle leave one by one, older cows followed by their adult daughters with their own weaned offspring clustered round them. Cattle from one household mingle with others to be herded together during the day. The small stock (goats and sheep) leave next and, after cows have gone, the calves are let out to be herded separately.

The family herds are thus commonly divided during the day into three separate grazing units of adult cattle, young calves together with sick or injured animals, and small stock. Each group may be pooled with the equivalent unit from other families using the same boma or based nearby, making the most of available herding labour. Depending on the location and time of year stock may walk only a few kilometres, or may spend most of the day on the move. Adult cattle are herded the greatest distance to grazing (and to water; chapter 8). The herd is commonly led by a belled ox, the sound of the bell keeping the animals together. A particularly greedy animal is often chosen, so that it will keep searching for new grazing and keep the herd on the move (Spencer pers. commun.). Special areas of grazing are reserved near the boma for calf use (chapters 8, 9). Small stock can make use of areas that are of limited value to cattle. When pasture and water conditions are good, herding may mean little more than the presence of the herder. The herdboys may spend long idle days talking, playing, and seeking out wild bee hives for their honey, edible fruits from wild plants, or good straight stems for herding sticks and play spears. On the other hand, herding may mean complex and energetic direction and retrieval of stock in steep and rocky terrain where animals could easily injure themselves, or controlling hundreds of thirsty cattle for hours in the sun while other herds are watered before them. These tasks need the help of older youths or men of the warrior or elder age-sets. Goats may be taken to browse in ravines or woodland thickets along watercourses, where leopards, buffalo or elephant can be a danger to stock and herder alike.

After the herds have left the women and girls first clean their gourds with burning charcoal from aromatic woods that give Maasai milk its characteristic and locally prized flavour. Other household utensils may be cleaned with cow's urine if water is not easy to come by. The women gather fresh dung from the corral and mix it with mud and cow's urine to plaster any parts of the house that need repair. The house is a traditional construction with a mud and dung plaster over an oblong or oval frame of poles and a hard mud floor. It is divided into a central living and cooking space, two or three sleeping compartments with raised platforms of poles and hides, and a partitioned area for young stock. The interior is dark and smoky, and approached through a low tunnel entrance. The structure discourages flies and mosquitoes and maintains a relatively even interior temperature despite hot days and cold nights.

Women walk to fetch water from sources up to 5 km away in gourds, metal or plastic containers holding a few gallons each. Some bomas may have donkeys to carry the water. On other days women walk similar distances to fetch back headloads of firewood. They carry their babies with them though older children may be left with co-wives or grandmothers in the boma. Any spare time is spent sitting in the shade of the hut wall or stockade, apparently untroubled by the clouds of flies that thrive on the concentration of dung in the boma. This is the time for tending small children, talking, and making the beadwork ornamentation for gourds, skin garments, jewellery and belts for themselves and their warrior sons.

Warrior youths and elders may be helping with any difficult or long distance herding that is to be done, but often the young men are away from the parental boma, living and travelling in warrior groups (see section on age-sets below). Elders meet up to talk with other herdowners in their own or neighbouring bomas, discussing matters of stock management, marriage settlements, disputes and compensation payments, and the education and initiation of their sons. They may leave for a day's journey to a nearby trading centre, or for longer trips to arrange stock sales and grain purchases.

In the late afternoon and evening the herds return. Usually the calves get back first to graze for a while on rich manured pasture close by the boma (chapters 6, 8). Herdowners take up position outside the stockade for the satisfying experience of watching their animals come home. The common grazing herd divides up as it nears the boma, and each household's animals enter by their owner's personal gate. The calves are followed later by the small stock and finally the adult cattle. Smaller calves are penned up to be let out one by one for the milking. The camp becomes a turmoil of lowing and bustling as calves try to rejoin their mothers while women attempt to control them so as to milk the cows first. Eventually the boma settles into calm as all the cows are milked, calves suckle, and cattle lie down to ruminate. Calves and small stock are shut in their respective corrals for the night. Women prepare the evening meal of milk or porridge, and children gather in each hut to be fed. Children vacate the hut when elders or warriors enter. The herdowner is served with any of his visitors, warriors commonly taking their milk in a group apart.

After the evening meal there is talking and storytelling. A wife may sing for her husband: if he is moved he may pledge her an animal to be held in trust for her sons. Warriors visiting a cluster of bomas may gather to dance together, or split up to flirt with unmarried girls and grass widows. Spencer (pers. commun.) points out that they have to be very discreet. Other elders are very watchful concerning the comings and goings of warriors (*murran*), especially to huts where they have no business. They can go to the huts of mothers-of-moran (see below), to their own sisters or to the huts of the wives of their own married age-mates. These last are the only grass widows with whom the *murran* may associate. Children curl up to sleep in the hut of their mother or foster-mother. Eventually everyone retires to shared sleeping quarters in the hut of wife or mother or, for visitors, the hut of an age-mate's wife: in this case the husband should sleep elsewhere (Spencer pers. commun.). In some cases visitors use a vacant hut or the hut of an older woman whose children are grown. Here they are billeted several to a sleeping platform. The hut is dark except for the fire that burns low through the night. There is the occasional commotion of a dog trying to break in to steal water from the small store in the hut, or young stock trying to break out to get to their mothers. Often lion and hyenas move and call nearby at night.

Within the household there is thus a marked division of labour (Talle 1988). Women are responsible for milking (2–3 hours a day), fetching water and wood (1.5–2 hours a day), cooking, child care, house building and maintenance. Spencer (1988:14–15) points out that women also have charge of the cattle at night and must be first to attend any calving or commotion in the yard. It is the junior wife by the elder's gate who is responsible for opening it up at night to any late visitor. Young boys and girls carry out the herding tasks; a few children attend school. Around puberty girls assume more of a woman's work role while boys graduate to more demanding stock herding work and eventually (after circumcision) the military and social pursuits of the junior warriors (see below). Men marry from their mid-twenties on to girls of 13–16 years old. Married men control livestock management, grazing and marketing policies, and political and religious affairs. Adult men have a considerably lighter physical work load than do women, though warriors and elders carry out difficult herding and watering work that is too complex or physically demanding for younger herders.

Wealth is reflected in the size of the herd and the family rather than in material possessions (Arhem 1981a). Individual households have little in the way of equipment: handmade (usually homemade) milk gourds bound and stoppered with leather; leather bags, skirts and belts; purchased or inherited knives and spears; alminium cooking pots, cups and spoons. Cloth is worn by men and women, blankets by elders. Women make and use bead jewellery. The wealthiest households in NCA may also possess a torch, radio, trousers, shirts and shoes, but cash is typically converted into grain purchases and stock holdings rather than goods.

(c) The seasonal cycle

The NCA Maasai are transhumant: they show a regular movement of their herds between dry and wet season pastures. Maasai transhumance centres on a permanent homestead based in a drought refuge area with permanent water and lasting pastures. In NCA such homesteads are typically sited in highland or intermediate altitude areas (the Crater Highlands or the Gol mountains), though elsewhere they may be associated with lowlying permanent swamp or lake areas such as Amboseli in Kenya.

As the dry season progresses the available forage within a day's return trek from the permanent boma deteriorates. The Maasai may set fire to the dry, coarse long grass. Burning releases nutrients that stimulate a flush of new growth, and destroys dormant or free-living stages of livestock parasites that have accumulated in the heavily used dry season refuge highland areas. At such times herding becomes a difficult and exhausting task. The herds may leave before dawn and return after nightfall, trekking long distances each day through dust and ash to find forage. Their browse consumption rises as preferred grasses become hard to find, and they are watered on alternate days only. This may involve arduous treks down steep scarps or rocky clefts to permanent waterholes, with little opportunity to feed on watering days. Small springs nearer the homestead may be reserved for domestic use and for calves. With so much time spent on the move, and so little food, the herds rapidly lose condition. They produce little milk for human use, and the households come to rely largely on grain imported from outside NCA in exchange for stock, and to some extent on meat and blood from male stock. If the dry season is prolonged, weaker animals die; even young calves may have to be weaned. Cattle from highland permanent bomas may be moved, as conditions worsen, to temporary stock camps so as to graze glades in the high altitude forests.

As the rains begin to fall, the arid plains show local flushes of new growth. At first these are restricted to stormtracks but gradually, as the rains progress, they coalesce into thousands of square kilometres of nutrient-rich short grass sward. Depressions, gullies and seasonal streams fill with water. It is time to shift the herds from the dry season ranges higher up, where conditions are deteriorating for the livestock as the insect vector populations multiply, dormant and free-living stages of parasites emerge and increase, and the cold and wet foster both respiratory and hoof infections. The main herds are moved down to temporary stock camps on the plains, and part or all of the household moves with them. A few members of the family, old people and women with young children, may remain at the permanent boma with a small number of milking animals. Down on the plains stock thrive on the high quality, mineral rich, short grass pastures, and many calve at this time. Milk is plentiful and makes up most or all of the pastoral diet (chapter 10). Many families commonly return to the same wet season camp several years running, but it is by no means necessarily made up of the same families that have been sharing a dry season homestead. In the more arid parts of NCA the herds may be moved through a series of temporary stock camps, making use of any patches of forage in reach of surface water. As the rains come to an end, surface waters dry up and pastures are grazed out. Eventually herds are moved back towards the permanent homestead where they will last out the dry season with permanent water and lower quality but more durable long grass pastures of the intermediate and high altitude zones.

Transhumant patterns cut across administrative boundaries and regularly take NCA Maasai beyond the borders of the NCA. Conversely, in serious drought the Crater Highlands act as a refuge that draws Maasai from a wide surrounding area. We can only describe general patterns of movement here: these are very flexible in response to considerations of climate, disease, and local, national or even international politics. Ideal patterns of transhumance are also modified by administrative bans, wildlife interactions and security problems (chapters 8, 10). Conservation bans have denied the Maasai a number of wet and dry season pastures and watering points (Olduvai Gorge; Eastern Serengeti; Ngorongoro, Olmoti and Empakaai Craters; the Highlands Forest Reserve). The risk of contracting disease from wildebeest during their calving period blocks cattle from access to the short grass plains in the wet season (chapters 8, 9). At the time of our study, Maasai with permanent bomas in the Gol mountains kept their herds at the main homestead throughout the rains, using what had traditionally been dry season grazing, while wildebeest used the short grass plains. During the dry season, the Gol herds were deployed to temporary stock camps on the margins of the now dry and unproductive short grass plains, using the remnants of grazing left on outlying hills. They had to make frequent movements in a semi-nomadic rather than transhumant fashion. An ever greater proportion of NCA Maasai are having to withhold their herds from the short grass plains for increasing periods during the wet season as the wildebeest population grows, as the antelope migration covers an ever greater area of the plains and as their calving extends over a cumulatively longer period. This has implications for both nutrition and disease transmission in livestock and ultimately for Maasai welfare (chapters 8, 9, 10). A third constraint that modifies ideal patterns of range use in NCA is the problem of raiding and stocktheft by Sukuma and Barabaig, which has caused the Maasai to retreat from the southwestern borders of NCA (chapters 9, 10).

Maasai social organisation: implications for human and livestock ecology

Traditional systems of section, clan and age-set provide the framework for social and economic cooperation, particularly as represented by livestock management, transactions and redistribution. These social structures have been described by a number of authors (eg. Fosbrooke, 1948; Jacobs 1965, Spencer 1988). As pointed out in the most recent and authoritative account, there are at least seven incompatible descriptions of Maasai age-set formation and eleven different versions of the clan system (Spencer 1988:2, 7). Different accounts tend to be incompatible perhaps because they refer to different Maasai groups, which possess variations on a similar social structure (see below, also Berntsen 1979, Galaty 1982). A brief and generalised review of the main features of Maasai society that pertain to ecology and management is given here, as well as some features characteristic of the Maasai of Ngorongoro.

(a) Section

The pastoral Maasai are subdivided into a number of sections (*ol oshon*: Fig. 3.2), entities with both social and spatial meaning. Fosbrooke (1948) described Maasai sections as territorial groups functioning as units in war, in allegiance to a particular religious leader and in performance of age-set and other ceremonies. Similarly, Jacobs (1975) presented Maasai sections as corresponding to loose territorial areas and possessing autonomous political structures. Spencer (1988) sees the Maasai as a federation of largely self-contained tribal sections:

> Most families never move or marry out of their own tribal section, but there is broad approval of those that have gone further afield within the Maasai milieu. Distant marriages cement the ideal of a more powerful union to which [the sections] belong . . .
>
> Spencer 1988:18

Maasai from any one section do not graze their stock in another section's territory unless exceptional drought or disease conditions force them to seek special permission to do so. Modern attempts to register group ranch boundaries that infringe section borders have led to murderous disputes between rival and warring sections in Kenya (chapter 10; see also Galaty 1980). Within NCA, wet and dry season movements take place within section territories, but inevitably cut across present day administrative boundaries. The modern Maasai sections, their geographical locations and their relative importance are partly products of the wars, famines and epidemics of the 1890s and of the colonial period that followed (Fig. 3.2; see also later in this chapter, chapter 10 and Waller 1988).

Each section is divided into localities (*enkutot, enkutet* or *inkutot*). Some authors present these as corresponding to watershed basin divisions demarcated by physical barriers such as a waterless plain, a wooded scarp or belt of tsetse bush (e.g. Jacobs 1965). To some extent localities may represent ecologically self-contained systems covering wet and dry season grazing and permanent water supplies sufficient to support regular transhumance in most years. However, others see the section territory itself as the smallest unit that could be seen as usually self-contained in ecological terms (Nestel 1986). Spencer (1988:15) presents the locality as often corresponding to a water catchment area, but stresses its importance more as a level of social organisation than a geographical or ecological unit. The locality

> has no boundary, but there is a collective rapport and a sense of identity that [the boma] normally lacks . . . Collectively the local elders of an age-set, of a clan or of the community at large assume responsibility to resolve relevant issues . . . Any recent immigrant from another part of [the section] has the right to participate in this decision-making . . . migration between localities . . . keeps alive the sense that in their decisions they all represent the interests and unity of [the section] as a whole Spencer 1988:15–18

Within a locality, neighbourhood clusters of bomas control local grazing, and individual families may maintain priority rights over particular springs and wells.

(b) Clan

Clan relationships are inherited from the father. Individual families usually acknowledge and remember at most a lineage of three generations' depth, back to the father of the oldest living man, and clan membership embodies the idea of common origins further back in the father's line. Jacobs (1965) recorded seven clans and 68 subclans, though he suggested there were probably around 100 of the latter. It is probably misleading to attempt a rigid classification. For example, marriage generally takes place between members of different clans, but where a clan has grown to be very large there may be intermarriage between subclans (e.g. Spencer 1988:19).

Clans are not organised as local geographically cohesive groups, nor are there formal clan leaders. Members of individual clans are dispersed across different sections throughout Maasailand, and thus provide a potential network of influence and social obligation alternative to the section system. However, Spencer (pers. commun.) sees the clan as considerably less important among the Maasai than for example among Samburu or Turkana. Perhaps its most important expression among the Maasai, apart from its constraints on marriage, is that new immigrants to a locality rely on clansmen already living there to help them settle in. Particular localities may develop a concentration of a particular clan.

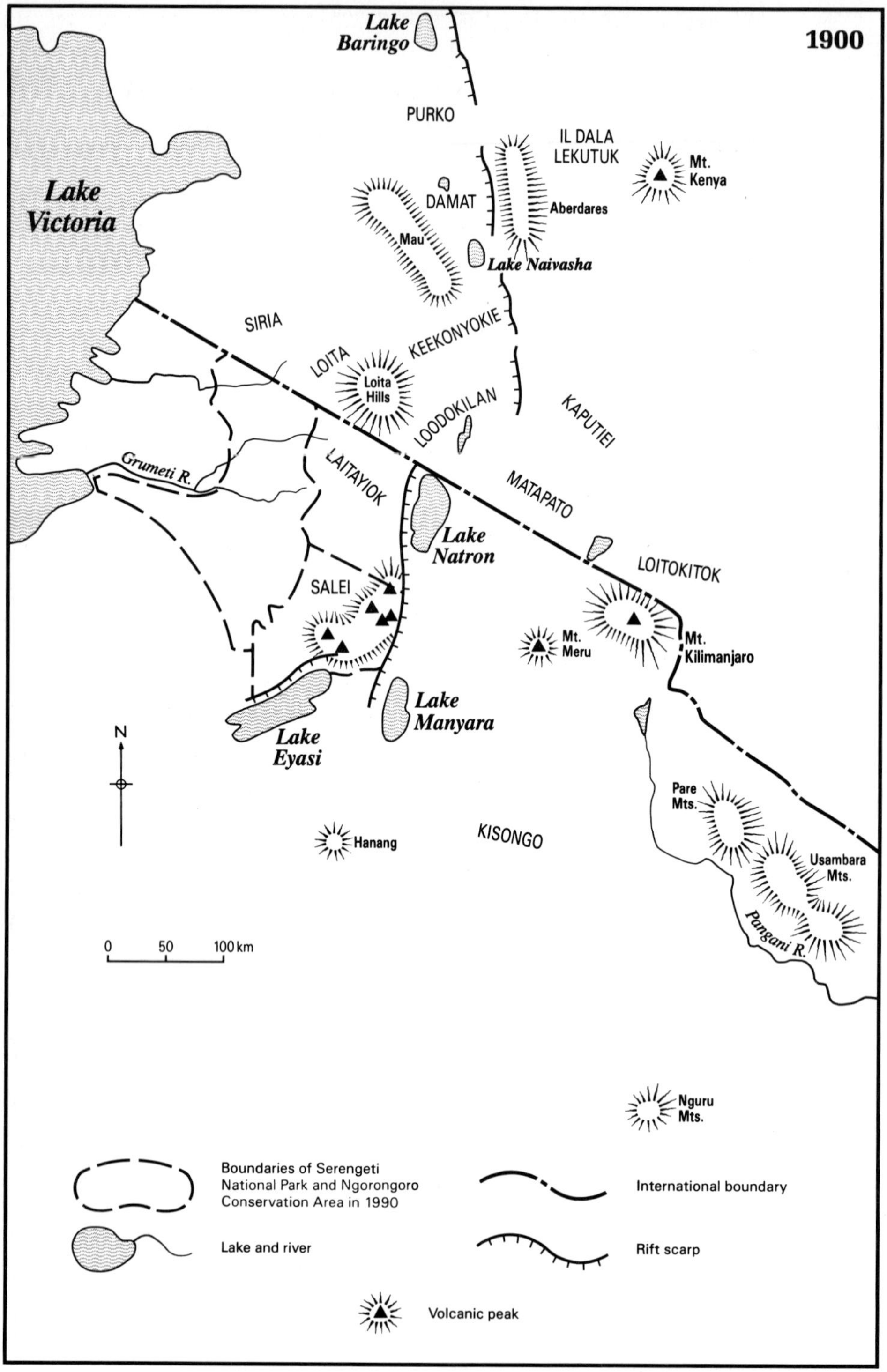

Fig. 3.2. Maasai sections in 1900 and 1990 (sources: Waller 1976, Nestel 1985, Jacobs 1965). 1990: 1. Serenget, 2. Purko, 3. Salei, 4. Kisongo (The Loitokitok (4a) make up a subsection of the Kisongo), 5. Laitayiok*, 6. Loita, 7. Siria, 8. Damat, 9. Keekonyokie, 10. Loodokilani, 11. Kaputiei, 12. Dalalaketok, 13. Matapato, 14. Sigirari*, 15. Moitanik, 16. Uasin Gishu. *The status of the Laitayiok section, its relationship to the Laitayiok clan, and its extent and distribution are not clear. Some

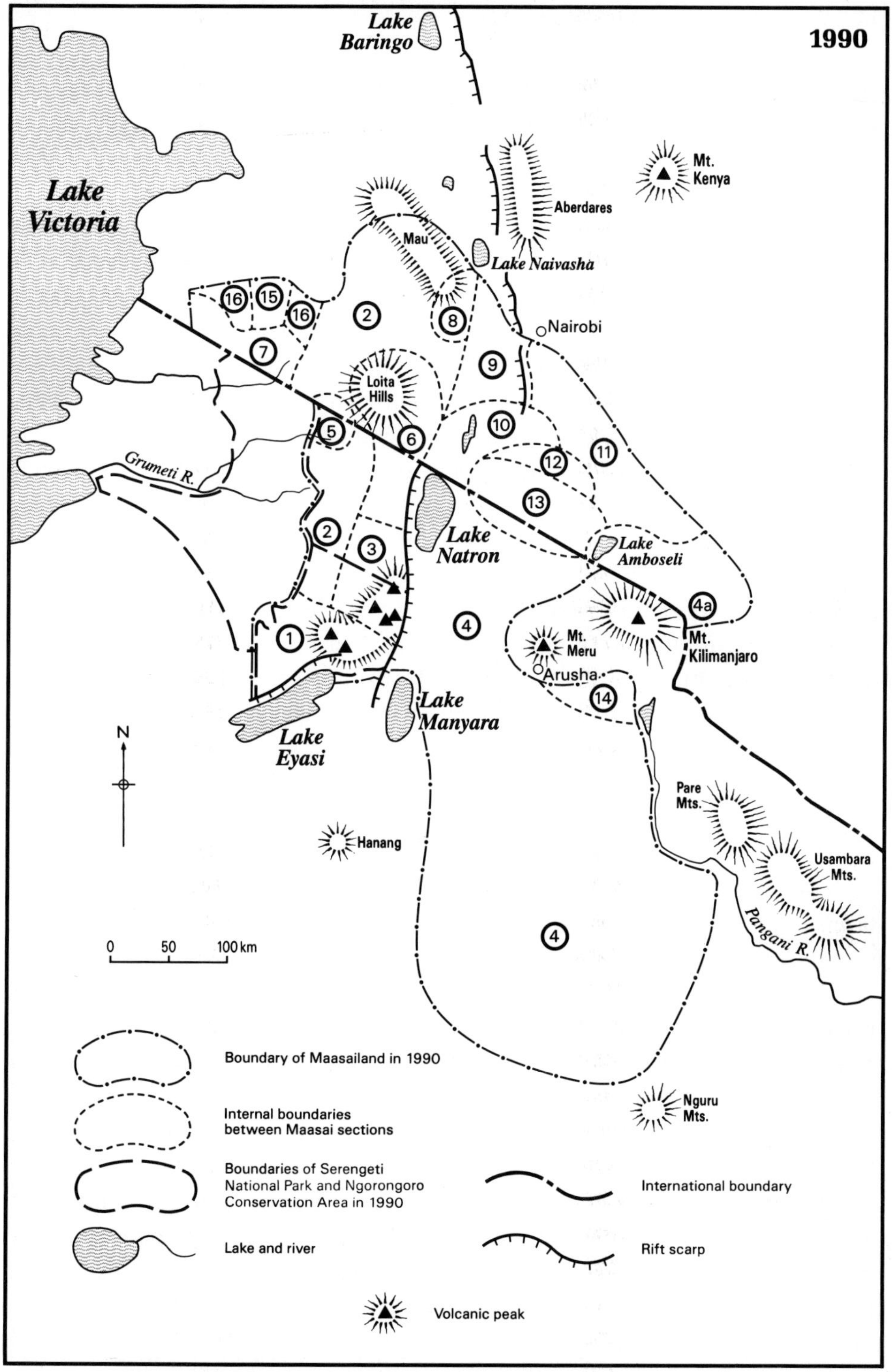

authors (eg Waller 1988) sees the section as destroyed and dispersed by the Iloikop Wars and surviving only as clan members scattered throughout Maasailand. Nestel (1985) includes it on her map of sections. Spencer (pers. commun.) makes a clear distinction between the Laitayiok clan and the Laitayiok section, which he states survives though in much reduced form. Spencer also queries the status of the Sigirari as a separate section, while Waller (1976) accepts them as such.

The role of clan membership in access to and control over livestock may have been overstressed by earlier authors (Fosbrooke 1948, Jacobs 1965). The Maasai household may be seen as a joint stock company, with the father as chief shareholder (Spencer 1988 quoting Merker 1904). It is stretching a point to claim that the clan 'owns jointly all cattle in possession of clansmen' (Fosbrooke 1948). Immediate family, rather than extended clan, help a man to meet his marriage settlement. To marry, a man must command enough cattle to make appropriate gifts to the girl's father (and build up the preliminary marriage debt that binds the father to him as stockfriend), to pay the bridewealth, and to provide the new wife with her allotted herd of one bull and eight heifers (see Spencer 1988 for a detailed account). The process of accumulating and presenting the stock to complete the marriage settlement takes place in stages over a long period. While in other pastoralist groups it is a major undertaking which depends on negotiating help from clansmen and stock friends (see e.g. Gulliver 1955 for the Turkana), Maasai marriage settlements involve far fewer animals and need less support. A Maasai boy gradually accumulates animals held for him in his mother's herd, whether allotted at the time of her marriage, born from allotted cows or acquired later as gifts from his father and other close relatives. Spencer (1988) gives an account of the formal occasions when such gifts are made among the Matapato Maasai. The household head has the power to decide most family stock transactions, with wives and sons being strongly interested parties. This stands in contrast to the situation in many other groups where stock transactions may be seen as clan rather than simply family business. In practice even within the Maasai family there are likely to be conflicts of interest, and the ideal of common ownership at family level is subject to manipulation by powerful individuals for their own ends. At the same time a man's reputation and ultimately his prospects and long-term security depend on the extent to which he is seen as a responsible manager of integrity and good judgment in handling stock and negotiating their transfers.

The cattle of the allotted herd and their offspring are for the wife to hold in trust for her sons. She is also assigned the care of some of her husband's hitherto unallocated cattle held in reserve for his future marriages. These cattle are hers to milk for the time being, but the husband decides on their management (e.g. which animals to castrate, which bullocks shall be exchanged for heifers, and whether any should be reassigned to a different wife with more mouths to feed). In time of need a married woman may be forced to beg animals from her natal household, rather than from her extended clan. Thus, households left destitute by drought or disease can call on long-term stock loans from their close family, and possibly but not commonly from clan members, to rebuild the nucleus of a milking and breeding herd. Stock loans commonly take the form of a cow with calf, which is left with the borrower for a specified number of calvings. When the cow is later returned, the maturing offspring become the property of the borrower.

Fosbrooke (1948) states that when clansmen of the husband come to beg cattle among the NCA Maasai, animals traditionally may be found from his own reserve, or begged from one of his wives and later repaid to her. If close kinsmen of a wife come begging, animals should traditionally be found from her own herd, or from the husband's reserve, but not from the herd of another wife. However, Spencer (1988) describes Matapato case studies that show the husband can do this, at some risk of opprobrium from family and community. He may even sell a wife's allotted cow in the best interests of the family as a whole. The one thing he cannot do is take a cow allotted to one wife and give it to another. Fosbrooke (1948) points out the implications of this system of rights over livestock for any attempt at introducing development schemes. For example, as individual women have complete control over the distribution of their milch cows' yield, no dairy projects could or should be implemented without their advice and consent (chapter 12).

(c) Age-set

The Maasai age-set system has local variants, as do the section and clan systems. Spencer (1988) gives a detailed analysis of the sequence of rituals, and the social and political importance of the age-set system in the Matapato Maasai. The age-set system has played an important historical role in the resilience and adaptability of Maasai society (Waller 1988). Fosbrooke (1948) gives an account of a Kisongo Maasai age-set cycle, but this particular case is in some respects atypical (Spencer pers. commun.). We describe only the main features, particularly those which have a bearing on human ecology, stock management and land use.

Put simply, a man's life is divided into a series of phases in which he and his peer group are progressively promoted from herdboy to warrior to elder. It is a sequence that spans a man's lifetime and that is of overwhelming importance in the changing roles it imposes, the powerful ceremonies through which it is expressed, and also in terms of the unfolding political, economic, social and ecological education and opportunities it opens up for the individual. Adjacent age-sets are locked into lifelong political and ritual opposition: the promotion of a junior age-set entails the ceding of privilege by the senior, and at the stage of opening a new warrior age-set this opposition is expressed in violent physical confrontation. This opposition results in alliance between members of alternate (next-but-one) age-sets, a feature that is particularly marked among the southern Maasai such as the Kisongo Maasai found in NCA (Spencer pers. commun.).

Briefly, boys are herders until around the age of puberty. Every fifteen years or so, depending on the strength of the emerging group relative to that of the reigning warrior age-set, each section holds a ceremony 'opening' a new age-set. This is followed by a period of years during which all boys of suitable age are circumcised. The sequence of events and the main ceremonies are shared within any one section. Rarely, as after the catastrophic years of the turn of the century (see below), several sections come together for a shared initiation. The circumcised boys spend some months as initiates. As they grow in numbers, physical strength and efficacy as a group they come progressively to challenge the established warriors more and more. The initiates are eventually promoted to become recruits in the new age-set of junior warriors (*il murran* or moran):

> Moran occupy a cherished position, associated with the reputation of the Maasai as a warrior people. Everyone in different ways is enchanted with the ideal of moranhood as a climax of male virility. Boys look forward to this period eagerly; elders hark back to it; girls look to the moran for lovers; young wives are suspected; mothers of moran dote on their sons' position; and the moran themselves bask in this limelight. They are held to excel . . . all others physically; and the symbol of their coveted position is a set of privileges . . . denied to boys.
> Spencer 1988:68

The period of recruitment into a *murran* age set is limited by pressures from either side: both from the age set immediately senior to them and, increasingly, from the uninitiated boys who are themselves potential recruits. The boys come to resist initiation because later recruits are relatively disadvantaged, facing a probably undistinguished career as the most junior members of the age-set, with a brief moranhood and an earlier transition to elderhood. The age-set immediately senior to the recruiting *murran* group is inevitably bent on limiting the numbers and power of the current *murran* and, at the same time, on promoting the interests of the uncircumcised boys for whom they will eventually act as patrons when a new cycle of circumcision and initiation begins. Once recruitment into a *murran* age-set is curtailed by the curse of the junior elder age-set immediately above, the older uncircumcised boys now have the prospect of becoming the most senior in the next *murran* age-set.

The new *murran* traditionally set up a warrior camp (*manyatta*) to protect their locality. During the formation of a *manyatta*, the *murran* stage a ritualised raid on their own parental homesteads and 'snatch away' married women – their own or other *murran*'s mothers – and cattle from the father's herd. The 'mothers-of-moran' set up hearths and huts in the *manyatta*, where each will provide a sort of maternal home base for the *murran*.

During their time as *murran* individuals with exceptional personal qualities emerge as natural leaders within a group which generally emphasises egalitarian solidarity, *manyatta* unity and loyalty, communal decision making, selflessness, generosity, and physical courage. The *murran* (together with their patrons from the next-but-one age-set above) select a respected and talented *murran* for the highly responsible, difficult and unwelcome position of manyatta spokesman (*olaigwenan*). He is then responsible for mediating between his own and other age-sets, establishing and maintaining *manyatta* standards and discipline, and liaising with the *laibon* (prophet or ritual expert) over religious and ceremonial matters. He chairs the political, ritual and military life of the *manyatta*. Other *murran* emerge as advocates or councillors who ensure the democratic nature of debate and decision making in the group.

During their time in the *manyatta* the young men compete in feats of physical prowess. Of these the hand-to-hand lion hunt is one that has captured popular imagination and is still commonly undertaken. Spencer (1988) describes individual hunts and local variations of the ritualised acts of bravery that they involve. The *murran* are exempt from regular herding, but are traditionally responsible for protecting the homesteads and livestock of their section. If any cattle are stolen or lost, *murran* are called on to retrieve them. They are also responsible for undertaking complex or long-distance stock movements. Any cattle theft implies raiding by neighbouring sections. The *manyatta* is then roused to pursue and punish the thieves, recapture the cattle and (at least in the past) to raid others to replenish its own section's herds.

The *murran* spend much of their time travelling, visiting distant bomas, helping with herding, and feasting in the bush. Spencer (1988) sees this mobility as maintaining a dynamic communications network that keeps the *manyatta* well informed as to the deployment of bomas and any threats from other sections or tribes. *Murran* may briefly visit elders' bomas, and can then take milk from the gourds set aside for the elders (but not from those of young children). The elders treat visiting *murran* with an uneasy mix of tolerance and suspicion: lifting milk can easily extend into seducing wives and pilfering small stock. *Murran* ask for animals which should be given under obligation in recognition of their status and role. The *murran* are expensive to feed: they live on milk, meat and blood, and still hold frequent traditional ox-feasts in the bush, away from *manyatta* and boma.

Murran visit young unmarried girls towards whom they act as protectors, friends and often lovers. The *manyatta* ideal emphasises comradeship and sharing among *murran* and also among the girls, discouraging exclusive relationships and any undue closeness between couples. Strong friendships emerge but these relationships rarely lead to marriage. Girls are married at the behest of their father, and usually become wives of an age-set senior to the *murran*. An illegitimate child will jeopardise a suitable marriage to an established and trustworthy herdowner. However, mutual liking and respect established in the *manyatta* can be the basis for lifelong friendships and reciprocal aid between men and women (Spencer 1988:114).

The climax of moranhood comes with the *Eunoto* ceremony which finally establishes the age-set. Spencer describes the process for the Matapato Maasai, where after the *Eunoto* ceremony there is a formal disbanding of the *manyatta*.

While the recruiting phase of the age-set is closed there is a period during which the next emerging age-set must come to assert itself both physically and socially before its own initiation can begin in turn. This progresses to the level of a prolonged demonstration, and eventually a dancing festival. The boys begin to muster like warriors for a raid, and finally their sponsors (members of the junior elder age-set who curtailed recruitment into the current *murran* group) lift their curse and ceremonially open a new age-set.

The timetable is now set for current *murran* to graduate. Over the next couple of years they are forced to cede their privileges to the emerging age-set. By their late thirties or mid-forties they have become junior elders. By now they are expected to have married, settled and to concern themselves with herd and family management and the education of their sons. The *Olngesher* ceremony, which is of particular importance among the Kisongo, completes their transition to elderhood and gives the age-set its formal name.

The actual sequence of the age-set cycle varies from one section to another: Fosbrooke (1948) describes a system whereby youths spend about seven years as junior *murran*, seven years as senior *murran*, and seven years as junior elder before progressing to senior elder status. This implies a total of around 14 years as *murran*. However, Fosbrooke's account describes the only case in recent history when Kisongo followed the pattern found in northern sections. According to Spencer (1988):

> A new age-set is established about once every fifteen years, and it is given a name common to all Maasai at the *olnegesher* ceremony, when its members pass the final hurdle to elderhood. Before that point is reached there are certain conventions concerning the organisation of the moran that vary betwen tribal sections. Among the Kisonko and related sections in the south, for instance, young men remain active moran for an extended period and there is a major transition from one ageset to the next about every fifteen years. In . . .

> most northern Maasai tribal sections there is an elaboration of this basic pattern in which each age-set is divided into two successive sub-age-sets. These are the 'right-hand circumcision' and the 'left-hand circumcision', whose names vary between tribal sections . . . As opposed to the Kisonko pattern, there are *two* successive age cycles during each fifteen year age-set period . . . until their *olnegesher* ceremony, which binds them formally into a unified age-set.
>
> Spencer 1988:94–5

The older age-sets with greater accumulated experience control the younger, and the elders of the locality in consultation with age-set spokesmen form the ultimate decision-making body. Men who emerged during their *murran* years as exceptionally wise, level-headed and even-handed are particularly influential, as are those whose later stock management and political skills build up large cattle holdings. The agemate relationships developed during the *murran* years are often the basis for the choice of boma colleagues. They allow alternatives in times of local or regional disaster, or when circumstances make continued coexistence with current boma colleagues difficult. On an individual level, the progression through the age-grade system is not just an education and a career but rather a shared and all-absorbing process in which

> the richness of ceremony and its associated beliefs forms the background of an unfolding emotional experience, which perhaps accounts for its tenacity and certainly is expressed vividly by the [Maasai] themselves Spencer 1988:7

(d) Implications of social structure for human and livestock ecology in NCA

Maasai social structures have evolved alongside the land use patterns of subsistence stock-rearing in arid and semi-arid rangelands. Environmental conditions do not dictate specific social structures, but successful and lasting social structures do develop so as to buffer the problems posed by the environment. Everyday life and social relationships are expressed through the idiom of herding and cattle. The wealth of descriptive terms for different kinds and conditions of stock becomes a vocabulary of respect and affection for kinsmen and stockfriends. The word that is used for herding stock (*a-irrita*) also means looking after a family: it

> implies the deft use of a stick to keep the herd firmly under control, and it is in this sense that [an elder] is expected 'to look after' his family . . . The metaphor of herding even works its way into their style of debating when the speakers ply their sticks to marshal

> together the strands and counterstrands of their arguments as though they were cattle to be manoeuvred. Spencer 1988:14, 18

Chapters 8 and 9 summarise the ways in which livestock management by subsistence pastoralists is adapted to the special conditions they face. Here we briefly discuss the way that the age-set system, and the section, clan and boma structures are inextricably bound up with livestock management.

> A boy's upbringing is geared towards the unremitting care of stock. From the time he can barely walk, he may try his hand at herding, clutching an upraised stalk as he tumbles towards the smallest stock as if to round them up. When he is first left by himself in charge of calves near the village, he has to learn to master a situation that constantly tends to get out of hand, as the calves spread out in search of grazing. His father or his older brothers teach him to control the herd and to respond to any situation that may arise, developing an awareness of the opportunities and hazards of the bush. The more he can be trusted, the greater his responsibility. As he becomes more involved with mature stock further afield, he is left to interpret the broad directions given to him in the morning, and to make his own judgments during the day.
>
> The skill of the herdboy is confined to this daily experience, and does not extend to decisions concerning migration or survival during a serious drought, which clearly demand the experience of an older man. Even so, this wider understanding grows out of the intimate knowledge of cattle, goats and sheep acquired during boyhood.
> Spencer 1988:51

During their progression from herdboy, to member of a roaming warrior peer group, to elder, Maasai men accumulate a wealth of information first on livestock and then on layout and potential variation of forage, water, minerals and disease risk of any area within a wide radius, under different conditions from year to year and season to season. Their impressive store of environmental knowledge is illustrated by the fact that the Maasai identify several hundred rangeland plants on a folk classification that corresponds closely to Western scientific species classification, but which goes further to distinguish ecotypes of individual species (cf. Heine *et al.* 1988; Homewood 1990, see also Niamir 1990). Individual plants are known for their properties as forage, fuel for different types of fire, building materials for different construction purposes, and materials for making utensils and containers, rope and string, weapons and herding sticks.

During their time as *murran* the young warriors establish a strong age-set identity. The relationships they form with their agemates mature into stock friendships that are later the basis of settlement site choice and risk avoidance mechanisms. The *manyatta* ideals of sharing and of democratic debate and decision-making later come of age in the handling of communal grazing and water resources, and the settling of community affairs through the elders' council. The *murran* become familiar with a wide geographical area, knowledge useful in later years of stock management when decisions as to long distance migrations may be necessary. The progression from herdboy to warrior to elder allows for maturing herdowners progressively to acquire skills, responsibility and eventually power in animal husbandry, herd management, military and finally political fields. Traditionally the age-set system ensured the maintenance of a mobile force to police grazing lands and fend off cattle raids (as well as carrying out such raids to increase the family herds). While its military role is to some extent obsolete, the importance of the age-set organisation as an education, and in establishing social values and networks, as well as its intense personal importance to the individual, are as real as ever. The age-set system also still provides a built-in mechanism for successive generations to adapt and respond to changing conditions, whether ecological, socioeconomic or political (e.g. Waller 1988), through an institution of formalised opposition to preceding age-sets and with the formalised support of older and strongly influential sets (Spencer 1988).

Section and locality systems maintain a form of communal land tenure which gives individual herdowners a degree of territorial security as well as the chance to shift grazing area in case of local drought or epidemic. In the short term, flexible boma composition (in terms of both numbers and kinship) allows rapid response to current local constraints and opportunities for stock management. Family obligations make some provision for redistribution of stock to those with legitimate needs. They establish a system of wealth store, credit, investment and insurance against risk where livestock are the main and most reliable currency.

(e) New social structures: Maasai villages in NCA

All the social structures described so far in this chapter retain fundamental importance in the life of Maasai in Ngorongoro. However, new frameworks of organisation have been imposed by outside agencies and to some extent these operate alongside the traditional systems. Since the mid-1970s the Maasai population of NCA, and of the rest of Tanzania, has been organised for administrative purposes into 'villages' (chapter 10). Each 'village' comprises a loose cluster of individual homesteads or bomas scattered over a wide area. These 'villages' are the pastoralist version of the Tanzanian *ujamaa* socialist settlements. They represent a political and administrative unit imposed by the process officially termed 'villagisation'. They are a Tanzanian phenomenon, and do not correspond to the Maasai villages described by Spencer (1988) for the Matapato Maasai in Kenya, where the term is applied to what we call a homestead or boma here. In NCA, 'village' centres were established by setting up dispensary, primary school and/or stock dip in or near previously existing trading centres, and in some cases by forcing homesteads to move closer. Overall, these 'villages' have had little lasting impact on patterns of settlement and seasonal movement, nor do they correspond with traditional economic or leadership structures. Individual families still live in widely dispersed bomas. Seasonal movements crosscut village boundaries and different families using the same village in the dry season may move to different wet season pastures, each associated with quite separate alternative villages (see chapter 8 for patterns of transhumance). Alongside the imposed village structure, the traditional social systems of section, clan, age-set and boma, still govern NCA Maasai access to resources and form the basis of their risk avoidance strategies and of their efficient livestock management in an unpredictable environment. The new villages represent the official structure through which education, health services, local government, law, and political representation outside NCA should all operate. However, their effectiveness is severely constrained by the internal administrative and legal structures of the NCAA (chapter 4).

Maasai origins and the prehistory of pastoralism in NCA

Losopuko l'Ol Tatua

(Maasai name for Ngorongoro – literally 'Highland pasture of the IlTatua')

The previous chapter summarised evidence for the occupation of NCA by hominids over some 3.5 million years. For almost all of this time people were gatherer-hunters, living off a variety of wild animal and plant species, as do many Hadza and Dorobo today in the vicinity of NCA. With the advent of peoples possessing domesticated crops and animals, cultivation and livestock husbandry came to dominate the Rift Valley. Here we look briefly at the origins of pastoralist peoples in East Africa, the emergence of Maa-speaking peoples, and the history of the Maasai over the last few centuries, leading up to their presence in Ngorongoro today.

(a) Archaeological and linguistic evidence for the origins of pastoralism in East Africa

The earliest remains of domestic stock in East Africa date back to around 3500 years BP, from north Kenyan sites left by herders of sheep and goats. These people also cultivated sorghum and millet, crops of indigenous African origin domesticated from ancestral wild cereal species in the Ethiopian highlands (Robertshaw and Collett 1983a; Phillipson 1977, 1985). With the economic advantage conferred by farming and herding they seem to have spread gradually throughout East Africa as far as the Southern Rift during the pastoral Neolithic (the first millennium BC). The most recent account is given in Robertshaw (in press).

Pastoralist peoples are identified in the archaeological record by a combination of one or more types of evidence such as remains of domesticated stock, characteristic pottery and other artefacts including stone bowls and small querns, and site location by ecoclimatic zone and soil type (Robertshaw and Collett 1983a). The earliest pastoral neolithic sites in present day Maasailand date back to around 2700 BP. The Serengeti has stone bowl sites of a largely hunting people with later evidence of domestic stock (Bower 1973, Bower *et al.* 1977, Bower and Nelson 1978, Robertshaw and Collett 1983 a, b, Collett 1987). Ngorongoro Crater has a number of stone bowl culture sites which probably date from within the last 2000 years (Fosbrooke 1972) but have artefacts comparable with those from south Kenya, dated at 2800 BP (Leakey, M. 1966). Ehret (1974) suggests the earliest signs of pastoralism in the area of Ngorongoro are around 2000 BP. A rough date of 2000–2500 BP seems appropriate for the emergence of pastoralism in NCA.

Linguistic evidence has contributed a great deal to unravelling the origins of East African peoples including the Maasai (Sutton 1974, Ehret 1974). All East African languages are derived from four distinct groups: Khoisan, Cushitic, Bantu and Nilotic (Greenberg 1963). All these linguistic groups have modern representatives in the vicinity of NCA, a phenomenon not found elsewhere in East Africa. The Nilotic linguistic group has three branches, of which the Plains Nilotes include the Maasai, together with other Maa-speaking peoples like the Samburu as well as the Karamojong-Teso speaking peoples like the Turkana. A second branch of the Nilotic group, the Highland Nilotes, include the Datog or Barabaig who live immediately to the south and southwest of NCA, as well as the numerous Kalenjin-speaking peoples of Kenya. Present-day descendants of the Cushitic linguistic group include the Southern Cushitic Iraqw language spoken by the people of the Mbulu plateau bordering the Ngorongoro Highlands, while modern Eastern Cushitic languages include Galla, Somali and Rendille. Present-day Bantu-speaking groups around NCA include the Sonjo and the Chagga who both have close links with the Maasai. The Khoisan linguistic group is still represented in East Africa by the Sandawe language and perhaps that of the Hadza who live to the south of NCA.

(b) The prehistory of pastoralism in East Africa

The interpretation presented here is largely based on the synthesis of archaeological and linguistic findings made by Ehret (1974), Sutton (1974) and Phillipson (1985). Recent work queries the timescale of events and the nature and extent of cultural influences Ehret has inferred from linguistic data (e.g. Lamphear 1988) as well as the relative importance of Nilotic and Bantu elements (e.g. Oliver 1982) but the broad patterns and sequences are not in doubt.

Before the advent of farming and herding the inhabitants of East Africa were gatherer-hunters speaking a Khoisan language. The earliest East African herders and farmers to appear in the archaeological record are thought to have spoken a Southern Cushitic language. From both archaeological and linguistic evidence they are thought to have originated in or near the Ethiopian highlands. As they spread southwards through East Africa during the first millennium BC they interacted with the Khoisan-speaking gatherer-hunters already living there. With the mingling of physical ancestry went a mingling of subsistence practices, and of languages, which left few remnants of the Khoisan languages or of the exclusively gathering and hunting way of life in East Africa, though the Hadza people retain strong elements of both. By 2000 BP Southern Cushitic speaking farmers and herders dominated Kenya and northern Tanzania, centred on the Rift Valley but stretching from Lake Victoria to the Indian Ocean and from the Kenya highlands to central Tanzania.

From linguistic, cultural and archaeological evidence, the first millennium AD was a period of interaction, assimilation and mingling as well as differentiation of new groups. While the Southern Cushitic communities were coming to dominate East Africa, the Nilotic linguistic group was already differentiating into three subgroups: the Plains, Highland and River–Lake Nilote language clusters. The ancestral Nilotes had a probable area of origin between Lake Turkana and the Ethiopian highlands, are thought to have kept and milked cattle, and may have possessed a cycling age-set system. During the first millennium AD groups of cattle-keeping ancestral Highland Nilotes spread southward and westward from their northern cradleland. During this period ancestral Plains Nilotes are thought to have been differentiating into the Maasaian group and the Karamojong-Teso group, ancestral to the Turkana among others. At the same time Bantu-speaking peoples spread into East Africa from the west and southwest, bringing new subsistence practices and cultural values centred on cultivation with planting of root crops and bananas. This spread may have been aided by their introduction or rapid adoption and use of ironworking, and their consequent efficiency in clearing and digging. The first millennium AD is often referred to as the pastoral iron age: by 1000 AD, some (Phillipson 1985) or possibly all (Sutton 1974) of the pastoral groups in East Africa had adopted iron implements. This period saw the emergence of strong specialisation into herding and farming, as well as archaeological evidence of considerable intercommunication throughout the region.

During this period, people of the Highland Nilotic linguistic group spread through much of present-day Maasailand, absorbing the Southern Cushitic communities that had preceded them. By around 500 AD the Highland Nilotic Dadog of northern Tanzania had differentiated from the related Kalenjin in west-central Kenya. From around 1000 AD onwards the Maasaians began to spread southwards through Kenya. By 1000–1500 AD the Dadog had occupied a large area of north and central Tanzania including the Crater Highlands. Oral history related by present inhabitants of the area tells how cultivators were chased out of the Loita Hills into Tanzania by the Dadog or Iltatua people prior to 1400–1600 AD (Jacobs in Odner 1972). By this date Highland Nilote Kalenjin groups had absorbed most of the South Cushitic speaking peoples of west-central Kenya, and probably extended as far as the Uasingishu Plains around present-day Eldoret.

(c) Pastoralists in Ngorongoro 1500–1850 AD

The Dadog (also called Watog, Datog, IlTatua or Tatwa) may have made some of the burial sites in the Crater (Fosbrooke 1962). They are thought to have had areas of cultivation on the Crater Highlands, and the present centuries-old mosaic of forest, shrub and grassland, for example on the slopes of Empakaai, may be due to their occupation and cultivating activities (Fosbrooke, pers. commun.; Frame 1982). Engaruka, on the eastern border of the Rift Wall, shows clearly a sixteenth-century system of irrigated cultivation using locally concentrated runoff (Sutton 1978, 1984). It has been tentatively attributed to the IlTatua, and provides evidence for large scale settled agricultural and pastoral communities in the past (Anderson 1988:242).

Between the sixteenth and eighteenth centuries the IlTatua seem to have been displaced or assimilated by a number of other groups from most of their wide territories. During this period the Maa-speaking peoples expanded, their influence extending from Lake Turkana to the north where the Maa-speaking Samburu border on the Turkana and Rendille, southward throughout the Rift Valley area and modern Tanzanian Maasailand. As part of this expansion, Maa-speaking peoples swept the remnants of the IlTatua from Ngorongoro, perhaps as early as 1600 AD (Tomikawa 1970). Anacleti (1975, 1978) also describes the Maasai driving Bantu cultivators away from 'rich cultivable land' probably near Ngorongoro highlands in about 1700 AD. These dates are much earlier than is generally accepted: Fosbrooke (1948) suggested about 1850 for the Maasai occupation of the Crater Highlands, and his chronology is still commonly quoted (eg Borgerhoff-Mulder *et al.* 1989). The Maasai themselves, as opposed to the wider category of Maa-speaking peoples, are generally thought to have undergone much of their rapid expansion during the nineteenth century (Waller 1979, 1990:84–5). The Maasai still refer to the Crater Highlands as *Losopuko l'Ol Tatwa* (the highland or dry season pastures of the Tatua). The modern remnants of the IlTatua, the Barabaig or Datog, now occupy much reduced territories south and southwest of NCA.

By the nineteenth century, Plains Nilotes dominated the Rift Valley from north Kenya to central Tanzania as far south as Gogo country around Dodoma, and the Maasai themselves had territories stretching from the Uasingishu Plains and the Laikipia Plateau in north-central Kenya through Ngorongoro to central Tanzania. On the fringes of the area that has come to be called Maasailand, peripheral Maa-speaking peoples bordered on other Nilotic and Bantu-speaking groups, with the Samburu adjacent to the Turkana and Rendille, the Il Chamus neighbouring Kalenjin peoples around Baringo, and the Arusha and Il Parakuyo coming up against Bantu peoples respectively to the east and south of Tanzania Maasailand.

(d) Interrelations of Maasai and other East African peoples

Eitingo isotuatin pukin neashunye ole nkaputi

When you marry, in-laws dominate your life (Maasai saying)

While these events left people with languages of specifically Bantu or Nilotic origin predominating in particular areas, these do not correlate with any separate physical ancestry or any single origin of cultural or subsistence practice. For example, the Southern Cushitic languages have all but disappeared in East Africa today, but this linguistic group made a considerable contribution both to the physical ancestry and to the cultural and subsistence practices of modern East African peoples. They contributed loanwords to the vocabulary of many modern East African languages, not only to cattle-keeping groups like the Maasai, but also vocabulary concerning herding, milking and other animal husbandry practices to some modern groups that are primarily Bantu-speaking cultivators. They also probably bequeathed a number of cultural practices such as circumcision and clitoridectomy to a wide range of groups throughout East Africa, and directly or indirectly to the modern Maasai. The Maasai language belongs to the Plains Nilote linguistic group, but the people speaking it represent a fusion of Plains Nilote together with Highland Nilote, Cushitic and Bantu influences in their physical ancestry, culture and subsistence practices, and loanwords from all these groups enrich the Maasai language.

The popular notion of the Maasai embodies an appearance and way of life so distinct from the cultivating peoples of East Africa that it is tempting to assume they form a self-contained ethnic unit with a separate origin. This is misleading. In many cases people that are now or were until recently identified as belonging to separate tribes not only share strong common elements in their ancestry but still show considerable movement from one group to another. The Maasai people show close and dynamic links not only of barter, but of language, intermarriage and social structure with a number of other Maa-speaking groups whose subsistence depends primarily on farming or hunting rather than cattle-keeping. These groups have until recently been labelled as quite separate tribes, but current understanding of their history and origins, as well as of the continuous drift of families into and out of herd-owning status, shows them to be inextricably bound up with the Maasai pastoralist system. The Dorobo (hunters), the Ilumbwa and the Nguruman, Nkuruman or Ilkurrman (cultivators) were originally taken as distinct tribes (Jacobs 1965), but these terms are now seen as describing subsistence categories rather than ethnic units (Spencer 1973:199–206, Berntsen 1979, Galaty 1982, Waller 1985). Stockless Maasai may settle with such groups, sometimes as a temporary measure to build up the capital to acquire new stock (Berntsen 1979), though these days destitute Maasai are more likely to seek wage labour in the townships (Spencer 1988). Terms such as Iloikop and Wakwavi are also markers of social distance, applied reciprocally between various groups of Maa-speaking peoples (Galaty 1982, Rigby 1985), rather than denoting distinct tribes as has commonly been assumed (Jacobs 1965). Together, all these groups form a wider political, economic and social network of interdependent and related people of whom the pastoralist Maasai have long made up only one aspect.

The processes of interaction between ancestral groups and their spread through East Africa probably took place largely by peaceful assimilation, but sometimes they must have involved more violent competition and conflict for prized resources. With the specialisation into herding and farming that took place in the Pastoral Iron Age (Robertshaw and Collett 1983a, b) certain areas would have acquired particular importance for cattle-keeping groups. For example, the Crater Highlands have a theoretical carrying capacity two to five times greater than that of the surrounding dry plains and woodlands (Pratt and Gwynne 1977), as do swamp areas such as Olbalbal. Such drought refuge areas form the core of pastoralist transhumant systems in sub-Saharan Africa. Their significance is not only for the vital dry season grazing and water resources they contain but also for the strategic control they give over surrounding wet season pastures and major stock routes (Waller 1979). Maasailand has for centuries included a number of such drought refuges: the Loita Hills, Nguruman

Escarpment and Ngong Hills as well as Baringo and Amboseli swamps in Kenya; the Loliondo Hills, Ngorongoro Highlands, Mts Kilimanjaro, Meru and Monduli in Tanzania. The permanent flows of the Pangani and Tarangire Rivers and Manyara streams have acted as temporary refuges until grazing was finished (Waller 1979). Such areas have long formed the focus of competition and conflict between rival cattle-keeping groups, whether between Maasai and non-Maasai, or between warring Maasai sections.

(e) The Maasai in Ngorongoro 1850–1910

The nineteenth-century history of the Maasai is known from oral history, from the accounts of contemporary observers (Merker 1904, Hollis 1905), and most recently from the detailed and vivid reconstruction of events in Maasailand from the late nineteenth to the early twentieth century that Waller presents (1976, 1979, 1985, 1988, 1990). It was a turbulent time in terms of intra-tribal strife and ecological perturbations. Various Maa-speaking sections allied together under Supeet and later his son Mbatian, each in turn a *laibon* of exceptional political influence. The allied sections fought against weaker sections for the control of grazing lands. The systematic military operations of the allied groups assumed the status of warfare rather than small scale raiding. By 1880 some previously distinct Maasai groups had been destroyed in these 'Iloikop Wars' (see below).

Following on these wars the European colonization of Africa introduced a series of previously unknown pests and pathogens, which had drastic effects on people with no prior exposure or resistance (Ford 1971, Kjekshus 1977; 'ecological imperialism': Crosby 1986). The famines and epidemics of this time had such an extreme ecological impact on the natural and human environment in Tanzania that this period is central to an understanding of NCA ecology, whether of vegetation, livestock, wildlife or human populations. In many parts of Tanzania, the removal of many of the able-bodied men to labour on colonial plantations or to fight in colonial wars combined with disease to bring about the disintegration of indigenous communities and production systems (Iliffe 1979). This latter effect was of limited importance in Maasailand (though see Read and Chapman 1982) compared to the impact of disease. Rinderpest – a disease of ungulates – swept from Sudan to West and South Africa as the great cattle plague of the 1890s. Though known in Europe since the time of Charlemagne, it had never before penetrated Africa south of the Sahara. It had a catastrophic impact, killing 90% of the cattle and decimating many susceptible wild ungulate populations on the way (Ford 1971, Scott 1985; see also chapters 3, 7, 9). The rinderpest cattle plague was accompanied by a disastrous epidemic of smallpox (Waller 1988:79). To the Maasai these events were of such cataclysmic importance that they are still collectively remembered and termed *Emutai* – meaning complete destruction (Waller 1988).

Ecological repercussions are still evident. There is some debate about the extent of vegetation change (e.g. Waller 1988:104, Waller 1990) but there is reason to think that the decimation of ungulates and pastoralist peoples across sub-Saharan Africa allowed the regrowth of bush in huge areas previously kept clear by grazing and burning (Ford 1971). This in turn provided the habitat for tsetse vectors of trypanosomiasis to spread, and in many places the fly populations and consequent disease foci persist, denying cattle-keeping people reoccupance of the land. Perhaps more importantly, the decimation of human, livestock and wildlife populations disrupted previously established patterns of disease transmission between human and animal populations. This weakened the immunological defences both of people and of animals, as these often rely on early exposure to and challenge by a disease while the individual is still protected by maternal immune factors. When transmission was reestablished the effect was disastrous (Ford 1971). Whether because new strains passaged through different species altered to show a new unexpected virulence, or whether because the chain of immunological defence previously maintained from generation to generation had been broken, trypanosomiasis became a serious problem in some parts of Maasailand (Waller 1990). Herders' keen environmental knowledge of potential disease areas and risks was rendered useless, as was their management of rangelands and herds to minimise such risks. In areas such as western Narok, the Maasai eventually found themselves confined to pockets of safe grazing while large areas were given over to the advance of tsetse and trypanosomiasis, and were ultimately designated as wildlife areas like the present-day Maasai Mara (Waller 1990:83). Virulent strains of rinderpest in cattle around the Serengeti and Ngorongoro continued to depress both livestock and wildebeest populations through the first half of the twentieth century, killing many yearlings (see chapters 7, 9 for more recent events).

The net effect of the initial epidemics on a pastoralist people, with no knowledge of these new diseases, their prevention or cure, and no system of famine relief adequate to meet the unprecedented disruption, was one of total calamity. Early travellers in Tanzanian Maasailand in general (and Ngorongoro in particular) saw and recorded some of the effects of the first rinderpest outbreak:

> The ravages on the Masai herds were catastrophic and caused severe famine among a people who were deprived of their daily food supply. Abandoned villages were almost without exception the only trace I found of the Masai people. All of their cattle have been wiped

> out and the surviving people have taken refuge with the agricultural tribes at the edge of the steppe.
>
> Stuhlmann, quoted in Kjekshus 1977

> Large numbers of those woeful creatures who now populate Masailand congregated around the thorn fence of our camp. There were skeleton-like women with the madness of starvation in their sunken eyes, children looking more like frogs than human beings, warriors who could hardly crawl on all fours, and apathetic languishing elders.
>
> Baumann 1894, of a boma in Ngorongoro Crater, quoted in Kjekshus 1977

Baumann estimated that fully two thirds of the Maasai died in the famine and epidemics, and that their entire way of life was destroyed. Waller (1988) casts some doubt on Baumann's estimate, but shows how the drastic stock losses and smallpox epidemics had a continuing effect on the Maasai through their disruption of traditional processes of stock loan and redistribution.

The stock raiding that followed between different Maasai sections escalated to the status of civil war between the Purko to the north and the Loita, Kisongo and Arusha to the south. Maasailand became polarised between two of Mbatian's sons, the *laibon* Olenana and his Purko followers, who originally dominated north and central Maasailand, and Olenana's brother Senteu, a rival *laibon* supported by Maasai groups south of the border (especially the Loita, together with the Kisongo and Arusha sections). Waller (1979; 1988:80–2, 107) describes how the relative importance of the different sections and their influence over different areas changed during this time. By 1901 the Kisongo and Arusha Maasai had also turned against the Loita, who were pushed westward to Ngorongoro and Serengeti with the help of the British. The Purko, formerly based in central-northern Kenya, now occupied much of Maasailand north of the Kenya border. The Kisongo, originally the southernmost section, now straddled the border and occupied much of NCA. The smaller Serenget and Salei sections moved eastwards to the Serengeti. All these sections are still represented in NCA and the Kisongo remain the dominant Maasai section there today.

The period 1895–1905 was one of recovery and internal reorganisation, with Maasai in British-administered territories forming some degree of alliance with the colonial regime, and rebuilding their herds partly through stock payments in return for their role in pacification raids on other tribes (Waller 1976). Weakened by natural calamities and intersection raiding, and undermined by German punitive raids, Maasai south of the border fared worse. Read and

Chapman (1982) give a vivid semi-fictional account of the life of a Tanzanian Maasai warrior and mercenary living in and around Ngorongoro at this time. Senteu and his followers eventually surrendered to Olenana and the Purko. In 1903 the first *Eunoto* ceremony was held since *Emutai*, shared between the victorious and the defeated sections. During the 'Maasai Moves' of the next couple of decades the Purko and other smaller sections were moved from the northern half of Kenya Maasailand. Laikipia, Naivasha, Nakuru, Uas Nkishu and other Maasai were concentrated in the south of Kenya Maasailand by the colonial administration.

Modern Maasai sections are partly the result of these events and of the colonial period that followed:

> As families rebuilt their herds and moved back into their own areas, the Maasai sections were recreated in their present form. The spatial organisation which emerged after 1900 differed in some respects from what had gone before. New sections, like Loitokitok, appeared while others, like Laitayiok, had effectively disappeared as separate entities. Sections like Siria had developed a closer cohesion as a result of their experiences, while Kisongo was beginning to split up. The main casualty was the wider system of territorial alliances which had, for instance, linked Purko, Keekonyukie, Damat and IlDalalkutuk together . . . during the 'Iloikop Wars'. After 1900, this level of organisation atrophied and was superseded by the colonial administrative structure of section/location. Two general trends are observable: one towards a greater degree of localisation and a stronger emphasis on the section itself as the maximal effective level of allegiance and activity; the other towards the concentration of resources and power in the hands of a few dominant sections . . . The Purko are the obvious example . . . Waller 1988

Sandford (1983) believes that after rinderpest had moved on, the cattle populations increased to their previous numbers within 20 years. Certainly substantial recovery could have taken place in this time (Dahl and Hjort 1976) but human population increases would have needed a very much greater time period. Waller (1988:77, 1990:93) concludes that much of Maasailand was left uninhabited in the 1890s. No accurate human or livestock or indeed wildlife population estimates are possible for conditions at the advent of colonial administration in Ngorongoro. All that can be said with confidence is that human, livestock and wildlife populations were all severely reduced and disrupted, vegetation was affected in major ways, and that recovery is still taking place (chapters 6, 7, 9). Early observations in the area cannot be taken as representing any sort of stable ecological baseline, but give a significant picture of the background to subsequent development of legal and political structures through the colonial and post Independence period (chapter 4).

The Maasai, the NCAA and the nation

Currently there are thought to be some 23 000 Maasai in NCA (chapter 10) with around 140 000 cattle and 140 000 small stock (chapter 8). Pastoralist numbers are inherently difficult to census, as are those of their stock. The Maasai move in and out of the villages and administrative zones of NCA, and to and fro across the borders with surrounding areas, on a seasonal and annual basis. Boma composition is continually changing, and individuals or families move into and out of pastoralism depending on their success with herding and other enterprises. They hold no written records of births, deaths, marriages or migrations. The land where they live is hard of access, particularly to officials with limited time and transport for their survey, and the motivation of the pastoralists themselves to comply with any census is lukewarm at best. Livestock numbers can change dramatically over a few months or years as a result of deaths from drought or epidemic disease, migration, and natural increase (chapter 8). To make matters worse, Maasailand straddles the Kenya/ Tanzania border (Fig. 3.2).

Bearing in mind all these problems, recent estimates put the Kenya Maasai pastoralist population at around 180 000 (Evangelou 1984) and Tanzanian Maasai at around 100 000. NCA probably holds a considerable proportion of all Maasai, and one-fifth to one-quarter of all Tanzanian Maasai. Of the three main pastoralist groupings in East Africa, the Maasai cluster (including the Maasai themselves, the Samburu, and the related more agropastoral Il Chamus and Il Parakuyo) represents the largest, and its members have among the higher cattle *per capita* ratios (Jahnke 1982).

The Maasai of NCA are almost all pastoralists. There is little opportunity or incentive within NCA for wage labour: Arhem (1981a) estimated a total of 75 persons engaged in such labour out of the 18 000 Maasai in NCA at the time of his survey. Few Maasai are able to command administrative or other posts in NCA (chapter 4). Women have two minor but regular and widespread cash activities: the manufacture and sale of honey beer, and the collection and sale of resin. The latter activity was prohibited by the NCAA in 1980 (Arhem 1981a). With a declining resource base (see chapter 10), there is a shift not so much from a livestock-based economy to a cash economy, as to a growing number of poor Maasai who must seek sources of ready cash however short term. Some young Maasai seek cash in return for displaying their traditional dress, dancing, and other attractive aspects of their culture to tourists. Some may also see involvement in poaching as an easy source of cash.

There is a tendency for governments to see mobile populations of pastoralists as problem people who choose to evade rather than participate in the process of national development. Officials not uncommonly portray Maasai as backward, living in a primitive way, possessing considerable wealth in stock but unwilling to join in the national economy, and characteristically aloof. Early travellers' tales represented the Maasai as continuously waging war and 'dipping their spears in blood', and this image has persisted, as has the assumption that Maasai are isolated from the rest of the nation. Earlier sections of this chapter have set out the way in which the Maasai maintain close social and economic links with other Maa-speaking farming and hunting groups. There has also long been an interdependence between Maasai pastoralists and neighbouring non-Maa speaking cultures, and this continues today. There is a long history of bartering stock for grain, and of intermarriage, shared settlement and cultivation with the Sonjo, a small farming group to the north of NCA (Berntsen 1979). The border with the more aggressive pastoralist Barabaig and agro-pastoralist WaSukuma to the south and south west of NCA has long been and remains an area of skirmishing and cattle raiding (chapters 8, 10). To the east of NCA the Iraqw or WaMbulu people represent a market for livestock and a source of grain (chapter 10; and perhaps on occasion a source of raided cattle as well). NCA Maasai maintain informal cross-border trading routes into Kenya and show rapid response to the relative market conditions in the neighbouring nations (chapters 8, 10). Over the last few decades the dearth of transport and communications in NCA, and the high gate fees imposed on all vehicles, have created a state of isolation for the NCA Maasai. Recent developments have allowed an influx of tourists but have made transport and trade no easier for the ordinary Tanzanian, whether pastoralist or other. This is a quirk of the circumstances surrounding the development of NCA, rather than any deliberate choice by the Maasai (chapters 10, 11).

Summary and conclusion

NCA has a sizeable population of Maasai with a traditional pastoralist lifestyle. Their livestock exploit the same resources of pasture, water and minerals that are used by the wildlife, and share the same diseases with them. The culture of the Maasai, and their social and economic structures, make the most of potentially rich but unpredictable rangelands, and minimise the risks of periodic drought and disease. The last few hundred years have seen periods of powerful military alliances alternating with drastic decimation and disintegration of Maasai groups, and periods of relative affluence alternating with catastrophic epidemics, stock losses and famine migration. Maasai culture and traditional Maasai social structures are strikingly robust, adaptable and resilient to change. They survived the vicissitudes of decimation of the Maasai population, dramatic stock losses, and alienation of territory. Spencer (1988) attributes much of the enduring strength of Maasai traditions to the rich, all-absorbing, and fulfilling nature of their social system, particularly the age-set system. Part of this enormous resilience must also be attributed to their land use and stock management, and part to the close economic and social ties the Maasai share with other East African pastoralist, cultivator and gatherer-hunter peoples. The NCA Maasai are currently to some extent isolated by the way the NCA functions, but traditional networks operating across the borders of the NCA are as important as ever in trade, marriage, migration and more generally in the survival and success of the Maasai in semi-arid rangelands.

4

History, politics and perceptions in Ngorongoro

Meibok nkishu enkanyit Respect does not safeguard cattle
(Maasai proverb: Waller 1979)

The richness and diversity of its natural resources mean that NCA is of importance to many different groups, including pastoralists, potential cultivators, wildlife conservationists, archaeologists and the tourist industry. The history of the area is a saga of tussles between rival groups seeking to pursue subsistence, economic, cultural, academic or leisure interests. At the moment the two main protagonists are the pastoralists, wanting more land use rights and greater security, and the conservationists wanting to protect the rangelands and their wildlife populations from human and domestic stock impacts. The background to current management conflicts must be seen not only in terms of the biological and ecological components of the system, but also in terms of the perceptions and powers of different user groups and in terms of the legal and political structures through which they work. It is important to realise that supposedly objective scientific arguments can be, and have been, misused to justify political ends rather than to support biologically sound management measures (cf. Sandford 1983:15). Before dealing with biological questions and ecological research relevant to NCA management (chapter 5 *et sequen.*) it is necessary to outline the development of the legal and political structures that now govern management of the area. This chapter goes on to examine the conservation values as currently seen by local, national and international communities respectively. These differing perceptions of ecological, subsistence and conservation crises form the background to current management conflicts and complete the context for the design of our research.

History of conservation management in Ngorongoro

(a) The emergence of conservation in Ngorongoro/Serengeti (1910–1959)

The present conflicts over management go back a long way. The German Government gazetted the NCA forests in 1914 for watershed conservation. Wildlife legislation dealing largely with the hunting of game animals, and providing for their protection, was laid down in outline in Tanzania in 1921 (Serengeti Committee of Enquiry 1957; Kitomari 1985). In 1928 a Complete Game Reserve was declared comprising the Ngorongoro Crater as defined by the rim, but excluding two German settlers' farms on the Crater floor. All hunting except that especially prescribed was prohibited in the Reserve. In 1929 part of the Western Serengeti was added to this closed reserve and the boundaries were greatly extended in 1930. In 1937 all hunting of lion, cheetah, leopard, giraffe, rhinoceros, buffalo, roan antelope, hyaena and wild dog was prohibited (Serengeti Committee of Enquiry 1957). The boundaries were redefined in 1939 to include Serengeti, Loliondo areas and most of present day NCA, apart from the forests. Initially gazetted by the Germans, the Forest Reserve was ratified and demarcated by the British and administered under a separate ordinance.

The initial hunting bans together with minor restrictions on settlement construction, stock movements and range management had little impact on the Maasai of the Serengeti–Ngorongoro area at the time (Legislative Council of Tanganyika 1956, Ole Saibull 1978, Arhem 1985a, b). However, later they formed the basis of major curtailment of human activities.

In 1940 the Game Ordinance replaced the earlier Game Preservation Ordinance. This empowered the Government to create National Parks. The entire Serengeti Closed Reserve of that time, which included most of NCA, was declared a National Park (without any consultation with traditional residents). This legislation restricted entry to and residence in the park area, but excepted those born there or with 'traditional rights' from such restriction. In 1948 a new National Park Ordinance was passed, and in 1951 the Serengeti/Ngorongoro area became a park, with much stricter legislation. For example, burning 'either wilfully or negligently' became a criminal offence (though this proved impossible to police), and an amendment in 1954 withdrew the right to cultivate.

Attempts to enforce the ban on cultivation brought about the first major overt conflict between wildlife conservation and human interests. Relative to the scale of pastoralist activities there was little cultivation within the Ngorongoro/Serengeti area. However, because of the dependence of pastoralists on dry season dietary supplements (see chapter 10) those few inhabitants who were primarily cultivators and suppliers of grain found support from the pastoralist majority (who were also part-time cultivators of subsistence crops). The residents of the Park wanted cultivation to be allowed, subject to control of its extent by the pastoralists themselves, and further wanted guarantees of unrestricted grazing access throughout the Ngorongoro highlands.

Increasingly strict conceptions of what a national park should be, and the incongruity of a growing and developing human community in Tanzania's only park, led the administration to conclude that 'the continued presence of the Maasai and their stock within a National Park was irreconcilable with the purpose of the Park' (Fosbrooke 1962). In 1956 the Tanganyika Government proposed a modification of the park boundaries, releasing much of the Ngorongoro area from the Park. This created a furore among European conservationists and Professor Pearsall was commissioned to report on the area for the Fauna Preservation Society of London (Pearsall 1957). Ensuing discussions decided that all Maasai rights in the Western Serengeti should be extinguished and that the eastern area, including the Crater Highlands, should be excluded from the park and administered by a separate Conservation Unit of the Government.

During and after the enquiry, the demands of the cultivator and pastoralist residents were taken up by TANU as a grievance against the colonial administration represented by the Park Trustees. In 1959 a compromise was reached with the Ngorongoro Conservation Area Ordinance. This separated the 12 000 km^2 Serengeti National Park from the 8292 km^2 Ngorongoro Conservation Area (NCA). The NCA was to be administered by the Ngorongoro Conservation Unit, charged with conserving and developing the natural resources as well as allowing for human use compatible with wildlife conservation. This provided for extensive but controlled Maasai grazing rights, settlement and small scale cultivation at the discretion of the Conservation Unit. Meanwhile, the 1959 National Parks Ordinance forbade human interests in the Serengeti. Current boundaries do not correspond with the zones censused by Grant (1954), and this census was preoccupied with distinguishing between families that had been resident prior to 1940 from later immigrants. However, using a minimum estimate (Grant 1954:13; Western Serengeti zone) it is possible to deduce that around 1000 Maasai inhabiting the Serengeti at the time were forced to move, together with over 25 000 cattle and 23 000 small stock and donkeys. This represented about one-tenth of the Maasai population of the Serengeti/Ngorongoro area. They were offered rights and services within NCA as compensation.

(b) Ngorongoro Conservation Area: 1960–1989

Ngorongoro Conservation Area Authority received grants for the construction of compensatory water developments to accommodate Maasai vacating the western Serengeti. A management plan was formulated in 1960 and revised in 1962. This management plan drew its direction from a speech of the Governor to the Maasai District Council outlining Government intentions over the Ngorongoro Area:

> Another matter which closely concerns the Masai is the new scheme for the protection of the Ngorongoro Crater. I should like to make it clear to you all that it is the intention of the Government to develop the Crater in the interests of the people who use it. At the same time the Government intends to protect the game animals of the area, but should there be any conflict between the interests of the game and the human inhabitants, those of the latter must take precedence. The Government is ready to start work on increasing the waters and improving the grazing ranges of the Crater and the country around it; for your part you must take care to fulfill the agreements into which you have entered to keep the countryside in good heart. You must not destroy the forests, nor may you graze your cattle in areas which have been closed under any controlled grazing scheme; at the same time you must be certain to follow veterinary instructions designed to prevent disease.

Government foresaw the need for integrated development and management when it stated that the conservation unit would

> cover the whole Ngorongoro Division so that development may in due course embrace the highlands and the adjacent plains together.

The first management plan described the objectives of conservation:

> As the Ngorongoro conservation area is not only the home of the Masai resident therein but is also a source of water for neighbouring areas, an asset of national value and an area of international interest, the natural resources (including water, soil, flora, fauna, and domestic animals) must be conserved and developed in such a way that they may provide a maximum sustained yield of products for the benefit of the humans dependent thereon without causing deterioration in the habitat and so maintaining the area's unique tourist attraction, aesthetic value and scientific interest.

These objectives were approved by the Minister concerned and provide the foundations for management of the area. It is noteworthy that this first set of objectives specifically mentioned the Maasai and their domestic animals, and stated the resources should be developed for the benefit of dependent human populations. The qualifying clause of not causing habitat deterioration set the limits on development practices.

The initial NCA administrative and decision-making body of four Maasai elders and four expatriate officers under the chairmanship of the colonial district officer rapidly collapsed (Arhem 1985b). An advisory board was set up in 1961 (without resident Maasai representation) to act as a forum for negotiation between interest groups. A series of management plans were drawn up for the Ngorongoro Conservation Area between 1960 and 1966 (Arhem 1985b:33); the third and most detailed (Dirschl 1966) was not officially accepted or implemented. By then there was a move by the controlling Ministry to give Ngorongoro and Empakaai Craters National Park status while the remainder of NCA was to be given over to livestock and agricultural development under the Maasai Rangeland Commission (Moris 1981). In 1969 the conflict surfaced with the Minister for Agriculture (who was also responsible for wildlife conservation), the MP, and the Regional Commissioner all supporting the degazetting of some 85% of the Conservation Area for ranching and wheat farming purposes. They were opposed by the Conservator of Ngorongoro, the Director of Natural Resources within the Ministry and the international conservation lobby which brought enormous pressure to bear. Ole Saibull (1978) gives his view of the politics, the personalities and personal interests involved. The public debate was finally curtailed by presidential intervention. Ngorongoro Conservation Area became the responsibility of the new Ministry of Natural Resources and Tourism (established in 1970). This new administrative arrangement emphasised conservation at the expense of human interests. However, at the same time the newly launched USAID Range Project registered Ngorongoro as one of the first four ranching associations designated for range and livestock development (Moris 1981:104–5), with dips, water developments and planned tenure changes.

In 1975 a new Conservation Ordinance restated the responsibility of the Ngorongoro Conservation Unit, now the NCA Authority, (see below) to safeguard and promote the pastoralist interests of the Maasai residents as well as preserving and developing the natural resources. At the same time the 1975 Ordinance prohibited all cultivation within NCA, although it had previously been allowed at the discretion of the administration. During the 1970s a conflict of interest emerged between the USAID range development plans and the national villagisation programme. Ngorongoro, along with all other putative ranching associations except Monduli, was never granted land rights. In 1978 the Tanzanian Government decided to disband the ranching associations, and in 1979 USAID ceased funding the defunct Maasai Range Management and Livestock Improvement Project.

In 1979 NCA became a World Heritage Site and in 1982 was made the Ngorongoro Biosphere Reserve in recognition of its international conservation value – not least as a site where man and wildlife have existed in harmony for so long (IUCN Directory 1987). The NCAA commissioned a new management plan with UNESCO financial support (Institute of Resource Assessment 1982). It was not officially endorsed, and has been superseded by a current IUCN/NCAA study. In the interim there have again been suggestions to impose more rigid zonation and control over pastoralists (e.g. Chausi 1985, Ole Kuwai 1981). The spectre of a mass expulsion of all Maasai from NCA persists (see e.g. Kitomari 1985, Kaiza-Boshe 1988). The suggested Serengeti Regional Conservation Strategy puts NCA as the first on the list of a series of planned buffer zones seen as essential to wildlife conservation and apparently designated for Game Reserve status (Malpas and Perkin 1986:47). Although some internal documents suggest that a considerably more favourable attitude has developed towards the NCA pastoralists (e.g. Kayera 1985), the present conservation strategy retains a negative view of the Maasai in many ways. For example, traditional pastoralism is seen as an anachronism bound to give way to commercial ranching; groundless assumptions are made about an explosive increase of the livestock population and the serious nature of its environmental impacts, particularly as concerns alleged overuse of the short grass plains (Malpas and Perkin 1986). Ministry wildlife officials interviewed in 1989 see Maasai residents as being nuclei of real and potential poaching activity. This negative attitude continues to be expressed in concrete action. For example, in 1986–87 and 1988–89 the conservation administration hunted out and destroyed small patches of maize, tobacco, potato and bean cultivation (e.g. Makacha and Ole Sayalel 1987). In what was described as a major land use problem in 1987, 666 people were arrested for the cultivation of 528 ha, causing a deterioration in NCAA/Maasai relations. A more efficient and people-oriented extension service would not have allowed such a situation to develop. The NCAA reports do not mention the real crux of the problem – the difficulty of purchasing grain (see chapter 10).

NCA: Legal and administrative structures

NCA is situated totally within the newly formed Loliondo District, previously part of Maasai District of Arusha Region in north Tanzania. The area is in fact a complete *taarafa* or subdistrict, composed of three *kata* or wards. Tanzania has a regional form of government, with most executive and decision-making power devolved to the districts themselves. Regional and national governments are, in theory, advisory and technical centres. If NCA were to function as most other inhabited areas of Tanzania, then the people, their stock, their infrastructure and their social and political development would have come under the district capital, now at Loliondo, previously at

Monduli. There would have been a nucleus of development and maintenance staff at *taarafa* and *kata* and even village level, and functioning locally elected political leaders and committees at these levels.

However, the NCA does not function as other areas of the country do. It is governed by a special Act of Parliament, the Ngorongoro Conservation Area Ordinance (Amendments) Act of 1975, which is a modification of the earlier Ngorongoro Conservation Area Ordinance of 1959. The functioning of these Acts is discussed in NCAA management plans of 1962, 1966 and 1983, and in more detail by Fimbo (1981) and Forster and Malecela (1988) in reviews commissioned for prospective management plans. The 1975 Act set up the NCA Authority (NCAA) as a parastatal institution with a board of directors and a conservator, to be appointed by the President of the Republic of Tanzania. With Ministerial approval, the Conservator can make legally binding Orders and Rules to control and regulate all activities within the NCA, including the lives of the resident pastoralist Maasai.

Other laws governing the lives of Tanzanian people are subsidiary to those of the NCA within the Conservation Area. The Regional, District and Subdistrict level government and political functionaries are therefore relatively powerless.

The NCAA Ordinances and Act have provided for a minimum of communication between the Authority and neighbouring government in that the local member of Parliament (for the district) or the District Council Chairman and the Regional Commissioner are to be members of the Board of Directors. For the last 20 years there has been no requirement for local representation on the Board from among the resident Maasai people. The Board today has a higher technical component with senior representatives from Livestock, Wildlife, District and Region. The Board rarely meets. It was completely bypassed in the case of the 1990 Commission of Enquiry. Current tourist lodge developments have been planned and executed in defiance of the Board's decisions (Chapter 11). This does not lead to efficient administration.

The Villages Act of 1975 in Tanzania gives recognition and power to lawfully established *Ujamaa* villages, which conform to socialist principles, Several villages have been created in NCA, by the use of considerable powers of coercion and persuasion (chapters 3, 10; Ndagala 1982, Arhem 1985a, b). While such villages have elected officials, they have no real power or function under the present NCAA organisation. Such elected leaders are often the same as the traditional leaders in Maasai society. Traditional leaders have not according to custom been rulers or decision makers for the community, but rather spokesmen for an age-set or for religious matters. (Chapter 11 discusses the problems of making such representatives responsible for management and development decisions in a society where traditionally management decisions are based on community discussion and concensus. There is a particular problem where an age-set system and strong division of labour between men and women, as well as the existence of strong clans and stock friendships, may lead to some groups monopolising channels of communication to the detriment of others' interests.)

Internal organisation within NCAA is largely concerned with self administration, tourism and especially natural resource conservation. There is no major headquarters section concerned with community services or development. In 1975 the whole area was divided into four administrative zones: Ngorongoro, Endulen/Kakesio, Empaakai and Kitete. These have little relevance for pastoralist development, being concerned largely with wildlife protection. Veterinary and pasture improvement services are minor, water maintenance services are decreasing and there is an uneasy sharing of health and education functions with the District (see Arhem 1981a, b, 1985a).

Not surprisingly, local political feeling is strongly against the present centralisation of power within the NCAA, which deprives resident people of the rights of expression and criticism they would have elsewhere in Tanzania (Ole Parkipuny 1981, 1983). There is a major imbalance in NCAAs management of conservation and tourism versus their input to veterinary, health, water and community development services (Institute of Resource Assessment 1982, Cobb 1989).

Perceptions

Political decisions, the laws that result and the resistance they arouse are based on and influenced by perceptions that differ dramatically between different participants. These conflicts of perception therefore generate the management conflicts that underlie our research. Chapter 2 outlined the natural resources of the Ngorongoro Conservation Area. Having sketched the legal and political structures within which these resources are managed we now go on to review them in terms of their values to the international, national and local communities and to the land use objectives of conservation, tourism and human subsistence.

Perceptions of NCA values involve appreciation of the costs entailed as much as awareness of the benefits derived. In Africa the benefits of wildlife conservation are usually enjoyed by a very different group from that which bears its costs (Bell 1987). The foreign visitor enjoys the aesthetic experience and the scientific opportunities at rather little (and entirely voluntary) financial cost, as does the international community through books and films. The national government bears the political and financial costs, but enjoys the national prestige and international political leverage associated with the conservation area. Government and private sector tourist trade and associated services share most of the revenues. The local rural population may traditionally have derived both subsistence and a high quality of life from the natural resources, but commonly suffers land alienation, restrictions on resource use, and even damage to land and property as a result of conservation measures.

(a) Conservation values to the international community

Together with the adjacent (and ecologically inseparable) Serengeti National Park, NCA has the greatest concentration of large mammals anywhere in the world and is Tanzania's – and probably Africa's – best known conservation area. Both Serengeti and Ngorongoro became World Heritage sites at the same time. Ngorongoro warranted World Heritage Site status as a natural heritage with outstanding aesthetic and scientific natural resources, and as a cultural heritage with archaeological sites of outstanding universal value for history, anthropology and palaeontology (Institute of Resource Assessment 1982). Many individual features are of striking value but their diversity, juxtaposition and unspoilt nature make them of extraordinary worth in international terms. The World Heritage Convention pays special attention to biological communities and landscapes that result from the long-term harmonious interaction of man and nature. The 1982 management plan emphasised that Ngorongoro can only be considered natural if pastoral man is allowed as an integral part of the natural ecosystem. The anthropogenic factors are particularly important in the case of the vegetation and floristic values, where it is the diversity and distribution of vegetation, rather than the presence of endemic, rare or threatened species that underlie their ecological interest. The greatest value of the vegetation is in its role as a habitat for the fauna. Its diversity and productivity make for the diversity and abundance of the wildlife, which are perhaps the best known Ngorongoro features in the international community. Again, it is not the presence of endemic or (with the exception of black rhino) rare or threatened species which constitutes the value of the wildlife resource. It is their diversity, the abundance of the plains game herds shared with the Serengeti, the scale of their migrations across the Serengeti/Ngorongoro ecosystem, their spectacular predators and their presence against a background of such scenic grandeur that makes them invaluable both aesthetically and for scientific research.

In addition to the physical and biotic values of NCA, the unique archaeological and palaeontological resources described in chapter 2 have been (and continue to be) fundamental to our current understanding of human evolution, and are recognised as such by the international community.

Maasai culture has been widely publicised in the international media through both scholarly and popular works (Spencer 1988, Jacobs 1975, Fosbrooke 1972, Waller 1979, 1988, Beckwith and Ole Saitoti 1980, Read and Chapman 1982). It has become something of a symbol of pastoralist culture and has generated a certain awareness of, and respect for, the pastoralist way of life and system of values. Pastoralist land use patterns are beginning to be recognised as ecologically valid and sustainable under many circumstances and livestock experts from developed nations are beginning to learn from their methods (chapters 8, 9). In this way the international community acknowledges the worth of Maasai culture in itself.

However, international appreciation of these values has not until recently been felt to any great extent in Tanzania in general or NCA in particular. The wild fluctuations in foreign tourism interests in NCA are discussed in chapter 11. A number of conservation organisations have provided funding for research or conservation, without having any major impact on wildlife, human or environmental conditions. Revenue from the many books and films about the area and about Maasai more generally tends not to percolate back to NCA. Over the last few years however, the international feeling for the value of Ngorongoro and the potential problems facing the area have led to a large-scale IUCN funded land use policy and strategy project addressing conservation, tourism, and pastoralist subsistence.

(b) Value to the nation

Among Tanzanian National Parks, Ngorongoro continues to attract by far the largest single number of visitors (over 25% of total park visits even with tourism at its lowest ebb – Jamhuri ya Muungano 1985), including a high proportion of foreign visitors, and is thus in gross terms a top earner among wildlife areas (see chapter 11). NCA also has an important role to play for Tanzanian visitors. It comes fourth in the league table for non-paying visits (e.g. by students) despite its relative isolation, lack of student accommodation, and comparative inaccessibility by public transport. For all this evidence of national interest in NCA, it is necessary to consider the opportunity costs the nation entails in conserving Ngorongoro. Tanzania has a greater proportion of her land surface gazetted as conservation estate than any other tropical nation (25%). Twelve National Parks and 15 game reserves cover some 15% of the country's 900 000 km^2 (IUCN Directory 1987), a proportion far higher than the 4% average for sub-Saharan Africa (Bell 1987). The majority of Tanzanian conservation areas occupy land with more than 750 mm rain p.a., which is potentially suitable for rainfed agriculture. Tanzania has a population of over 22 million, increasing at an estimated 2.8% p.a., doubling in less than 30 years (August 1988 census). Land with good soil, adequate grazing and fuelwood, no tsetse and rainfall sufficient to allow cultivation is already in short supply. The national conservation estate contains a significant proportion of the land of highest agricultural potential, and of this the Ngorongoro Highlands is among the most fertile. The ambivalence within Tanzania over dedicating all of this resource to conservation is obvious from the repeated attempts both at grassroots and at ministerial levels to open it up for agriculture. National perceptions of NCA are coloured by the extent to which wildlife is seen to pay its own way. Put bluntly, despite its potential NCA has for many years not been a major earner of foreign exchange, nor even able to pay for its own conservation administration from the proceeds of wildlife tourism. Today it generates a foreign exchange surplus, but the East African tourist industry is subject to major fluctuations and also entails social and ecological costs that have not as yet been taken into account (chapter 11).

(c) Values to the local community

It is precisely those biological values that make NCA such a haven for the diverse and abundant wildlife populations, that make it an invaluable grazing resource for the Maasai. The diverse vegetation, the permanent and seasonal water sources, and the salt licks are vital to both ungulate wildlife and domestic stock. Collett (1987) uses archaeological evidence as well as the present-day close coexistence of Maasai and wildlife to bear out the idea that Maasai pastoralism exists by virtue of the sustainable use of those same resources that sustain the wildlife. He sees the Maasai as the best conservers of their own wildlife. It is certainly the case that many of the great savannah wildlife areas of East Africa – among the greatest in the world – are on land alienated from the Maasai over the last 50 years (Western 1971:3, Sindiga 1984). Collett (1987) and Ole Parkipuny (1981) maintain that it is only with the advent of imposed conservation restrictions that conflicts between Maasai and wildlife land use have arisen. Population growth since the low levels following rinderpest must have exacerbated any increased conflict.

Resistance to the ways in which conservation administration has restricted human activities does not mean that Maasai are indifferent to the remarkable natural values of NCA. Maasai are keenly aware of the aesthetic values of their homeland (Ole Parkipuny, 1981, Ole Saitoti, 1986, Read and Chapman 1982). In addition to this appreciation of its beauty the Maasai have a long-standing philosophy of moral responsibility towards their land and its wildlife (Ole Parkipuny 1981, 1983). They traditionally hunted only for ritual purposes or in times of famine (chapters 3, 10).

What of the costs to the Ngorongoro Maasai of conservation in NCA? Ole Saibull (1978) and Arhem (1985a, b) give slightly different interpretations of

Maasai perceptions of conservation restrictions in the Serengeti/Ngorongoro area. Both agree that until 1954 restrictions interfered little with the Maasai inhabitants, although by this period the whole Serengeti/Ngorongoro area had been defined as conservation estate and there were minor and rarely enforced restrictions on settlement, stock movements, range management, cultivation and hunting. Both authors agree that the 1954 total ban on cultivation brought about the first major conflict. Arhem (1985a, b) sees events from then on as a progressive restriction of Maasai rights in decision making, area of access and forms of land use, with a steady growth of Maasai resentment against the NCAA.

By contrast Ole Saibull (1978) presents a picture of a Maasai community concerned that their future rights depended only on administrative promises (while wildlife rights in NCA were protected by law), but which at the same time 'felt privileged by being under the umbrella of the comparatively well-off Authority'. Ole Saibull (1978) saw the Maasai as more than willing to put up with conservation restrictions in return for the range, veterinary and water developments initially made available, as well as remission of District Council taxes (a proportion of gate revenues now substituted for these). This picture is coloured by his experience as Conservator of NCA and by the fact that in the 1960s and early 1970s there was little other development in Tanzanian Maasailand. The Maasai were perennially uncertain about their own long-term status within NCA. This uncertainty was made worse by the ever-changing administrative structures. During the 15 years 1962–1977 NCA came under the shifting authority of three Ministries, four Ministers, six Regional Commissioners, four Area Commissioners, two MPs, four Directors of National Parks and three Conservators of Ngorongoro. The Maasai need for the products of cultivation to make pastoral life viable within NCA is indisputable (chapter 10). However, Ole Saibull sees the socioeconomic advantages accruing to NCA Maasai to be such that Maasai support of cultivation rights within NCA was from the start a largely political move to forestall further restrictions, force a legal statement of their rights and finally rule out the possibility of total expulsion.

Either way, the eviction of all Serengeti Maasai to NCA and also of the Maasai inhabitants of the Ngorongoro Crater Floor to the Crater Rim following the 1959 Ordinances meant considerable costs in practical, aesthetic and emotional terms. Subsequent grazing restrictions, progressive breakdown of NCA facilities and the renewed total cultivation ban all meant growing socioeconomic problems (see chapters 8, 9 and especially 10 for an evaluation).

Throughout the history of NCA, Maasai representatives have contrasted Maasai tolerance of, and established coexistence with, wildlife and natural values, with the conservationist insistence on exclusion of humans for the sake of wildlife (Maasai elders Ole Surupe and Ole Pose, quoted in Ole Saibull 1978, and in Serengeti Committee of Enquiry 1957; Ole Parkipuny 1981, 1983). This tolerance of wildlife can be eroded by political illwill. In both NCA and Amboseli, the Maasai *ilmurran* responded to exclusion for conservation ends by deliberately hunting those rare species whose presence constitutes the main conservation value and tourist attraction of the National Park (chapters 7, 11; see also Lindsay 1987).

(d) Perceptions of conservation and subsistence problems

So far this chapter has outlined the history of NCA, its significance to different groups, and the costs that conservation entails for each. The NCA has unparallelled conservation value, but is precariously balanced both in financial terms and in terms of national and local perceptions of opportunity costs. In this final section we complete the background to the research by outlining the different perceptions of subsistence and conservation problems within the area.

Maasai see the NCA not only as their homeland, with all the attendant emotional and spiritual implications, but also as a rich grazing resource. They have few complaints about range condition in NCA. However there is no doubt but that their standard of living has fallen dramatically over the last few decades (chapters 8, 10). Their perception is that problems of human subsistence arise largely as a result of administrative restrictions on their ability to make full use of the NCA. The ban on even small-scale cultivation, exclusion from critical grazing and watering areas, and the ban on burning (with its attendant disease and range management problems – chapter 6) are seen as major issues (Arhem 1981a, b). Livestock marketing and grain trade conditions exacerbate the situation (chapters 8, 10). By contrast, the NCAA perception is that Maasai subsistence problems arise as a direct result of an outmoded and inefficient way of life and of resource use, coupled with an inexorably increasing population. The general feeling is that if the Maasai cannot make a satisfactory living under current circumstances in NCA they should move elsewhere (see chapter 11, also Arhem 1985a, b). There is also an assumption that traditional pastoralism must inevitably develop towards commercial ranching and dairying enterprises, and high population and stock densities, inherently incompatible with conservation (Ole Saibull 1978, Makacha and Frame 1986). This is discussed further in chapters 8 to 11.

There are major differences in the perception of conservation problems. The NCAA expresses concern that the Maasai both cause specific conservation crises, and also have an overall detrimental effect on the environment, which must lead to adverse effects on the wildlife. The most obvious example of a specific crisis is the near-extinction of black rhino in the area. The NCAA accuse the Maasai of organised poaching. Conversely the Maasai accuse the NCAA employees of using their antipoaching patrols, arms and vehicles as opportunity and equipment to poach rhinos and dispose of the currently highly valuable rhino horn. The issue is discussed in detail in chapter 7. A second conservation crisis as perceived by the Authority is the decline of woodland and forest areas (Kaihula 1983, Kikula 1981) which is seen as an outcome of Maasai activity. On the other hand, vegetation ecologists invoke long-term change from causes other than human activity, for example senescence of single-age stands, changing groundwater levels, and growing wild ungulate numbers. If anything there is a phase of bushland invasion with regeneration of woody species in many edge areas of the plains (NEMP 1989, Chamshama *et al.* 1989), perhaps attributable to lower elephant numbers, and less fire, as much as to changing livestock pressure. These issues are analysed in detail in chapter 6.

The NCAA attitude overall is that the presence of the Maasai is detrimental to NCA environment and wildlife populations. The original official description of the Serengeti National Park and Ngorongoro Conservation Area (Legislative Council of Tanganyika, 1956) stated clearly that Maasai interests were to be excluded from the Park not on the basis of any supposed damage, but because of the political problems envisaged in a multiple land use future, and because of the possible eventual incompatibility of Maasai and conservation interests. However, the subsequent Serengeti Committee of Enquiry (1957) took a different tone. Professor Frank Pearsall together with Pasture Research Officer T. Robson gave subjective evidence to the effect that serious damage to the environment was already being caused by the Maasai through burning, overgrazing, trampling around water points, and tree cutting. It was felt that human and stock numbers would inevitably rise, that damage by Maasai would increase, and competition for grazing and water would intensify. The original decision to exclude Maasai from the 12 000 km^2 Serengeti National Park and to place substantial restrictions on their use of NCA was justified on these grounds. The ecological damage argument is regularly repeated (Ole Saibull 1978; Ole Kuwai 1981; Chausi 1985; Makacha and Frame 1986). There are rare statements against this conventional wisdom (e.g. Branagan 1974). In NCA, as for most sub-Saharan semi-arid areas, longitudinal quantitative data are few, and primary productivity fluctuates from year to year in a way likely to mask any long-term trend. Subsequent chapters discuss the problems of assessing trends in vegetation and environment, and evaluate the statements of the 1957 Serengeti Committee of Enquiry and later authors on Maasai impacts on NCA environment and wildlife.

Background to research

When NCA was declared a World Heritage Site UNESCO commissioned a new management plan, to be drafted by relevant departments of the University of Dar es Salaam. A wealth of descriptive material on the natural resources of NCA together with detailed ecological studies of vegetation and wildlife studies were already available, but there was virtually no baseline information on the pastoralist inhabitants. In order to evaluate the different perceptions of subsistence and conservation problems outlined above and to make a useful contribution to future policy and management we needed an understanding of the ecology of the Maasai and their livestock in NCA. While our initial brief was to collect baseline ecological data, specific questions emerged. Firstly, what could ecological studies tell us about the sustainability of Maasai pastoralism in NCA? What was the impact of pastoralism on conservation values, in terms of environment, of wildlife populations and of 'naturalness'? Secondly, what could such studies reveal about the problems of pastoral subsistence in NCA? To what extent were such problems a result of 'natural' ecological factors of the physical and biotic environment, a result of poorly adapted methods of land use and stock management, or a result of imposed management constraints? The next six chapters outline our choice of research methods and review our own and others' studies of range, wildlife, livestock and human subsistence ecology in the light of these questions. This will form a basis for our concluding chapters which discuss possible management alternatives and their likely outcomes.

5

Management-oriented research in NCA

Menyanyukie esajati oloilelee katukul Even the shank of the cow is worth something
(You can always build on what you have – Maasai proverb: Waller 1979)

Over the years a great deal of research effort has gone into investigating the ecology of the Serengeti/Ngoronogoro area. Much of this effort has been directed towards natural history and single-species ecology. Little has been of immediate management relevance and only recently have studies addressed community and system ecology. Our aim was to add selectively to the material already available to make it applicable to management problems. This chapter looks firstly at the central issues behind our study. The background of scientific knowledge which was used as the springboard for the 1982 management plan is reviewed. The chapter than outlines the way in which our own study objectives and sites were chosen, and research on range, livestock and human ecology planned, to complement and make relevant the earlier material. Since our study the Ngorongoro Ecological Monitoring Programme has been revitalised with funding and direction from World Conservation International. The Ngorongoro Conservation and Development Project (NCDP), a joint venture between NCAA, IUCN and the Tanzanian Ministry of Lands, Natural Resources and Tourism, has coordinated a series of consultancies leading up to a new management plan for NCA. We draw on these sources throughout the book.

Management and research

(a) Central issues in NCA

The crucial management policy questions in NCA revolve mainly round the future of Maasai land use practices, and to a lesser degree the economic and conservation possibilities of alternative forms of wildlife utilisation in the area. The questions of Maasai land use centre on a few recurrent issues:

1. What is the impact of Maasai pastoralism on rangelands, on wildlife populations, and on conservation values?

2. Is Maasai pastoralism environmentally sustainable in its current form?
3. Are problems of Maasai subsistence in NCA the product of environmental constraints, ecologically inappropriate land use, or imposed administrative and conservation constraints?

These questions lead on to others which cannot yet be tested, but where comparative material can be used to evaluate the probability of alternative future developments:

1. Will NCA Maasai pastoralism continue in its present form?
2. If Maasai pastoralism is likely to change, will it develop naturally into intensive livestock ranching and dairying enterprises inherently less compatible with conservation values?
3. If not, to what extent will the different possible future forms of Maasai pastoralism be compatible with conservation values?
4. What population changes are likely to take place and what ecological changes are these likely to involve?
5. Could technical interventions help the tradeoff between Maasai subsistence production and conservation in NCA?
6. Could administrative and conservation restrictions be modified so as to alleviate Maasai subsistence problems without endangering conservation values?
7. Are there forms of wildlife utilisation whereby conservation could pay its own way in Ngorongoro to a greater extent than at present?

(b) Past studies

Despite the original Conservator's holistic approach (Fosbrooke 1972) research within the Serengeti/Ngorongoro area was, until the present study, focused on the dynamics and interactions of wildlife populations and on their socioecology. The Serengeti/Ngorongoro area was seen as a gigantic 'natural' laboratory for general biological and predominantly wildlife research. It was felt that here natural history could be documented, and theories of behaviour and community ecology tested, without the interference of people (despite the long history of human use of the area, and the continued Maasai presence in NCA). Earlier books on the area are milestones in wildlife biology (Schaller 1972, Kruuk 1972, Sinclair 1977, etc.) and the Serengeti Research Institute (SRI) produced an outstanding series of tropical ecology research papers. In most of these works NCA was seen as a minor section of subsidiary importance within the Serengeti Ecological Unit (SEU). The factors determining primary production; grazing, browsing and fire effects; long-term influence on both animal and plant populations of epidemic disease in the primary consumers; competition and facilitation among the herbivore populations have all been discussed in a major synthesis of research on the SEU (Sinclair and Norton-Griffiths 1979).

The Serengeti Park administration was perennially sceptical of the value of much of this research to Park management issues (Sinclair and Norton-Griffiths 1979). Despite the later research interest in such management-related issues as the impact of burning and of large mammals on woodland regeneration (Pellew 1983), and settlement on the Park boundaries (Kurji 1981) research tended to avoid the management problem-solving approach. Single-species studies tended to predominate over community or ecosystem ecology. Some areas, such as the forests of the Crater Highlands, were ignored altogether. Virtually nothing was known of the Maasai and their livestock, and this was assumed to be the business of the NCAA rather than of the research institute. A number of workers (see chapter 6) advocated or set up studies of rangeland problems in NCA, but unlike the single-species studies in SNP none of these were completed.

It was only in the 1980s that pastoralist ecology began to be a subject of general interest. Over the last few years the West has given wide publicity to pastoralist peoples, both in popular portrayals of a 'noble savage' lifestyle (Ole Saitoti 1986, Read and Chapman 1982) and in the exposure given to drought, famine and disaster in the Sahel and Ethiopia, as well as in factual documentaries on pastoralist life. Scientific interest in biological and management aspects of pastoralism has shown a rapid expansion (Dahl and Hjort 1976, Sandford 1982, 1983, Simpson and Evangelou 1984, Hill 1985, Coughenour, McNaughton and Wallace 1985). However, this is a recent development. The massive growth of research in SEU in the 1960s and 1970s did not apply to pastoralist and livestock ecology.

The need for management-oriented research in Serengeti/Ngorongoro is now more widely recognised for a number of reasons. The large mammal populations and habitat zones there have shown themselves to be anything but stable. Social and economic pressures for development (or at least for a more secure subsistence) have intensified.

Study design

(a) Study focus

The 1980s NCA management plan study aimed to investigate the ecological facts underlying different perceptions of management problems (chapter 4), and to inform management decisions to be made on these issues, rather than to study every aspect of Maasai ecology. Enough material was already available from the wealth of past research to make it necessary only to calibrate current data on many aspects of rangeland and wildlife. However, livestock and human populations and their interrelations with range and wildlife were poorly understood in ecological terms. Our research therefore concentrated on NCA pastoralists and their domestic stock: population dynamics; utilisation patterns; production patterns; subsistence and stress; sustainability and degradation in different zones. In order to answer or even allow informed speculation on the questions outlined in the previous section, baseline data were needed on the following:

1. Past and current dynamics of and main factors affecting human, livestock and wildlife populations.
2. Natural resource utilisation by people, livestock and wildlife.
3. Pastoral stock production.
4. Human subsistence (household and individual consumption).
5. Disease and/or nutritional stress in human, livestock and wildlife populations.
6. Signs of environmental degradation and evidence as to their origin and importance.
7. The relationship of each of the above to patterns of rangeland type, productivity and condition.

Specific research questions are given in the sections on range, livestock and human ecology below.

(b) Study sites and schedule

The basic criterion in choice of study sites was to represent a cross-section of the diversity of habitats found in NCA, and of the equally great diversity of ecological and economic problems associated with them. Detailed studies of geology and vegetation already available for NCA formed the basis for the initial choice of three areas spanning the altitudinal range from the plains at 1500 m around Ol Doinyo Ogol (Gol Mountains) to the volcanic highlands that rise to around 3000 m. Discussions with local Maasai range and veterinary staff of the NCAA suggested appropriate villages within each zone. A preliminary visit to these villages during April 1981 identified three bomas each:

1. willing to allow intensive study of their stock and management practices;
2. of average size, neither so large as to be unmanageable nor so small as to be unrepresentative, and of average wealth (see chapters 8, 10 for evaluation of their representative nature);
3. relatively accessible by vehicle.

The final choice (Fig. 5.1) was made on our behalf by the Maasai of each village, and resulted in intensive study sites at Oldumgom/Nasera bomas in Ndureta village in the Gol Mountains; Ilmesigio boma cluster in Oloirobi village and Sendui boma cluster in Alaililai village. Additional information came from other bomas in the same villages, and from NCAA records.

The Gol mountains together with their surrounding short grass plains represent the hot, arid end of the habitat spectrum in NCA. Some 1000 Maasai, 5200 cattle and 4100 small stock (Arhem, 1981a, b) share the area with a large proportion of the Serengeti migratory wildlife population during several months of the rainy season. Tickborne diseases are rare but malignant catarrhal fever (MCF), scarce water sources and access to grain supplies are major problems and herdowners shift frequently to temporary camps within a 30–40 km radius (chapter 8).

Ilmesigio, the westernmost part of the Oloirobi 'village', comprises a cluster of bomas spread over 6 km^2 of Lemagrut mountain. At an altitude of 2300 m it typifies the ecotone between lower altitude tall grass pastures and higher altitude woodlands and montane grasslands (see chapters 2, 6 and Table 2.3). This moderate rainfall and temperature zone has a medium population and high livestock density and major tickborne disease problems (Rodgers and Homewood 1986). Ilmesigio herds show transhumance to the eastern edge of the Serengeti plain at the foot of Lemagrut on a more regular and restricted pattern than seen in Gol.

Sendui, situated at 2700–3000 m on the plateau ridge between the volcanic peaks of Olmoti and Empakaai, represents the high altitude, cold climate zone of NCA. The population has grown rapidly over the last ten years. Past observers suggested that grazing pressure has led to spread of unpalatable species, particularly the tussock grass *Eleusine jaegeri* (chapter 6). With porous soils and a ridgetop position no surface water is available: villagers and livestock depend on a gravity feed pipe from Olmoti to a trough in the Embulbul depression 7 km from the settlement. Before this water point was constructed, dry season water supplies depended on access to the Olmoti and Empakaai Craters (Figs. 2.3, 5.1), both now restricted conservation areas. Sendui herds show transhumance: in the case of our study herds, there were seasonal movements from November to April to Lera and Loipukie at the western foot of the Crater Highlands (Fig. 5.1).

These three very different areas represent the major ecological zones of NCA which are used by pastoralists and illustrate the main constraints operating in each.

An initial field period of six weeks (July–August 1981) was followed by visits in December 1981, May 1982, December 1982, July–August 1983, July–August 1989 and September 1990.

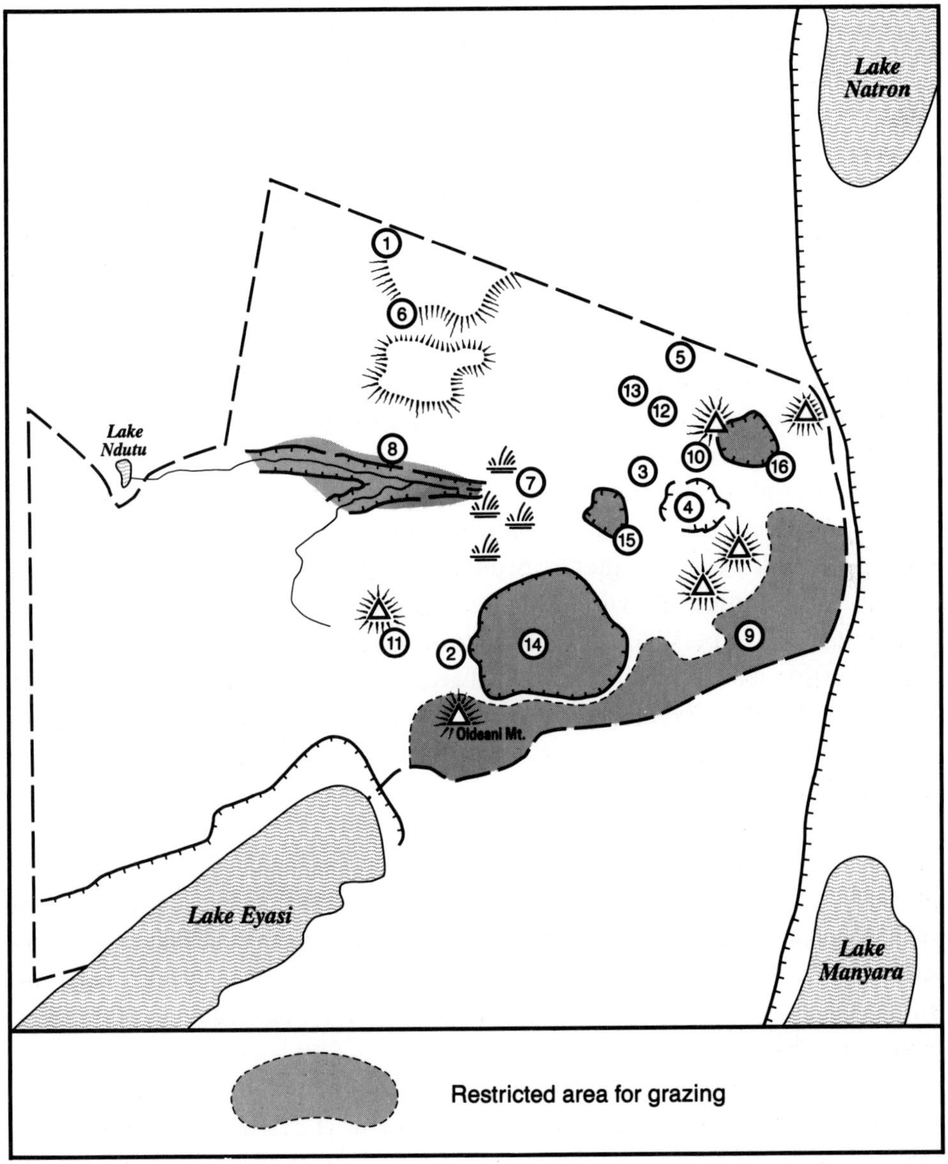

Fig. 5.1. Main study sites and place names. 1. Oldumgom, 2. Oloirobi, 3. Alaililai, 4. Embulbul Depression, 5. Loipukie, 6. Nasera (Andrea Lesian), 7. Olbalbal Swamp, 8. Olduvai Gorge, 9. Northern Highlands Forest Reserve, 10. Sendui (Ole Senguyan), 11. Ilmesigio (Ole Lekando), 12. Lera, 13. Naibor Ajijik, 14. Ngorongoro Crater, 15. Olmoti Crater, 16. Empakaai Crater. (See also inside cover.)

Study methods

(a) Range

The point of investigating range resources in this study was to gain some understanding of the impact of Maasai pastoralism on the rangelands. Allegations of overgrazing and woodland clearance implied harmful effects of Maasai pastoralism on plant species composition and abundance, vegetation structure, and primary production, and on degradation effects such as erosion in NCA (Ole Kuwai 1980:6, Kaihula 1983, Kikula 1981). Our study needed to investigate whether or not such effects were taking place. It aimed to evaluate the extent to which such land use is sustainable, and under what conditions. It had to evaluate the effect of Maasai pastoralism on primary production and on wildlife utilisation of the same shared range resources.

Basic descriptions were available for both Serengeti and NCA vegetation, as well as empirical evidence and theoretical papers on vegetation dynamics and response to utilisation (chapter 6). Studies were available on the importance of different communities to wildlife populations and on some aspects of utilisation patterns (chapters 6, 7). Observations over a two-year period are not enough to describe trends in a semi-arid area (Sandford 1983, Homewood and Rodgers 1987). Ideally, long-term vegetation plots and exclosures would have been used to investigate the suggestion that specific areas are undergoing degradation as a result of overgrazing. Plots established two decades earlier by Frame had been destroyed or lost, and this line of research was not open to us. Studies from long-term plots in the eastern Serengeti, ecologically continuous with NCA, provide important evidence on the issues of overgrazing, stability and resilience of these grasslands. Erosion features were quantified by Ecosystems Ltd (1980) and King (1980, 1982). Here we outline the methods whereby our study could add to the body of information already available and link it more immediately to management.

Despite the considerable background knowledge on vegetation there was no specific information available on NCA range as a resource for pastoralist livestock nor on other aspects of human use of the vegetation (e.g. for fuelwood). Our first aim was to produce a brief quantitative description of the habitat types and plant production characteristics of each study area, and to quantify seasonal variation in primary production. This then provided a basis for comparison between seasons and sites on livestock performance and human subsistence patterns, and allowed wider extrapolation through established reference data on rangeland variables. Our second aim was to map patterns of human and stock utilisation, these being central to distinguishing pastoralist impacts as opposed to those of wildlife and natural climatic factors.

We needed rapid description and evaluation of the quality and quantity of range resources in each area. Field reconnaissance together with small scale aerial photo analysis identified major communities (photo series 1958–1983, 1:30 000–1:70 000). Study bomas were marked on aerial photographs and the area of each habitat type within an 8 km grazing radius determined, this representing the normal maximum accessible area suggested by Brown (1971), Western and Dunne (1979), etc. Topographic maps at 1:50 000 were used to plot livestock grazing routes, water sources and boma sites, and the principal communities used by stock were described in terms of structure, cover, grass height and greenness.

A simple two-step transect technique (Riney 1982) was used to assess the vegetation quantitatively for ground cover, leaf table height, and greenness. Percentage grass cover and mean grass height allowed the calculation of a forage volume index, e.g.

$$20\%\ \text{cover} \times 10\,\text{cm high grass} = 0.02\,\text{m}^3\ \text{grass/m}^2$$

$$50\%\ \text{cover} \times 30\,\text{cm high grass} = 0.15\,\text{m}^3\ \text{grass/m}^2$$

Representative samples of grass were analysed for crude protein (see Homewood, Rodgers and Arhem 1987, for details of all techniques).

(b) Livestock

Despite the earlier lack of research interest in NCA livestock, information was available from a number of sources. NCAA holds periodic censuses, reported as numbers of each stock species associated with each named boma in each named village. Ecosystems Ltd (1980) had also carried out aerial censuses covering among other things estimates of the numbers and distribution of stock in the whole of Arusha Region. NCAA files contained patchy records on stock marketing, veterinary and dip inputs. No synthesis had however been attempted to give a long-term picture of stock dynamics. The background data available gave no information on patterns of range use, on stock production characteristics, on constraints affecting that production nor, finally, on the relationship between stock holdings, their management, their productivity and Maasai subsistence. The focus of data collection in our study was therefore to fill in these gaps. This meant herd counts were needed to give a baseline from which to measure dynamics over the study period (as well as to check against other sources); monitoring of births, deaths and transfers; monitoring of patterns of range use, activity patterns, feeding patterns and pasture selection; measurement of production in terms of stock condition, milk and calves; and finally identification of the effects of climate, range variables, disease or management practice.

Within each study boma a register of named identified cattle was built up and the performance and fate of individual cattle monitored over a two-year period, together with rangeland parameters that might affect or be affected by livestock populations.

Pastoralists are generally thought of as reluctant to allow outside attempts to estimate stock numbers. However, we were not concentrating on the potentially sensitive issue of ownership and size of stock holdings but rather on the 'access herd' (the subsistence herd available at that place and time – Dahl and Hjort 1976). This avoids the complexities of type and degree of rights accruing to different persons over individual animals while allowing detailed monitoring of fertility, mortality and transfers by sale, exchange, gift or loan. The counts thus represent a true picture of the subsistence herds available at the time, and historical information allows this to be related to a longer-term perspective.

Our figures for herd sizes were derived from four different data sources:

1. Repeated gate counts as herds entered or left the boma each day.
2. Independent accounts of individual herds by different members of the households associated with each boma.
3. Opportunistic counts of herds encountered during the day.
4. NCA ground census counts carried out in 1980 and reported by name for each boma.

Cattle condition was scored on criteria established for range quality zebu (Pullan 1978).

Cattle population dynamics were monitored in considerable detail by recording at each study visit for individually named animals the births, deaths and transfers that had taken place since the last sample. This information was used to calculate fertility and mortality during the study by totalling the number of cow-months for the relevant age/sex class(es), converting this to cow-years and dividing the number of events by the result. In addition to this individual histories of named adult cows were recorded from interviews with their owners giving age, total number of calves borne, the sex and the fate of each (death/sale/slaughter/exchanged/retained, etc.) for all adult cows in the study herds. This allowed calculation of historical fertility and mortality rates over a period covering approximately the previous ten years.

Cattle biomass density is of clear importance in evaluating local impacts. It was estimated for each study site in the following way. The position of each boma within a radius of 16 km of the reference sites was plotted. It was assumed that cattle from each boma graze mainly within a circle of radius 8 km (chapter 8). Cattle from 16 km away would have no effective overlap; 1 km away high overlap; 8 km away intermediate overlap and so on. Potential percentage overlap was calculable from simple geometric principles. By totalling these values a combined potential percentage overlap was obtained for each site, as a basis for estimating local biomass density of stock. For example, for a study area with only the reference corral and no others present in the 16 km radius the basic utilisation would be 100%. With two extra corrals immediately adjacent utilisation plus overlap would total 300%. An additional corral at 8 km distance would give a further 39.1% overlap.

Mean livestock numbers for bomas in each study area came from the 1980 NCA stock census (see chapter 8 for evaluation). Average weight for range condition small East African zebu cattle was taken as 180 kg (e.g. Peden 1987, Ecosystems Ltd 1980); liveweight data for sheep, goats and donkeys were 15 kg, 15 kg and 100 kg respectively. Livestock biomass for an average boma could then be calculated and by multiplying by the overlap factor the actual biomass for the grazing circle of the reference boma determined. Biomass density was calculated for the usable habitat of the grazing circle (chapters 6, 8) and allowance made for areas usable for a short period only (e.g. Gol plains).

General information on range occupance, activity patterns and energy expenditure came from systematic activity sampling methods (Rollinson *et al.* 1956, Lewis 1977, Altmann 1974, Coppock, Ellis and Swift 1986) interpreted in the light of additional opportunistic observation and discussion with the Maasai. Cattle herds were followed for one or more complete day's activity during each sample, from leaving the boma in the morning to returning at night. At 15 minute intervals the observer would record range conditions and herd position, as well as sample animal condition and activity for a scan sample of ten individuals. Activities were summed per hour and per day as percentages of time outside the boma. This allowed comparison of differences between sites and seasons.

Feeding activity records specified component (browse or graze), habit (tussock, turf or herb) and plant species wherever possible. Food plant intake at Sendui was compared with item availability in the range, using data from transects for the area as a whole as well as a series of line intercept point cover values at the actual grazing area. Transects involved ten metre tapes laid across the sward with cover type being noted at each ten centimetre point. Freshly collected cattle faeces were dried and analysed to allow calculation of diet protein values from published regression values for pastoralist cattle in East African rangelands (Arman, Hopcraft and McDonald 1975, Bredon and Marshall 1962).

Our measurement of milk production relied entirely on the goodwill and cooperation of the women in the boma. We asked each one to name her milking cows and to let us weigh her milking gourd before and after each cow was milked, using a Salter spring balance weighing up to 5 kg in 20 g graduations. This gave complete records for morning and evening milkings for several consecutive days in each boma during each study visit without causing disruption or inconvenience. As with many pastoralist groups, the Maasai allow the calf to suckle to get the cow to let down her milk. Unless it is nearly weaned the calf is then allowed to continue suckling while the mother is milked. The milk yields measured therefore represent not total output but only that part which is taken by the Maasai for their own use (see Grandin 1988 for a study of how this proportion may vary). The form of the data allowed direct measurement of variation in the amount of milk available to different households (see below). In addition to milk yield, chemical composition of milk was analysed for comparison with figures on pastoralist milk from elsewhere and to allow rough estimation of nutritional value (see chapter 7 and Homewood, Rodgers and Arhem 1987).

(c) Human subsistence

NCAA had census information on human population numbers (chapter 10) as well as files on attendance at clinics and schools, and trading figures for livestock auctions and shops. Ole Parkipuny, former MP for the area, has been an articulate and energetic spokesman on Maasai needs and problems. As part of the studies commissioned by UNESCO in preparation for the 1983 management plan, Dr Kai Arhem undertook a socioeconomic survey of all the NCA villages (Arhem 1981a). This outlined the current state of water supplies, housing, health, school and livestock services together with availability of consumer goods, transport and communications, and administration and law enforcement structures. Maasai perceptions of ecological, socioeconomic and development problems were also investigated (chapter 4; see also Arhem 1981a, b). However, there was no information on Maasai land use and stock management, nor on the relation between patterns of livestock production and Maasai subsistence, both crucial to management questions.

During the first six weeks of our study (July–August 1981), Dr Arhem collaborated with us in a survey of the Maasai food system which then formed the baseline for the following two years' information on production and consumption (Arhem, Homewood and Rodgers 1981, Homewood, Rodgers and Arhem 1987). The membership of the three study boma households was elucidated in detail and the food intake of their ten component houses studied over a total of 26 meal-days. Arhem chose the house as the unit of observation rather than the individual or the household for practical reasons. It was difficult to estimate individual intake accurately as meals were often taken in other houses. It was easier to determine the amount of food going in to and coming out of a particular house and the residents and visitors eating there at any one time (Arhem, Homewood and Rodgers 1981), and so to calculate average intakes per adult equivalent. The number of cows in milk for each house was recorded during each study visit as well as the current house membership. Estimation of food entering the house relied on milk yield measurements together with our knowledge of numbers of lactating cows, on everyday measures of grain and flour rations shown us by the woman of the house, and on observing the distribution to all boma households of meat from slaughtered animals. These figures allowed the relationship between milk availability and the consumption of purchased grain to be established. In addition to the food survey Arhem recorded simple work diaries allowing estimates of the work patterns, activity budgets and energy expenditure of NCA Maasai of different ages and sexes.

It was thus possible in later samples to extrapolate from our figures on milk yields and numbers of residents to the contribution of milk to the diet, in terms of absolute amounts and their relative adequacy in different areas and seasons. Taking this in conjunction with estimates of energy requirements, the dietary deficit and/or the need for supplementary grain or meat in the diet could then be estimated for different study sites under different conditions.

Summary and conclusion

This chapter outlines the rationale for research design in a management-oriented study, where priorities and constraints dictate that wherever possible established sources should be calibrated and put to full use rather than duplicating research effort. To a great extent the methods presented here allow us to tackle the management questions that stimulated the research. However, it was clear to us that an even more objective understanding of the system could have been gained, and more clearcut management decisions ultimately be made, had long-term monitoring information been available since the inception of the NCU and the Serengeti Research Institute. Such monitoring would concentrate not on the single-species, large mammal studies for which NCA and Serengeti are renowned, but on regular assessment of the soil, water and vegetation resources whose long-term trends are now the crux of management debates. It is a pity that the human population and its livestock were not thought important enough to warrant research monitoring from an early date, despite their known contribution to shaping the ecosystem and the persistent debate over their current impacts. Hopefully systematic records will be kept from now on.

6

Range resources

Eishorua opa Enkai inkishu o nkujit, mikior intokitin neishoo iyiook Enkai
God gave us cattle and grass, we do not separate the things God gave us
(Maasai proverb: Mol 1978)

All grazing animals in the NCA, both wild and domestic, are ultimately dependent on the quantity, quality and distribution of primary production in the rangelands. Some 23 000 NCA Maasai are dependent on a quarter of a million livestock, themselves using some 3–4000 km^2 of rangelands. The NCAA say that the pastoralists' herds are over-utilising the range resources, causing increasingly severe habitat degradation (Ole Kuwai 1981). Pastoralists say the changing pattern and intensity of grassland use by wild herbivores, and the increasing prohibition of livestock grazing in key areas, means they can no longer practice successful and sustainable subsistence pastoralism. This sums up the opposing sides of the conservation argument in NCA. Clearly the key to understanding much of this land use controversy lies in the status and dynamics of the range resources. Parallel conflicts prevail over the forested areas in NCA (Struhsaker *et al.* 1989, Chamshama, Kerkhof and Singunda, 1989).

This chapter sets out the main facts and debates on productivity, utilisation and dynamics of rangeland and forests in NCA. It reviews NCA grazing land productivity in the context of studies for Serengeti, East African and other African rangelands in order to emphasise the special nature of the NCA resources. The dynamics of long-term change, central to any evaluation of impacts, are then considered. NCA presents a microcosm where the interacting stresses and disturbances that affect savannas throughout the tropics may be seen in operation. These disturbances are first outlined in general terms and their importance is then analysed for individual study areas. The situation in NCA can best be understood in the light of recent work on stability, resilience and long-term trends of grassland ecosystems. General theories of savanna dynamics are illustrated with individual study areas. Consideration of both productivity and dynamics leads to the central management issues of pastoralist impacts on rangeland and the sustainability of those impacts. This chapter ends with a section considering forest and woodland status, dynamics and causes of change.

Grassland production

(a) General principles

Primary production in tropical savannas has received considerable attention and is the subject of increasing concern. Regional studies (Le Houérou and Hoste 1977) including East and Southern Africa (Deshmukh 1984) are available, as well as overviews for the African continent (Coe, Cumming and Phillipson 1976) and recent reviews have produced cross-continental syntheses (Frost *et al.* 1986). All these reviews centre on the availability of water and nutrients as major determinants of tropical rangeland production (Table 6.1; Bell 1982).

NCA Crater Highlands are an example of production on nutrient-rich soils where water is not limiting. Any one of a number of different forest, woodland or grassland vegetation types may dominate, or there may be a mosaic of several, depending on secondary factors such as altitude, fire, browsing and human land use. The short grass plains of NCA are an example of nutrient-rich soils limited by water availability. They respond to a pulse of soil moisture with a brief but large production of very high quality pasture. Such areas are able to

Table 6.1 *Interaction of nutrient and water availability determining tropical grassland production (after Bell 1982)*

	Nutrients	
Water	High	Low
High	High plant biomass High quality forage High animal biomass High animal diversity e.g. Crater Highlands and Rift Valley wall forest/grassland mosaic	High plant biomass Poor quality forage Low animal biomass Low animal diversity Fires common e.g. miombo of Southern Tanzania
Low	Water-limited but seasonally high quantity of high quality forage Seasonally high animal biomass and diversity e.g. NCA short grass plains, Serengeti and Amboseli	Low plant production Low/medium quality forage With increasing water and plant production, quality declines because nutrients become ever more limiting, e.g. West African Sahel

support a seasonally very high biomass and diversity of grazing herbivores (chapter 7). Within limits, grazing has a considerable effect in stimulating even higher primary production (McNaughton 1979:83, 85). The eroded hill slopes of the ancient Gol range may be a possible exception to the generally nutrient-rich soils of NCA.

East African sites show a steep increase of grassland production with increasing rainfall (Deshmukh 1984), due to the nutrient-rich soils. The North African/Mediterranean zone shows a virtually identical pattern, but by contrast in the infertile Sahelo–Soudanian pastures nutrients become limiting even at fairly low rainfall levels (Breman and de Wit 1983) and primary production shows only a gradual increase with rainfall (Le Houérou and Hoste 1977). McNaughton (1985) reviews comparable productivity/rainfall regressions that have been produced by several researchers for various Serengeti sites (Table 6.2). All are broadly in agreement. The relationship established by Deshmukh (1984) is based on a range of areas with altitudinal and other conditions more likely to be representative of NCA habitats than are data for Serengeti plains

Table 6.2 *Aboveground primary production (AGP in g/m²) related to precipitation (P_t in mm)*

Site	Period	Grassland type: Short	Grassland type: Medium	Grassland type: All types	Source
SEU (control/no grazing)	Annual	$0.69P_t - 102$ (Short and Medium)			1
SEU (actual/grazed)	Annual	$0.96P_t + 68$ (Short and Medium)			1
SEU	Annual			$0.34P_t - 122$	1
SEU (wet season)	Short term	$2.11P_t - 54$	$1.14P_t - 23$		1
SEU	Annual	$1.03P_t - 132$ (Short and Medium)			2
SEU	Short term	$1.29P_t - 249$	$0.77P_t - 20$		3
EAR (peak biomass)	Variable			$0.85P_t - 19.58$	4
N.Afr/Med (linear)				$0.87P_t - 41.49$	5
Sahel (linear)				$0.26P_t - 10.54$	5

Notes:
Figures all represent equations of the form $AGP = bP_t + c$ where b = slope and c = intercept.
SEU = Serengeti Ecological Unit, EAR = East African rangelands.
Source: 1. McNaughton 1985; 2. Braun 1971, 1973; 3. Sinclair 1975; 4. Deshmukh 1984; 5. Le Houérou and Hoste 1977.

grasslands alone. Spatial and temporal patterns of production for the drier areas are very patchy and follow stormtracks, local showers and water table height (McNaughton 1979:55–6). With increasing altitude and rainfall, and decreasing evapotranspiration, more uniform primary production is found. The exact quantities predicted by the regressions are of less interest than the overall conclusion that all NCA rangeland types are highly productive by any standards (McNaughton 1985), and particularly by contrast with the Sahelian grasslands or those elsewhere in Tanzania (Table 6.1; Bell 1982).

(b) Primary production of study sites

Site differences in terms of availability of habitat types, cover values, forage volumes, and nutrient contents have been described elsewhere (Homewood, Rodgers and Arhem 1987). The main points are summarised here.

Table 6.3 shows the relative availability of major habitat types within an 8 km radius around each study site. Nasera and Oldumgom sites in Gol differ in proportion of wooded hill slope grazing land, Nasera having 18% and Oldumgom 64% of this habitat.

Table 6.4 shows our estimates of forage volume for the three main study

Table 6.3 *Availability of major habitat types*

	Gol		Sendui		Ilmesigio
	Nasera	Oldumgom	Sendui juu	Loipukie	
Short grass plains (SGA: valley)	77.7	23.2	15.8	—	—
SGA (hilltop)	4.3	12.6	—		
Hillslopes *Acacia/ Commiphora*	17.9	64.2	—	15.5	
Eleusine tussock grassland	—	—	16.7	} 41.2	} 42.2
Pennisetum/Themeda grassland			51.2		
Satureja herbland			4.9		
Ravine thicket			11.4	6.8	
Juniperus forest					6.4
Acacia lahai woodland					16.7
Hyparrhenia/Themeda grassland					34.7
Salei plains grassland				36.5	

Note:
Values are expressed as a percentage of 8 km radius grazing circle around reference boma. Each grazing circle has an area of almost 201 km^2.

sites, providing a comparative scale for the different pasture zones. Ilmesigio forage availability values are highest and Gol values lowest, except for the continually cropped short grass of the Embulbul floor at Sendui (chapter 8). In Gol the hill slope forage volumes for dry and wet seasons are nearly double those of the plains. These figures partly reflect the tussock form of hill slope grasses as opposed to the mat-like short grass association of the plains at Gol (as do cover values: Table 6.5). Ilmesigio at intermediate altitude has a range of montane and lower altitude pasture types. The *Acacia lahai* woodland zone shows a high grass cover (91%, mostly turf). Montane pastures have a lower average of 79% grass cover, with more tussock (22%–Table 6.5). This is mostly *Pennisetum sphacelatum* (= *P. schimperi*) but includes some *Eleusine*. The lower pastures have only 49% grass cover with more bare soil, typical of drier situations. Forage volume figures (Table 6.4) are highest in the montane tussock grassland area. The *A. lahai* and the lowland pastures have similar forage volumes.

Table 6.4 *Grass height, forage volume and greenness for different vegetation types (see text for details)*

	a Grass height (m)	*b* Cover value[1] (m^2)	$a \times b$ Grass volume (m^3/ha)	Greenness[2]
Gol				
Short grass – dry	0.03	0.62	186	2.7
– wet	0.07	0.56	392	5.0
Wooded slope – dry	0.08	0.34	272	3.8
– wet	0.16	0.38	668	4.7
Sendui				
Upland (dry)				
Eleusine	0.12	0.59	708	3.8
Pennisetum	0.10	0.76	760	3.5
Embulbul	0.02	0.53	106	2.4
Lowland (wet)				
Ridge	0.12	0.60	720	4.9
Valley	0.13	0.35	455	4.2
Ilmesigio				
Upland (dry)	0.12	0.85	1020	3.2
Acacia	0.08	0.91	728	2.2
Lowland (dry)	0.15	0.49	735	4.1

Notes:

[1] See Table 6.5.

[2] Mean of scores on a 1–5 scale (see text).

Table 6.5 *Ground layer cover values in different vegetation types (%)*

	Rock + soil	Molerat tip	Litter + dung	Total nonplant	Shrub + herb	Tussock grass	Turf	Total plant	Total grass
Gol									
Short grass									
dry $n=3$	25	0	10	35	3	0	62	65	62
wet $n=3$	31	0	5	36	7	0	57	64	56
Wooded slopes									
dry $n=3$	48	0	15	63	3	34	0	37	34
wet $n=3$	35	0	3	38	24	38	0	62	38
Sendui									
Upland (dry)									
Eleusine $n=10$	3	13	7	23	18	39	20	77	59
Pennisetum $n=6$	2	7	5	4	9	53	23	85	76
Embulbul $n=4$	0	34	5	39	8	0	53	61	53
Lowland (wet)									
ridge $n=3$	17	0	4	21	19	60	0	79	60
valley $n=2$	47	0	3	50	15	35	0	50	35
Ilmesigio (dry)									
Upland $n=6$	1	1	13	15	6	22	57	85	79
Acacia $n=4$	3	0	2	5	4	23	68	95	91
Lowland $n=4$	25	0	12	37	14	49	0	63	49

In Sendui sites mole rat activity results in significant loss of pasture, with low cover values and up to 50% bare soil tip in the Embulbul Depression (Table 6.5; Homewood, Rodgers and Arhem 1987). The tussock grasslands are predominantly *Eleusine*, with *Pennisetum sphacelatum* at slightly lower altitudes. *Eleusine* tussocks are much larger and more widely spaced, resulting in a rather lower cover value. Both tussock grassland types have similar short grass cover between tussocks as well as some herb component. Both have similar forage volumes, as do the ridges of the lower altitude wet season Loipukie site. The latter make preferred grazing for cattle (chapter 8).

The nutrient quality of grassland production is as important for a herbivore as is quantity. Dry standing grasses may have digestible nutrient values well below threshold values for the maintenance of condition and body weight (Sinclair 1975). There is a general correlation between grass greenness estimates and crude protein value and palatability (Western 1973, Grimsdell 1979). Greenness indices are given in Table 6.4. As expected, rainy season values are greener than dry season values, high altitudes greener than low, and turf greener than tussock. *Eleusine* values are complicated by the deliberate burning of individual large tussocks inducing palatable fresh green growth in the dry season (personal observation, Newbould 1961, Frame, Frame and Spillett 1975).

Crude protein (CP) values are given in Table 6.6 for a variety of forage species, components, seasons and study sites. All values are within the normal range for these pasture types, with grass green leaf CP values some 60% higher than stems, and dry leaf fractions appreciably lower than green leaf. All species sampled had leaf:stem weight ratios close to 1:1, apart from *Pennisetum sphacelatum* with its preponderance of stem. *Eleusine* tussock structure and protein varied greatly with age: as leaf was abundant and stem not used by stock, only leaf was taken for analysis. Short grass species had higher CP values than tussock species. *Themeda* tussocks at Sendui had relatively high values, perhaps in response to grazing pressure as well as climatic factors. Ilmesigio *Themeda* values were lower, and plants there were postmature at the time of study. Browse plants sampled had CP values considerably higher than grasses, as expected (Le Houérou 1980).

Mineral content is another important aspect of grassland quality. Recent studies have confirmed selection for mineral-rich pasture by grazers (McNaughton 1988, 1990; calcium selection by lactating wildebeest, Kreulen 1975, McNaughton 1990). Access to natural salt licks and to the mineral-rich pastures of the short grass plains are perceived as major issues by the Maasai, and quite subtle differences in availability and timing of mineral inputs may be critical (chapter 8).

Rangeland dynamics

(a) Fire and herbivore impacts: general principles

Fire and herbivores affect vegetation structure and levels of rangeland productivity otherwise intrinsically determined by plant-available moisture and nutrients. Fire is common in tropical grasslands wherever low grazing pressure allows the accumulation of dry plant matter. Fire leads to the loss of volatile compounds of nitrogen, carbon, and sulphur. It tends to destroy woody seedlings and sensitive species, particularly those lacking seed adaptations, belowground reserves and the capacity to sprout back. Rangeland systems such as the *Themeda triandra* fire climax grasslands and wooded grasslands of the Ngorongoro/Serengeti region, where fire has been a regular feature for centuries, have a correspondingly fire-adapted species composition. In such systems periodic burning enhances the production of good grazing. Temporary protection against fire will, in the absence of compensatory grazing, allow an accumulation of standing plant biomass of low nutritive value and an increase of fire-sensitive, often unpalatable herbaceous species. This makes adverse fire effects that much more serious if and when fire is

Table 6.6 *Forage crude protein values (%)*

	Dry season			Wet season		
	Leaf	Stem	L+S	Leaf	Stem	L+S
SGA + turf						
Gol	6.7	3.8	—	—	—	8.3
Sendui: ridge	—	—	6.5	—	—	—
Sendui: Embulbul	—	—	6.6	—	—	—
Ilmesigio	—	—	7.2	—	—	8.7
Hillsides						
Gol	6.0	4.1	—	—	—	7.9
Sendui: Loipukie	—	—	—	8.3	—	6.8
Ilmesigio	—	—	3.4	—	—	—
Montane grassland						
Sendui: *Eleusine*	4.0	—	—	6.0	—	—
Pennisetum	5.8	—	—	—	—	—
Ilmesigio: *Eleusine*	4.2	—	—	—	—	—
Pennisetum	5.2	4.2	—	—	—	—
Themeda	6.9	3.4	—	—	—	—
Browse (all)		11.2			12.3	

deliberately or accidentally allowed to return. Stopping burning can thus lead to unwanted side effects on productivity and species composition, as shown in a recent comparison of burnt Maasai Mara and unburnt Nairobi National Park grasslands (Boutton, Tieszen and Ibamba 1988a, b). Fire, however, is more than just another factor of ecological interest. It is a subject of management concern, and affects not only rangeland quality and quantity of fodder, but also patterns of succession, differential habitat use by wild and domestic stock, ectoparasites, disease levels, etc. As a potential management tool of the NCAA (not currently used) and as a traditional management tool of the pastoralists (currently prohibited) it has brought NCAA into conflict with ecologists and local people. Fire effects on range are discussed further in this chapter; on wildlife (chapter 7), on disease (chapter 9) and as a management input of significance (chapter 12).

Herbivore impacts consist of defoliation, trampling, and nutrient cycling. Large mammal, small mammal and invertebrate herbivores each have characteristic but to some extent overlapping impacts on rangeland vegetation (Sinclair 1975). Small mammals and invertebrates may have significant grazing effects in terms of competition for green biomass at times of low forage availability (Sinclair 1975) but their defoliation effects may be more significant for browse (Belsky 1984) than grass species.

Large herbivore grazing or defoliation effects have been the focus of recent interest in rangeland studies. Generalised models of plant herbivore population interactions (Caughley 1976) and of savanna response to disturbance (Noy-Meir 1982, Walker and Noy-Meir 1982) have implications for NCA rangeland dynamics. Noy-Meir (1975, 1978, 1982) has explored different possible patterns of pasture growth combined with different herbivore consumption responses. Plant growth shows an initial increase in response to grazing, up to a maximum productivity for both plant and animal populations. Some systems then show a progressive steady decline with heavier grazing. Others continue at a high level of productivity up to a critical threshold grazing pressure, at which there is a dramatic crash to a new much lower level. Noy-Meir suggests that savannas tend to correspond to the former and 'improved' or commercially managed pastures to the latter. The ultimate outcome depends on forage palatability and accessibility, and on the efficiency of the grazer. Green biomass in dwarf or prostrate forms, or underground storage organs of perennials, may form an ungrazeable reserve. Thus creeping perennials like *Cynodon*, and the relatively unpalatable *Eleusine* with its high silica content have relatively secure ungrazeable reserves, while erect, palatable perennials like *Themeda* have little protected reserve. The secondary toxic compounds in *Cynodon* young growth also protect against this species being grazed to the point of destruction.

Defoliation by large ungulates has been intensively studied for various Serengeti grassland communities by McNaughton (1979, 1983, 1985) and Belsky (1985, 1986a, 1986b, 1987). Grasses grow from basal meristems and by tillering while browse species grow from the apical meristem. These different patterns have specific implications for response to defoliation. A range of grassland species show initial stimulation of growth in response to defoliation. In the Serengeti, *Kyllinga* sedges (Coughenour, McNaughton and Wallace 1985) and *Andropogon greenwayii* (Belsky 1986a) have both been shown experimentally to respond to repeated clipping by increased tillering, and *A. greenwayii* may actually be dependent on grazers removing its dense canopy to allow tillering to take place (Belsky 1986a). Individual grass species show different degrees of stimulation by, and susceptibility to, defoliation. This depends on their belowground reserves, and their capacity to resprout. *Themeda triandra* is stimulated by moderate defoliation and needs dry season burning to ensure its continued survival (Braun 1973, Coughenour, McNaughton and Wallace 1985). Such burning increases palatability and availability and has been practiced by centuries of pastoralists. However, *Themeda* is unable to withstand heavy grazing.

(b) NCA lowland plains: grazing pressure and trampling effects

The plains short grass association (SGA) is represented by our Gol study site. On these shallow porous soils production is limited to periods of moisture availability. They have seasonally very high productivity and can then support very high animal biomasses, mostly of wildlife (see chapters 7, 8 and 9). Migratory wildebeest remove most green leaf in the wet season. The lengthy dry season with no further moisture means no further growth, so the SGA is almost unused by cattle.

The ecologically continuous Serengeti short grasslands are not thought to be overgrazed despite extremely high grazing densities (chapters 7, 9; McNaughton 1983, 1985). The short grass association species are classified as 'obligate grazophils'. On average about 66% above ground production (AGP) is consumed, and on occasion up to 94% AGP is taken without harmful effects. On the contrary, exclosure of grazers causes short grass areas dominated by *A. greenwayii* and *Sporobolus ioclados* to shift to a taller community dominated by *Pennisetum* species, which shade out the former dominants and have similar or lower green biomass despite their large forage volume (McNaughton 1985, Belsky 1986b).

In an aerial survey of erosion patterns in Arusha Region, NCA scored lowest for animal trail erosion, and lower than the regional average on gully and sheet erosion, while erosion fans mainly attributable to climate and landform were common in NCA (Ecosystems Ltd 1980, Cobb 1989). Overall, Ngorongoro District showed less severe erosion than the rest of Arusha Region, but NCA showed more areas of moderate or severe erosion than did Salei and Loliondo, the other two divisions of the District. The NCA plains showed more serious erosion than the highlands.

Following on from the aerial survey, landsat image analysis (King 1980, 1982) has been taken to suggest environmental degradation in NCA short grass plains on two counts. The first is a comparision of 1972/3 and 1978/9 wet season green biomass as estimated from the simple division ratio of Landsat bands 7 and 5. The lower value for 1979 is taken to indicate a trend of declining productivity resulting from overgrazing by wildebeest. Field checking showed some areas with the lowest reflectance ratio, e.g. around Olduvai, to have 25–75% bare ground with deep wildebeest tracks between residual perched tufts of grass. However, the extent to which the simple division ratio truly reflects green biomass is debatable (King 1982). More to the point, any inferred trend of degradation based on two spot measurements of regional green biomass in intrinsically variable semi-arid rangelands is highly suspect. The extensive research into Serengeti grassland production and dynamics cast doubt on this particular way of using remote sensing as evidence of declining productivity.

The second line of evidence is an increase of bare dust and moving dunes in the Angata Salei area from 77 ha in 1972 to 619 ha in 1978, representing an average annual increase of 100 ha from 1973. This is attributed to cattle overgrazing, partly as a result of compression by the wildebeest herds (King 1982). The suggestion is that while the adjacent and ecologically continuous Serengeti plains show no sign of overgrazing or environmental deterioration, the concentrated passage of the migratory herds through the NCA short grasslands together with the year round presence of livestock is enough to bring about deterioration here.

Trampling may damage individual grass plants leading to a change in species composition and/or ground cover, and may cause compaction of bare soil with the attendant adverse effects on raindrop impact, infiltration, plant-available moisture and soil structure. Belsky (1986a) presents evidence suggesting that some trampling actually stimulates rooting and spread of *Andropogon greenwayii* tillers. However, all grass species in Belsky's experiments were destroyed by simulated heavy trampling.

The fine volcanic dust soils of the Angata Salei are inherently unstable. King (1980) suggests that concern over loss of topsoil and associated fertility is unwarranted because 'the soils are too young to have ever formed a topsoil, and there seems to be little change of nutrient status with depth' and the windblown volcanic ash is rapidly redeposited. Belsky and Amundson (1986) have documented the successional sequence that the resulting moving dunes or barchans leave in their track. Total recovery takes some 60 years from the initial total destruction of vegetation cover caused by the invasion of the dune.

Further analysis of recent imagery and photography have not shown trends towards increased degradation of the plains grasslands in NCA (Perkin pers. commun. 1989) although the bare dust areas of erosion impact are still evident in Angata Salei. The main change to be detected is the now obvious bushland encroachment in the Endulen, Lemagrut, Ndutu and North Malanja areas. Chapters 7, 8 and 9 look at evidence on intensity and duration of use by different wild and domestic species and their relative trampling impacts.

(c) Intermediate altitude grasslands

The NCA intermediate altitude open grasslands between about 1700 m and 2300 m comprise medium and tall grasslands which make up over 20% of NCA vegetation cover (Table 2.2). These grasslands are supported by high soil moisture levels due to altitude and drainage, rather than by a higher rainfall due to the proximity of Lake Victoria as is the case for the medium grasslands of the western Serengeti. However, the dominant grassland species are the same as in the Western Serengeti (*Themeda triandra, Pennisetum* spp., *Hyparrhenia* spp.) and their response to environmental disturbance may be closely comparable. These grasslands are generally accepted to have been derived from denser woodland formations by fire, grazing and browsing (see below, also Lind and Morrison 1974). They are comparatively densely settled (e.g. around Ilmesigio: chapter 10) with high livestock densities (chapter 8).

In the medium and tall grasslands above-ground green biomass declines in the absence of grazing despite continued moisture availability. Above-ground biomass is mostly stem, and postflowering senescence brings rapid withdrawal of nutrients and loss of palatability. Grazing and/or burning reverses this process and stimulates new growth. The species dominating these medium and tall grasslands (e.g. *Themeda*) are less resilient in the face of heavy grazing than are the short grass communities (McNaughton 1983, 1985). In spite of this, neither residents nor NCAA see overgrazing as a problem in this zone, even with the locally high livestock densities.

Tussock species vary in production and palatability with moisture availability. The *Themeda–Hyparrhenia* pastures at lower altitudes in Ilmesigio are in the rainshadow of Oldeani Mountain and subject to higher temperatures. Dry season growth and hence potential utilisation is low. Higher altitude grasses draw on greater moisture stores and maintain some green leaf year-round. *Pennisetum sphacelatum* is heavily grazed, as shown by typical leaf table heights of 15–20 cm for this grass.

The main debate over the dynamics of the medium grass zones comes down to whether it is desirable to burn intermediate and lower altitude *Pennisetum–Themeda–Hyparrhenia* grasslands, and with what frequency and timing. Burning has been banned for decades (chapter 4) because of concern over destruction of woodland and forest edges. It is now generally accepted by savanna ecologists that an early dry season burn in the medium grasslands stimulates a flush of green growth, eliminates ectoparasites and encourages the palatable and productive *Themeda triandra*. Dirschl (1966) suggested protection of the forest and woodland edges against burning, together with 'pasture improvement' throughout the rest of this zone by the removal of coarse tussock grasses (largely *Pennisetum*) with ploughing, burning and seeding, followed by rotational grazing. Pasture improvement and rotational grazing arc unlikely to be practicable or desirable (see chapter 12). However, if fires can be set and controlled with sufficient care to protect the forest and woodland against further encroachment, this may represent the best compromise for both conservation and human land use. The same arguments apply to the Crater floor where controlled burning would improve the grasslands for wildlife (chapter 7; see also NEMP 1989). Large areas of intermediate altitude grasslands in the Kakesio–Osinoni zone are little used by cattle because of the combined deterrents of an increasing wildebeest presence (chapters 7, 8, 9) and WaSukuma raids (chapters 3, 10).

(d) Highland grasslands: the Eleusine *debate*

Highland grasslands are dominated by *Eleusine* tussocks, interspersed with mat-forming species like *Pennisetum clandestinum.* At lower altitudes *Eleusine* gives way to a *Pennisetum sphacelatum* tussock grassland, again derived from woodland formations (though a treeline would be expected at 3000 m in undisturbed vegetation). These areas are increasingly densely settled and residents perceive shortage of grazing land and of good forage to be problems (chapters 8, 10).

The main management debate in the highland grasslands centres around whether or not *Eleusine* tussock grass is spreading in response to heavy grazing by livestock. *Eleusine* tussocks are very long lived and can reach massive dimensions of 1 m radius and 2 m height. *Eleusine* concentrates silicon in the leaf tissue and so presents a tough and scabrid surface which deters many grazing animals. Cattle from dense *Eleusine* pastures show extreme wear on incisors and middle molariform teeth at a relatively early stage.

A number of studies have been initiated to investigate the *Eleusine* problem (Glover 1961, Newbould 1961, Frame 1976, NEMP 1987); unfortunately not one has been completed. Here we summarise the preliminary assumptions and theories of the different workers.

Glover (1961) and Frame (1976), following Pearsall (1957), suggest that *Eleusine jaegeri* has increased as a result of stock selectively overgrazing the palatable short mat-forming species, thus allowing the competitive spread of the relatively unpalatable *Eleusine*. NCAA holds this to be the case. However, no study has demonstrated this process. There are a number of papers demonstrating the effects of different control techniques, but none demonstrate *Eleusine* spread or its causation (O'Rourke, Frame and Terry 1975, O'Rourke, Terry and Frame, 1976). Branagan (1974) takes a different view from the conventional wisdom, seeing *Eleusine* spread as the result of undergrazing due to administrative constraints on burning. This is based on the fact that individual *Eleusine* tussocks are deliberately burnt by Maasai (personal observation; Branagan 1974) and the new growth of young *Eleusine* leaf is palatable. Newbould (1961) gives an emphatic statement on the basis of his several years' experience as range officer in the area, unfortunately unsupported by published data:

> [In the highland pastures] the present situation is the result of undergrazing . . . I consider that the importance of *Eleusine jaegeri* has been consistently overemphasised . . . I can find absolutely no evidence that it is increasing or that it results from overgrazing, and I think that it has existed in its present abundance for many human generations . . . It fulfills an essential role in binding the light volcanic dust soils of the Highland . . . The best compromise is to keep it down by burning . . . followed by intensive grazing . . . before the *Eleusine* has grown above . . . c.9 inches. Once burnt, grazing will keep it short . . . Young shoots of *Eleusine* are palatable and nutritious as fibre deposition and the formation of siliceous saw-edges only occurs after the shoot has exceeded c.9″ in length . . . short *Eleusine* is not seriously detrimental to its associated grasses. Newbould 1961

Eleusine tussock grassland in Olmoti Crater has remained at least superficially unchanged despite two decades' grazing ban on Maasai stock. Recent reports suggest *Eleusine* is spreading in areas from which stock are excluded, and where it can therefore not be attributed to pastoralist overgrazing (NEMP 1987). We suggest that the extent of *Eleusine* spread has not yet been satisfactorily established, and that such spread is unlikely to be due to overgrazing by Maasai stock (Homewood and Rodgers 1984, 1987). We suggest that the non-burning policy of the last twenty years has led to unpalatable, undergrazed *Eleusine* spreading at the expense of heavily grazed intervening turf species.

The dynamics of *Eleusine jaegeri* await a full scale study before these opposing views can be evaluated. The available information simply does not allow one to distinguish between the alternatives. This has not prevented a series of workers (including ourselves) from expressing strong feelings on the subject, and attempting to invest these with the authority of a well-founded conclusion. Beware!

Alongside the dearth of data on *Eleusine* dynamics there is an equally important lack of experiment with alternative land use effects. What would happen if burning were to be reintroduced? We suggest a decrease in *Eleusine* and a more productive sward overall.

The montane grasslands also include distinctive short grass associations. These, like the plains SGA, have species of high nutritional value, which react favourably to grazing pressure (McNaughton 1979). Montanc short grass pastures are on deeper soils with higher moisture availability meaning higher productivities. They are not subject to intense grazing by wild herbivores but support local populations of zebra, steinbok, reedbuck, etc. Some green growth is thus available for most of the year, and continued grazing by domestic stock (especially the locally high densities of sheep) stimulates further production of leaf.

(e) Nutrient recycling

Nutrient recycling is perhaps unlikely to be of crucial mineral importance in the NCA nutrient-rich soils, particularly in the short grasslands both montane and lowland. Dung or fertiliser applications have little effect on the arid plains short grass productivity through most of the growing season (Banyikwa 1976, Belsky 1985, 1986a, 1986b). McNaughton (1985) comments on the tendency of wildebeest to graze in the medium and tall grasslands during the dry season but to ruminate, rest and deposit dung in the shorter grass areas of their dry season habitat. He suggests this may have implications for nutrient stripping and concentration respectively in the two habitat types. Nutrient concentration through dung accumulation has a visible effect on vegetation growth both in abandoned and around current Maasai bomas (Stelfox 1986). The environmental implications of herbivore nutrient stripping and redeposition are examined in chapter 9.

(f) Stability and resilience in rangeland dynamics

Savannas are characterised by frequent major changes in range conditions due to fire, fluctuating water availability, major changes in herbivore numbers, and fluctuating grazing pressures. However, they are highly resilient, with a strong tendency to recover despite disturbance (Noy-Meir 1982, Walker and Noy-Meir 1982). This is attributed to high reproductive rates of savanna plants under stress conditions; increased growth rates of vegetation at low biomass; spatial heterogeneity which encourages herbivore migration and provides sources of recolonisation; underground reserves; dormancy mechanisms and the species selection flexibility of the herbivore community exploiting the multispecies plant biomass. A temporary trend towards increasing proportions of less palatable species may be part of a series of self-regulating feedback processes associated with a grazing system which has evolved over the past millions of years. Increasing grazing populations eventually lead to a decrease in preferred species and an increase in unpalatable species, leading in turn to a reduction in grazing populations and the subsequent recovery of the palatable species. Such inbuilt fluctuations are to be expected in semi-arid grazing systems (McNaughton 1979, Sandford 1983, Caughley, Shepherd and Short 1987), and recent work suggests that the Ngorongoro-Serengeti grasslands are particularly resilient and show rapid recovery even with very major changes in herbivore populations.

Belsky (1985, 1986a, b, c, 1987) has investigated several aspects of grassland change and resilience in the Serengeti Ecological Unit (SEU). Earlier studies expressed concern about overgrazing. McNaughton (1983) suggested perennial grassland was being converted to annual as a result of overgrazing in localised specific sites. Sinclair (1979) suggested an increase of herb and shrub relative to grass would be expected to have taken place throughout the 1960s and 1970s, associated with herbivore population increases (for example the sixfold increase of wildebeest from 250 000 to 1.5 million – chapter 7). Anderson and Talbot (1965) recorded the presence of 'plant indicators of overgrazing'. However, Belsky's follow up study of long term plots and site descriptions show no detectable change since the first detailed descriptions made 20 years previously. Vegetation mosaics and the majority of individual plants were shown to have remained stable over a ten-year period and there was no evidence of grassland deterioration.

More recently Belsky (1985, 1986a, b) has published studies of colonisation and successional change in the revegetation of experimental disturbances in short, short/medium, and tall grasslands of the Serengeti. In all sites species that regrew after disturbance had all been present originally and in most cases the most abundant were those that had previously been abundant. By one to three years post-disturbance all areas had returned to their previous species composition (Belsky 1986a, b). The frequency of disturbance, soil type and grazing pressure did not appear to influence regrowing species composition. Although repeated severe trampling was shown experimentally to kill all study species, Belsky (1985) suggests that contrary to the findings of McNaughton (1983), Banyikwa (1976) and others, the large mammal impacts of trampling, defoliation and nutrient recycling are not primarily responsible for the 'rich community pattern found in the Serengeti grasslands'. For example, mosaic and tall grasslands remained unchanged by exclosure, though short grassland species gave way to a taller community under these circumstances. Belsky's revegetation studies confirm the theoretically expected resilience and rapid recovery of the Ngorongoro/Serengeti grassland species after grazing and physical disturbance. The Ngorongoro/Serengeti grassland communities also appear to have largely resisted invasion by introduced plant species, unlike many other conservation areas (McDonald and Frame 1988).

Forest and woodland

Woodland and forest areas in NCA contribute to the water catchment function of the Crater Highlands. They are important to the Maasai for dry season grazing, for timber, fuel, resins, medicines and honey, and to wildlife as cover, shade and fodder. The plant species composition of the different forest woodland formations was summarised in Table 2.3.

(a) Woodland utilisation and dynamics

Chausi's (1985) estimates of the areas of NCA vegetation types, based on Herlocker and Dirschl's (1972) vegetation map (Fig. 2.5, Table 2.2), show that highland woodland makes up around 10% of NCA vegetation cover, and lowland woodland 9%. It is thought that most of NCA could support a climax woodland vegetation (other than high altitude communities at >3000 m, evergreen forest and the plains short grass association). The great majority of the potential woodland area has been converted to derived grassland by fire, browsing and physical impacts of large ungulates (Laws 1970, Field 1971, Eltringham 1980, Croze 1974a, Pellew 1983). Human and livestock impacts have contributed to this process. Anthropogenic influences on NCA woodland change include fire, pole cutting, firewood collection, charcoal burning, farming, pastoralism and bee keeping or wild honey collection (Kikula 1981).

Working from admittedly coarse resolution 1:500 000 Landsat images, Kikula (1981) suggested rapid loss of woody vegetation in specific areas in 1972–1975, followed by a slower rate or cessation of clearance in 1975–1979. His analysis suggested the loss of 67 km^2 woodland and 36 km^2 forest plus 5 km^2 bamboo in the first period, and 14 km^2 woodland with 5 km^2 bamboo in the second. The earlier rapid clearance is mostly attributed to cultivation pressure immediately before the 1975 cultivation ban. Kikula quoted independent studies as suggesting the regeneration of many major woody species. Detailed studies using 1982 photography and recent imagery now show areas of definite bush regrowth, including the *Acacia lahai* stands at intermediate altitudes and *Acacia tortilis* at lower altitudes in the west (Perkin pers. commun., 1989)

There are few estimates available for use of timber or fuelwood by any East African pastoralist group, including the Maasai (Chamshama, Kerkhof and Singunda 1989). Fuel use for pastoralists with a largely milk diet is low. For example, the IPAL study found low rates of consumption of wood for fuel among Gabra, Rendille, Boran and Samburu pastoralists in Northern Kenya (0.1 m^3/person/year). Both Fosbrooke (1972) and Arhem (1981b) estimate Maasai collection of firewood as negligible. In a recent survey for the NCDP, firewood was very tentatively estimated at around 1.7 m^3/person/year for intermediate altitude and highland bomas such as Ilmesigio and Nainokanoka. Firewood collection was found to rank low among Maasai priorities and there was a close correspondence between species preferred and those collected. Both findings suggest there is no shortage (Chamshama, Kerkhof and Singunda 1989). Some bomas also commonly use dried dung as fuel, which does not necessarily indicate a shortage of firewood. The survey identified few Masai settlements where fuelwood plantation would be likely to be necessary or successful, and advised against any project with the single purpose of increasing firewood availability for the Maasai. By contrast, Maasai involvement with natural forest management for the full range of purposes including fuelwood, timber, dry season grazing, and other uses such as medicinal plant collection or simple through passage, was seen as an immediate priority bringing considerable long-term conservation benefits (Chamshama, Kerkhof and Singunda 1989).

The NCAA and tourist hotels show a comparatively very heavy fuelwood use. The total energy bill for the tourist lodges went mostly on gas and diesel in 1987. Judging by the figures presented in Chamshama, Kerkhof and Singunda (1989:11, 17) the firewood and charcoal used by the Park village and the three lodges was nonetheless equivalent in amount to the firewood used by some 1900 Maasai households during this same period. Fuelwood for lodges and NCAA commonly comes from Lemagrut *Acacia lahai* as well as from past *Eucalyptus* plantation. Charcoal is commonly (and wastefully) used for space heating in lodge fireplaces designed for firewood (Chamshama, Kerkhof and Singunda 1989). Most NCAA employees use charcoal for cooking, or locally collected fuelwood if charcoal is hard to get.

Maasai boma construction and repair is infrequent compared to that of more nomadic groups. The IPAL study found that northern Kenya pastoralists had to build boma stockades anew with each move, and with an average six moves/year this amounted to a consumption of 1.5–3.0 m^3/person/year. By comparison the Maasai use considerably less timber for construction, given the longer life and regular re-use of individual bomas. Chamshama, Kerkhof and Singunda (1989:7) estimated a timber use for Maasai fence construction of around 0.27 m^3/*boma*/year plus 0.25 m^3/*house*/year. Maasai boma construction types vary from *Acacia* thorn fencing such as is found in Amboseli and in Gol to the cedar stockades common in Ilmesigio. Boma life in Amboseli is 7–10 years; in the Crater Highlands the posts of a cedar-built boma last for 50 years (Western and Dunne 1979, Chamshama, Kerkhof and Singunda 1989, see also chapters 8, 10). Cedar fencing takes considerable time and labour to make and an increasing proportion of NCA families are too poor to afford the long-term investment the cedar stockades represent. The Maasai use of timber and poles are probably negligible in terms of impact on NCA woody vegetation. Broad estimates of timber wood demand and supply suggest that the current cedar offtake from Southern Lemagrut forest is well within sustainable production capacity, and *Acacia lahai* stands are known to have shown measurable spread at Ilmesigio (see Chamshama, Kerkhof and Singunda 1989 for detail on both points).

Drawing on studies of similar formations in Serengeti National Park, Amboseli and other East African conservation areas it is clear that even without clearance by people, forest and even more markedly woodland may show substantial change. Amboseli lost a considerable proportion of its *Acacia xanthophloea* woodland owing to water table and associated salinity changes (Western and van Praet 1973). The same effect seems to have operated in the Lerai swamp woodland inside Ngorongoro Crater where many *A. xanthophloea* trees have died since the 1960s (Kaihula 1983). Between 1962 and 1972 Serengeti National Park is estimated to have lost 13% of its woody cover; locally, losses accounted for up to 50% cover (Norton-Griffiths 1979:314). Similar changes have taken place in Manyara National Park (Mwalyosi 1977, 1981, 1987). These changes can be due to the impact of non-anthropogenic fire, large and small ungulates browsing and causing physical damage, and water table changes. All these factors can be synergistic. For example, by retarding growth and 'escape' of individual seedlings and saplings into the next size class, browsers extend their period of vulnerability to fire. Impacts of browsers and physical damage by elephants together interact to produce a woodland system which oscillates between mature canopy and open regeneration/grassland phases (possibly as a stable limit cycle – Caughley 1976). It is possible that periodic removal of browsers (e.g. by the 1890s rinderpest epidemic) allows the escape of even-age stands. The periodic absence of fire (as a result of runs of low rainfall/low grassland production years) would have the same effect. There are signs that NCA may be going through one of these phases of bush and ultimately woodland regeneration as a result of an interplay of low elephant numbers, low fire incidence and changing livestock and wildebeest pressures. Woodland is growing back vigorously around abandoned NCA waterpoints where concerned conservationists had predicted lasting degradation (Cobb 1989). Pellew (1983) stresses that there is no attainable natural equilibrium as the system is continously in transition. Conservation management often aims to maintain a status quo and is ill-equipped to deal with such variability. Possible interventions to bring about particular management ends range from culling large mammal browsers and elephants to excluding fire (Pellew 1983 for the Serengeti National Park).

(b) Forest utilisation and dynamics

The closed forests (defined as > 80% canopy and understorey cover) comprise around 8% of NCA and are restricted to the Crater Highlands and eastern Rift Valley wall where they make up the Forest Reserve, plus stands of *Juniperus* and *Arundinaria* on Lemagrut and Oldeani respectively. They are of major importance as water catchments both for NCA and for adjacent agricultural areas. They were traditionally used for timber, fuelwood and medicines by both Maasai and WaMbulu, and this use continues illegally now (Struhsaker *et al.* 1989). As well as providing habitat for forest animals, they provide dry season grazing for immigrant wildlife and when permitted (rarely) for domestic stock. Forests, as do woodlands, contribute to the habitat diversity, species richness and ecological interest of the area.

The forests are now being mapped and described in some detail (NEMP 1989) revealing areas of significant loss. Extensive areas near the lower boundary lost more than 75% of the canopy cover in the past 15 years. The Karatu and Oldeani lower margins of the Forest Reserve border on Mbulu settled land, with considerable areas of bare soil and erosion, and show signs of illegal use for timber and fuelwood. Upper level NCA forests along the periphery of the Forest Reserve were subject to heavy clearing in the early 1970s (36 km^2 1972–75: Kikula 1981), possibly as a result of cultivation pressure immediately before the cultivation ban. Both King (1980, 1982) and Kikula (1981) suspected continued encroachment at forest edges on Lemagrut, Oldeani, west of Engaruka, etc. (but on a scale too small to show on Landsat image series). This suspected encroachment is thrown into doubt by more recent findings of forest, woodland and bush spread (Chamshama, Kerkhof and Singunda 1989). Cobb (1989) suggests there has been significant clearance of *Juniperus procera* forest on the highland slopes north of Nainokanoka, on the slopes of the Munge valley and of Olmoti Crater. If this is the case, it may have been a result of the construction of new bomas with the sudden rapid growth of the population in the 1970s (chapter 10). There is no evidence from current levels of use that such clearance is continuing (Chamshama, Kerkhof and Singunda 1989). A very general report on forest status reiterates past fears on pastoralist impacts (Struhsaker *et al.* 1989). However, the more detailed and quantified study by Chamshama, Kerkhof and Singunda (1989) sees pastoralist use as broadly sustainable and puts this in perspective against forest use by NCAA, tourist lodges and by the agricultural villages and estates around the border of NCA. The report concludes:

> Perhaps the greatest immediate threat to the forests of NCA arises from the agricultural communities composed of both peasant farmers and large estates . . .
> The forest produce needs of the Ngorongoro lodges and staff village are currently being met by the importation of fuel from Mangola Chini (due south of Oldeani Mountain) and by the illegal exploitation of NHFR adjacent to the staff village . . .

The impact of the NCA on the Mangola woodlands is severe and totally uncontrolled. The public image of a government organisation which has responsibility for the conservation of an 8000 km^2 area and ravages neighbouring areas for fuelwood is as extraordinary as it is unacceptable.

So far NCAA has found no effective means of protecting the outer margins of the Forest Reserve against increasing needs of local people for fuel and pole wood, nor has the problem had the attention it deserves. The lower forest edges are indisputably more damaged than those adjoining NCA rangelands, and the agricultural and estate fuelwood and construction timber needs here warrant plantation projects on or outside the NCA boundary.

The forest glades constitute another area of debate concerning NCA vegetation management. Several workers in NCA have subscribed to the idea that forest clearings, like the Crater Highlands forest edge mosaic of woodland and bush, are due to former agropastoralist impact (Dirschl 1966, Herlocker and Dirschl 1972, Frame 1976). The formation of these glades is attributed to a sequence of factors similar to those responsible for the forest glades in Western Mt. Kilimanjaro: predisposing soil and microclimate; dry season grazing and intermittent agropastoralist occupation, and finally burning and trampling which maintain and extend the clearings.

It has been suggested that use of these forest clearings for dry season grazing may cause compaction of the soil and changes in vegetation cover, which may contribute to reduced water infiltration and increase surface runoff, effects detrimental to the water catchment function of the forest reserve (Hamilton 1984). It is on these grounds together with concern over general forest destruction that dry season grazing has been banned in the Forest Reserve since the late 1960s. Permits, however, are given for grazing in the larger glades (e.g. Rotian), and an estimated 26 000 cattle were grazing within the Northern Highlands Forest Reserve in the dry season of 1988 (Struhsaker *et al.* 1989: the basis for this estimate is not given). NEMP (1989) draw attention to the great habitat and species diversity ensuing from such differential use, but caution against excessive pressure, especially from frequent fires. Chamshama, Kerkhof and Singunda (1989) see clarification of the forest grazing permit system as one of the main issues in NCA forest conservation.

Recommendations on the utilisation of the NCA forests, as for the montane forests of Mt. Kulal and other IPAL areas, stand in contrast to the traditional conservationist stance on subsistence use of forest resources (Chamshama, Kerkhof and Singunda 1989 and Synnott 1979, versus e.g. Struhsaker *et al.* 1989). Both Chamshama's and Synnott's reports outline types and extents of land use compatible with forest conservation aims. They recommend controlled removal of firewood and building poles together with the use of drought refuge grazing in the montane forest 'not as a privilege but as an accepted part of the land use plan'. They also make clear that wherever accustomed resources are restricted for conservation reasons, adequate provision must be made to furnish a substitute. Where this has not been done the effect is to accelerate environmental deterioration on surrounding resources. This general principle should be given careful consideration in the case of NCA.

Summary and conclusion

1. The diversity and high productivity of the rangelands are the key to the herbivore diversity and productivity and therefore the key to management in NCA. The main management debates centre on grazing resource availability and pastoralist impacts on range.

2. Rangelands show long-term fluctuations in vegetation structure and productivity due to fire, grazing, trampling and nutrient recycling by hervibores. These various factors show different degrees of importance in different zones.

3. Heavy grazing pressure has been suggested to be a problem in the lowland plains but the evidence of detailed vegetation studies shows this is not the case. Trampling has caused localised degradation in the plains and subsequent chapters investigate the extent to which wildlife and livestock respectively may be bringing this about. In the medium grasslands fire is a major factor and its use the subject of major management debate. There is also disagreement firstly as to whether unpalatable *Eleusine* tussock grass is spreading in the highland grasslands, and secondly as to whether this is caused by livestock overgrazing, or by bans on burning.

4. Savanna grasslands have evolved with fire, at both species and community levels. Stopping burning can adversely affect productivity and species composition and this appears to be the case in Ngorongoro. There is a need to reintroduce controlled burning programmes in the medium and tall grasslands of the intermediate and high altitude zones. Long-term pastoralist grazing and burning have helped shape the present day rangelands, and have a similar role in the future. Simple measures can protect forest edges from fire.

5. NCA rangelands like other savanna grasslands are unstable in terms of showing long-term fluctuation in response to changing rainfall, fire regime and herbivore populations. However, like other savanna grasslands they are very resilient in terms of their tolerance to disturbance and their tendency to return to original species composition and vegetation structure. There seems to be little cause for concern over environmental degradation. Isolated areas of change are due to several causes, and pastoralist impacts are by no means the sole or even the main agents of change.

6. NCA woodlands showed measurable losses of canopy cover during the 1970s, but measurable regeneration is now taking place. Much of the earlier loss was due to clearing for cultivation prior to the 1975 ban. Timber and fuelwood collection by Maasai pastoralists has a minimal effect. Other factors driving woodland dynamics include water table changes, browsing herbivores and physical damage by large herbivores (particularly elephant).

7. Forests are important for the conservation of soil, water and of biotic diversity in NCA. Pastoralist impacts on forest appear to be minimal and sustainable; pressure on forest resources by NCAA, tourist lodges and agriculturalists surrounding NCA is less so. The controlled use of forest glades for dry season grazing and of the forests for collection of medicines, timber and fuelwood should be planned and catered for.

8. The Ngorongoro/Serengeti system is seen by professional ecologists as a dynamic mosaic of vegetation types. It is forever changing, but retains diversity and illustrates the natural processes of plant–herbivore population interaction (as well as anthropogenic effects) that modern conservation areas seek to display.

9. There is a dearth of basic ecological information on the NCA range resources, especially in the montane grasslands. The recently reestablished Ngorongoro Ecological Monitoring Programme (NEMP) presents an opportunity for a stronger research input equivalent to the MacNaughton–Belsky studies of the short grass plains. This will need strong and continued scientific and financial support.

7

Wildlife

Metii oidipa oldoinyo ake oidipa otunokine ewueji nemedotunye.
It is only the mountains which do not move from their places.
(Animals, plants and people all migrate or change – Maasai saying: Mol 1978)

Bitir akenya
'The morning is like a warthog'
(It is unpredictable – Maasai saying:Kipury 1983)

So much has been written on the wildlife of the Ngorongoro and Serengeti ecosystems that we do not attempt to summarise it all here. Instead this chapter concentrates on the main patterns in the wildlife community, and on management themes. The term 'wildlife' has changed over the past few decades, from meaning large game animals worthy of sport hunting, to to the present connotation of whole communities of wild plants and animals linked by intricate patterns of ecological interdependence. The main features of these communities were outlined for NCA in chapter 2, and factors governing the plant resources were discussed in chapter 5. In this chapter we introduce the main animal components, and outline the major natural processes governing the interactions and dynamics of the wild animal populations of NCA – migration, competition, disease, predation and plant–herbivore interactions. We then go on to select specific management issues: wildlife management goals; the concept of naturalness and the extent and desirability of change; species and areas of particular conservation importance; subsistence hunting and ivory poaching. Later chapters compare the ecology of NCA wildlife with that of domestic stock, and put NCA wildlife populations in perspective against other joint pastoralist/wildlife grazing systems in East Africa (chapter 9). Wildlife tourism is dealt with in chapter 11.

The large mammal herbivore and attendant carnivore communities of Serengeti/NCA are unique among conservation areas in Africa in terms of the scale of species diversity, the population numbers involved (Table 7.1), and the range of taxa of conservation significance (East 1988). This array of species can be divided into forest and grassland communities. The grassland communities are biologically the most diverse and striking and hence the most studied; they are also those most closely interwoven with the ecology of Maasai and their livestock. The forest community is no less important in its contribution to the

Table 7.1 *Large mammal populations in the Serengeti/Ngorongoro system*

Species	Scientific name	SNP–NCA population	NCA population	Source (estimates date late 1970s–late 1980s)
Carnivores				
Lion	*Panthera leo*	2000–2400	200	6,7
Leopard	*Panthera pardus*	1000	?	7
Cheetah	*Acinonyx jubatus*	200–500	100?	7
Wild dog	*Lycaon pictus*	150–300	?	7
Spotted hyena	*Crocuta crocuta*	3000–5000?	1000?	7
Striped hyena	*Hyaena hyaena*	rare	rare	
Aardwolf	*Proteles cristatus*	rare	rare	
Jackal – Golden	*Canis aureus*			
– Side striped	*Canis adustus*			
– Blackbacked	*Canis mesomelas*			
Herbivores				
Elephant	*Loxodonta africana*	1800	300?	5
Rhinoceros	*Diceros bicornis*	< 100	50?	see text
Zebra	*Equus burchelli*	200 000	63 000	9,10,11
Hippopotamus	*Hippopotamus amphibius*	?	?	
Giraffe	*Giraffa camelopardalis*	7200	1000	8
Bushpig	*Potamochoerus porcus*	?	?	
Warthog	*Phacochoerus aethiopicus*	5000	?	6
Buffalo	*Syncerus caffer*	67 000	3500	9,11
Eland	*Taurotragus oryx*	20 000	5400	1,11

Bushbuck	*Tragelaphus scriptus*	1000–2000	1000 +	1
Waterbuck (2 subspp.)	*Kobus defassa*	3000	120 in Crater	1,2
Reedbuck	*Redunca redunca*	1700	?	1
Mountain reedbuck	*Redunca fulvorufula*	Very rare	< 200	1
Roan	*Hippotragus equinus*	100–300	0	1
Oryx	*Oryx beisa*	Absent	400?	1
Topi	*Damaliscus korrigum*	55 000	Very rare	1
Hartebeest	*Alcelaphus buselaphus*	10 000–20 000	350	1,2,11
Wildebeest	*Connochaetes taurinus*	1 300 000	8300–1 100 000 (seasonal)	1,2,3,10,11
Impala	*Aepyceros melampus*	120 000	3300 (none in crater)	1,2,11
Oribi	*Ourebia ourebi*	3000	?	1
Klipspringer	*Oreotragus oreotragus*	500	?100–200	1
Steinbuck	*Raphicerus campestris*	2500	Frequent?	1
Dikdik	*Madoqua kirki*	32 000	Common?	1
Grey duiker	*Sylvicapra grimmia*	2500	?	1
Blue duiker	*Cephalophus monticola*	Rare	Absent	1
Grant's gazelle	*Gazella granti*	52 000	6300	1,2,11
Thomson's gazelle	*Gazella thomsoni*	250 000	5800–150 000 (seasonal)	1,2,11
Primates				
Olive baboon	*Papio anubis*	Common	Common	
Blue monkey	*Cercopithecus mitis*	Rare	Common	
Vervet monkey	*Cercopithecus aethiops*	Common	Common	
Black and white colobus	*Colobus guereza*	Very restricted	Absent	

Source: 1. Rodgers and Swai in East 1988; 2. Estes and Small 1981; 3. Sinclair *et al.* 1985; 4. Rodgers 1981b; 5. Dublin and Douglas-Hamilton 1987; 6. Hanby and Bygott in Sinclair and Norton-Griffiths 1979; 7. Bertram in Sinclair and Norton-Griffiths 1979; 8. Grimsdell in Sinclair and Norton-Griffiths 1979; 9. Sinclair and Norton-Griffiths 1979; 10. Malpas and Perkin 1986; 11. Boshe 1988.

overall diversity and conservation values of NCA, but in biological terms is less unusual and less well studied. We concentrate mainly on the grassland communities.

Grassland herbivore community

In the three decades over which they have been studied, the animal populations of the Serengeti/NCA grasslands have undergone some major changes. These have been most conspicuous in the migratory herds that dominate the animal biomass (sections (*a*) and (*b*) below). The Serengeti/NCA wildlife community is one of the best documented in the world, but there are still conflicting ideas as to the major determinants of plant–herbivore interactions and population dynamics. The importance of competition in triggering population eruptions, and in limiting their extent, is still a matter for debate (section (*c*)). The relative roles of different plant–herbivore interactions in driving or stabilising woodland–grassland oscillations are also in dispute (section (*d*)).

(a) Community structure

Sinclair (1975) estimated the biomasses of large herbivores, small mammal herbivores and herbivorous invertebrates in the Serengeti grasslands as 22.9, 0.4 and 0.8 kg/ha respectively in the long grass areas and 17.9, 0.1, 0.3 kg/ha in the short grass areas. He estimated that they accounted respectively for around 19%, 1% and 8% offtake of plant production in the long grassland and 34%, 0.1% and 4% offtake in the short grassland. Fire and detritivores accounted for the remainder. There is no doubt that the large mammal herbivores dominate plant–herbivore interactions.

The wild ungulates total some three million head for the Serengeti Ecological Unit (Ecosystems Ltd 1980:5; Malpas and Perkin 1986). A wet season aerial census estimated 1.55 million large mammals in NCA (Ecosystems Ltd 1980:10). Of these an estimated 250 000 were permanent residents of NCA, and the rest migratory animals using NCA between December and April. Carnivores make up only around 7% of the large mammal biomass. The migratory herbivores dominate the large mammal community.

(b) Migratory herds

It is the migratory herds of the Serengeti/Ngorongoro system that have captured international imagination. The largely migratory wildebeest, Thompson's gazelle and zebra account for around 86% of the SEU large herbivores biomass (Serengeti Ecological Monitoring Programme 1987). Current NEMP figures together with the Ecosystems report (1980:163) suggest that the seasonal influx of migrants pushes NCA wildlife biomass up to an average 11 200 kg/km^2 from December to April, with local concentrations reaching far higher values. Migration allows considerably greater biomasses to accumulate than would be the case for resident populations (Sinclair and Fryxell 1985, Fryxell and Sinclair 1988a,b). The particular characteristics of the wildebeest – body size and gait (Pennycuick 1979), gregarious herding behaviour, resistance to malignant catarrhal fever (MCF: see below and chapter 9) – allow them to monopolise resources and dominate the system (McNaughton 1979, 1983). This predominance is at the expense of domestic livestock (see chapter 9).

Figure 7.1 shows the extent to which NCA is involved in the migration during the wet season (redrawn from Pennycuick 1975 and the April 1987 NEMP aerial census). The migration is described in detail by Pennycuick (1975), Inglis (1976) and Maddock (1979). Wildebeest are short grass grazers and concentrate on the short grass plains in the wet season where their movements follow sporadic localised rain and new grass growth. With the onset of the dry season they move west and north to the longer grass savannas of higher rainfall areas of the Serengeti and Mara. McNaughton (1988, 1990) explains range use by SEU migratory wildebeest in terms of nutrient and especially mineral requirements. Past analyses have invoked gereral forage qualities, avoidance of waterlogged, muddy, sticky substrates and of high predation risk areas as additional factors.

Figure 7.1 shows rather minor differences between the 1960–1970s distribution and that of April 1987. Although the data are not available in strictly comparable form, long-term aerial monitoring suggests that from the late 1970s a markedly greater proportion of the migratory herds have used the Angata Kiti and Salei plains of NCA as far east as Mosonik, some moving north with the dry season into Loliondo GCA and then west to the North Serengeti (Norton-Griffiths pers. commun.). This has affected forage resources for the pastoralist community around the Gol Mountains. With a sixfold increase in numbers (see below) and a greatly extended calving period, the wildebeest monopolise the short grass plains, and the Maasai in Gol frequently resort to thornbush barriers to keep the migrants off the adjacent critical hillside dry season grazing areas. In the past the cattle could be moved either westward into the Serengeti (mainly around Moru), or eastwards to the Highlands. The choice no longer exists.

In the south of NCA there has been an increase in the wildebeest use of the wooded grasslands joining the Endulen/Kakesio ridge. This may be the result of one or more factors, including Maasai withdrawal due to stock theft (chapters 8,9), woodland opening by elephant, changed patterns of grassland production or simply pressure of numbers. Together with the sixfold increase in

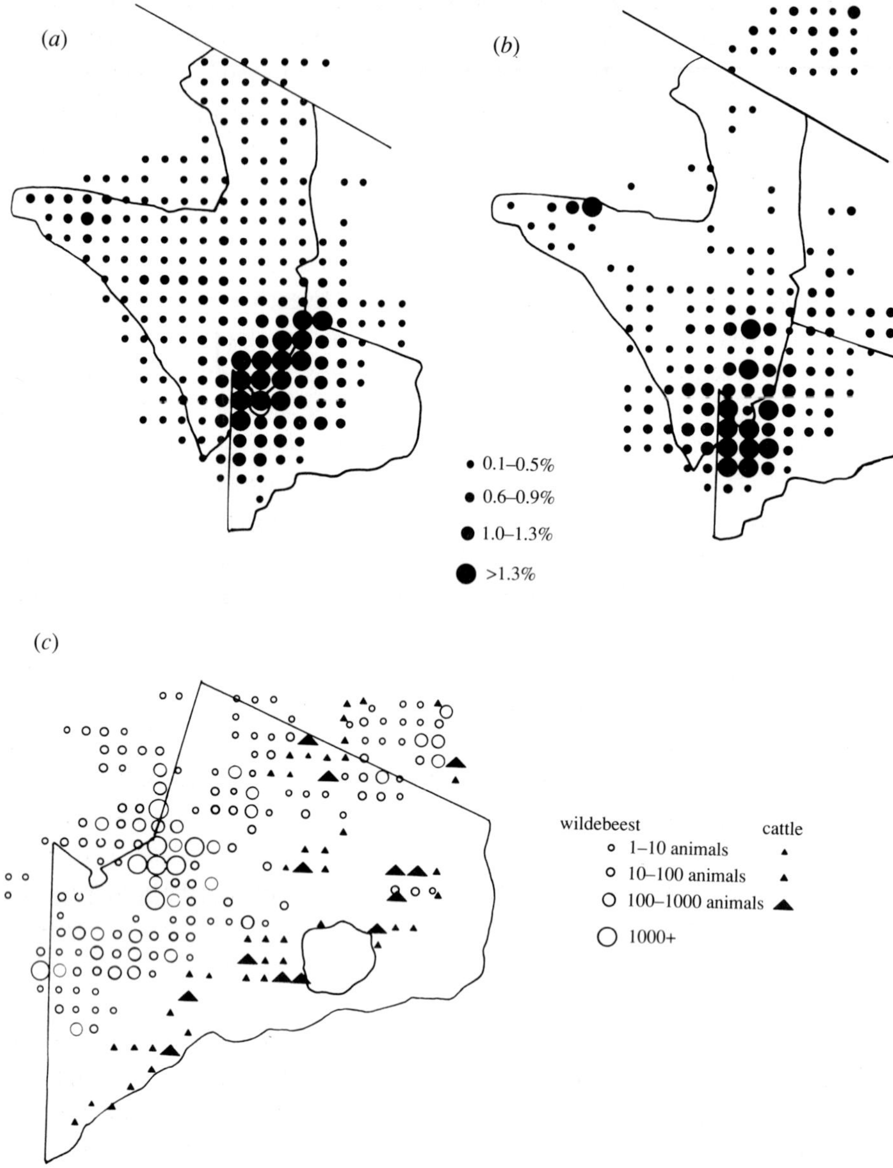

Fig. 7.1. Wet season wildebeest distribution derived from aerial systematic reconnaissance flight survey. (*a*) December–April period, 1960–1969 (Pennycuick 1975), (*b*) December–April period, 1969–1972 (Pennycuick 1975), (*c*) April 1987 NCA area only (NEMP). (*a*) and (*b*) relative density of wildebeest, (*c*) estimated numbers of wildebeest and cattle per 5 × 5 km² grid square.

wildebeest numbers these distribution changes have affected the availability of fodder for domestic animals and exacerbated pastoralist perceptions of wildlife conflict for the Gol/Ol Balbal and Endulen/Kakesio area (Ole Parkipuny 1981, 1983). The wildebeest occupation of the Kakesio/Osinoni wooded grasslands and woodlands has continued despite recent decreases in the total wildebeest population (Cobb 1989:72).

(c) Interspecific competition, facilitation and predation

Bell (1970) presented evidence for a grazing succession, whereby a sequence of species (zebra, wildebeest, Thompson's gazelle) with different grazing requirements pass through the Serengeti/NCA grasslands, each modifying the pasture to a condition favourable to the next species. Similar successional sequences have been suggested for Amboseli (Western 1973, 1975) and the Sudd (Mefit-Babtie 1983, Fryxell and Sinclair 1988a), domestic stock being included in the sequence in both cases. However, McNaughton (1988, 1990) explains species concentrations in terms of local soil and plant mineral composition. Sinclair and Norton-Griffiths (1982) point out that SEU population data for wildebeest, zebra and Thomson's gazelle for 1960–1980 do not show the linked increases in all three populations that grazing succession theory would predict. Instead the zebra population has remained constant, and the wildebeest increased from around 200 000 to 1.4 million (down to one million in a recent count – SEMP 1988 quoted in Cobb 1989:72). The Thomson's gazelle population showed a fall-off in the mid-seventies (Fig. 7.2). They interpret these changes as being more consistent with the migratory wildebeest being ultimately food limited, the Thomson's gazelle being limited by interspecific competition with wildebeest and the zebra population being controlled by predation. Wildebeest dry season mortality can be predicted from dry season food availability and is of an order to limit the wildebeest population overall (Sinclair, Dublin and Borner 1985). In the Mara, Sinclair (1985) found that a mixture of interspecific competition and predation influenced most of the large herbivore populations, with facilitation and intraspecific competition being less evident. Historically, these wild herbivore populations have been strongly affected by competitive and disease interactions with domestic herds. The extent to which such influences still affect the wild ungulates, particularly over the last few decades, is discussed in chapter 9. Recently, increasingly severe poaching in the Western Serengeti may have begun to affect population size (Perkin pers.commun. 1989).

(d) Plant–herbivore dynamics: changes in woody vegetation

The overall resilience of the grasslands in the face of grazing has already been discussed (chapter 6) as has the scale of change shown by the woodland areas. Several woodland areas in NCA have shown a decline in canopy cover, a decrease in total woodland extent, and/or a failure of regeneration over the last few decades. Anthropogenic impacts (e.g. pre-1975 clearing for cultivation in the Endulen/Kakesio zone) and the effects of changing groundwater levels (e.g. for the *Acacia xanthophloea* Lerai Forest in the Crater) are both involved. However, wild herbivores also play a large part in woodland change. The 'elephant problem' – large scale damage to woodland by high density elephant populations, and their control through culling – became a major management controversy in the 1960s and 1970s (Laws 1970, Eltringham 1980, Field 1971, Owen-Smith 1983). With the rise in elephant poaching in the 1970s the woodland damage aspect of the elephant problem lost urgency. However, NCA is sandwiched between two areas where large mammal damage to woodland has been studied in detail: the Seronera woodlands (Croze 1974 a,b, Pellew 1983) and Manyara (Mwalyosi 1977, 1981). Although not studied in depth in NCA, elephant, giraffe and other herbivore

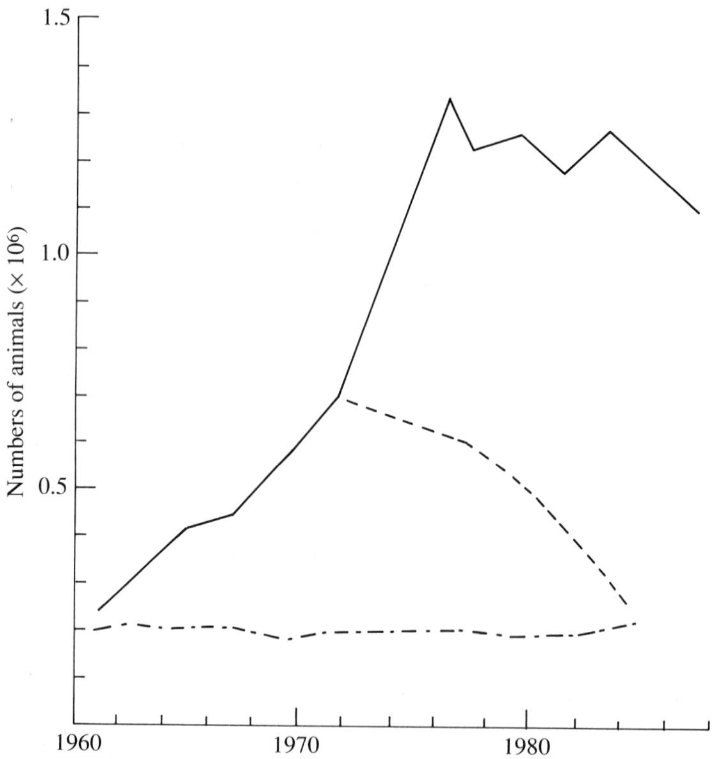

Fig. 7.2. Changes in main large wild herbivore populations of the Serengeti Ecological Unit including NCA (after Malpas and Perkin 1986). —— wildebeest, --- Thomson's gazelle, ·—· zebra.

damage is implicated in the decline of woodland around Olduvai and Lake Ndutu (where wildebeest cause considerable damage to young acacias), and in Lerai, Loitoktok Springs and Laiyani in the Crater (Kaihula 1983).

Pellew (1983) has modelled woodland dynamics in the Seronera area, and his discussion of the main factors, including both wildlife and past Maasai livestock effects, is relevant to NCA. The woodlands of Serengeti and by extension of NCA are seen as 'a fluctuating mosaic of change, in which each constituent cell of the mosaic will be in a different stage of the cycle'. Changes in wildlife populations make for major extrinsic perturbations. For example the system has been influenced by rinderpest decimating giraffe; ivory hunting decimating elephants; and low grazing pressure together with high rainfall allowing high standing grass biomass and exacerbating fire effects. Size class frequency data and tree girth increment rates suggest Lerai may have had a period of intensive regeneration at the turn of the century when elephant were heavily hunted for ivory (Spinage 1973, Kaihula 1983). The current crash in rhino numbers due to poaching is probably associated with a concurrent reduction in elephant in the 1970s and 1980s (see below). This would tie in with the possibility of a current period of woodland regeneration in NCA, as happened for example in Tsavo and in Uganda, and as is suggested for NCA by current analyses of aerial photographs and satellite imagery (see chapter 6). Pellew (1983) tentatively attributes the existence of older (pre-1888) stands in the Serengeti to the pre-Park and pre-rinderpest Maasai presence. He suggests that before the Maasai were excluded, higher grazing pressure in times of greater cattle numbers may have suppressed fire, encouraged bush encroachment and allowed a period of woodland regeneration. Again, similar effects of a pastoralist presence may apply in NCA at the moment. In addition to the large mammal impacts reviewed by Pellew, Belsky (1984) presents experimental evidence to suggest that the smaller browsers (Thomson's and Grant's gazelle, dikdik, impala) slow the growth of the smaller size classes of woody plants. Pellew (1983) stresses that the dynamic mosaic of successional stages of regeneration and mature woodland does not allow for the maintenance of any one specific equilibrium state, despite conservation management's desire to do so. Changing plant–herbivore equilibria in NCA/Serengeti do not indicate a need for control of the herbivore populations involved.

Forest wildlife populations

The forest and its wildlife are in a national and international context biologically less outstanding than the NCA grassland communities. However, they make an important contribution to the diversity and conservation values of the NCA as a whole, and the watershed protection role of the forest is vital not only for NCA water supply but also for the comparatively densely settled and farmed Mbulu areas to the east.

The NCA forest wildlife is species-poor (chapter 2) and has not been well studied. What little is known, apart from a basic species list, derives from observations on the forest margins and adjacent habitats. Of the large mammals, buffalo, rhino, elephant and bushbuck are present and waterbuck are found in the larger clearings. Giant forest hog are rumoured to exist in the Oldeani bamboo (an unconfirmed sighting in 1975; also Child 1965:89) – if these records are correct, this is their only locality in Tanzania. The persistent doubt is symptomatic of the lack of knowledge on NCA forests. Black and white colobus are mentioned by Fosbrooke (1972) but do not exist in NCA. A common guenon (Sykes or blue monkey) may be seen. The forest buffalo and elephant populations also use the Crater walls and floor (Rose 1975; Kiwia, Kabigumila pers. commun.). The forest wildlife populations come into contact with subsistence and commercial farming along the eastern and southern borders from Mangola/Oldeani/Karatu/Mbulumbulu to Lositete (Fig. 7.3). They cause local crop damage, and compensation or crop protection measures must be considered, especially where illegal hunting, pole cutting or firewood collection are to be discouraged.

It is likely that there was originally continuous forest cover from the Ngorongoro forest to Lake Manyara (Douglas-Hamilton 1972) allowing elephant to move between two dry season concentration areas. Tsetse clearing and subsequent Mbulu settlement in the 1940s cut this corridor and today only a fragmented forest strip remains from Kitete and Lositete down the Rift wall to tiny isolated groundwater patches north of Mto wa Mbu (Makacha and Frame 1977a,b). NCAA is not well equipped to conserve forest habitats effectively. There are plans for larger-scale irrigated agriculture in this area. Together with the notorious difficulties of trying to maintain isolated elephant groups amid other developments, these factors preclude rehabilitation of this migratory route (Fig. 7.3 and Rodgers 1981a). The NCA elephants are thus an isolated population, as settlement around Mangola has effectively severed the past southern linkage around Lake Eyasi to Yaida Chini.

Management policies and problems

(a) Wildlife administration

The wildlife resources of Tanzania belong to the government irrespective of whether they are on public or private land. Their management is entrusted to the Wildlife Division of the Ministry of Natural Resources. It controls two parastatals (Tanzania National Parks Authority and NCAA) each with its own ordinances and governing body. It also has the civil service staff administering Game Reserves (in which there is no settlement or cultivation, and limited wildlife exploitation is strictly controlled) and Game Controlled Areas (in which settlement and cultivation are allowed, but hunting is controlled). The wildlife of the Serengeti/NCA ecological unit while ostensibly under a single Ministry is controlled in practice by three distinct organisations each with its own legislation. It covers three separate regions, each with different wildlife policies and priorities (Arusha/NCA; Musoma/SNP and Shinyanga/Maswa GR); and when animals migrate to Kenya they come under the control of another country. Each organisation involved maintains its own infrastructure and administration, and so while in terms of GNP or per capita income expenditure on the management of the resource is high, there is often considerable overlap and wastage. In real terms, and with deteriorating national economies and inflation, actual expenditure on wildlife has greatly decreased during the 1980s. Boshe (1988), in a brief overview of wildlife status

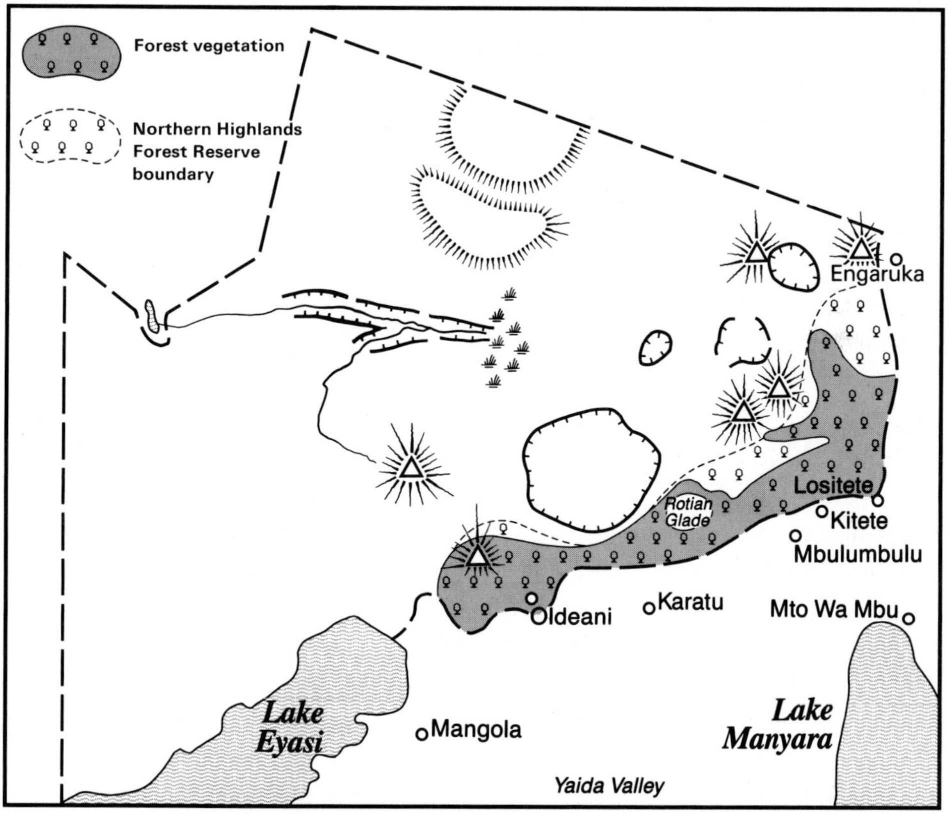

Fig. 7.3. Settlement along the borders of the Northern Highlands Forest Reserve.

for the IUCN study, was unable to identify any wildlife situation or species causing concern to management or biologists. This is in part due to the absence of any management objectives.

(b) Naturalness and change

The well-documented changes in plant and animal communities this century are viewed in different ways by different people. Some changes, such as the sixfold increase in wildebeest, are seen as desirable in wildlife management terms; others, such as the overall decline in woodland, as deleterious. The welfare of the wildlife resources of the NCA continues to be the main focus of attention of management authorities and conservationists in general. However, with NCA's history of incomplete management studies and unconfirmed management plans, there is a lack of accepted and clearly defined objectives for wildlife management. It is thus difficult to evaluate resource status, let alone suggest causal factors that may be threatening environment or wildlife conditions.

Shepherd and Caughley (1987) give a concise and entertaining review of the range and sequence of successive fashions in conservation philosophy and management, summarised below in approximate chronological order:

1. Preservation of scenery and 'nice' animals (and elimination of 'ugly species' – e.g. past shooting of wild dog and hyena in NCA)
2. Conservation of soil and plants (post-1930s US dustbowl)
3. Preservation of a single specific physical or biological state of the area (usually as first observed by Europeans)
4. Conservation of representative plant and animal associations
5. Conservation of biological diversity – species richness and variety of community types
6. Conservation of genetic variability
7. Conservation of biological processes – 'the resource is *wildness*'

New objectives are often added onto the old rather than replacing them. These conservation objectives differ in kind: most are about conserving states, but the last is about processes. The objectives may not be mutually compatible.

These various aims and the contradictions they enshrine are apparant in NCA conservation policies. The East African policy of non-interventionist or laissez-faire management prefers to monitor rather than regulate natural changes over large areas (Sinclair 1983b). This is in contrast to South African policies of fine-tuning manipulative management involving culling and reintroductions (Pienaar 1983). NCA wildlife management now involves a compromise between actively conserving species and physical components that could be radically affected by natural processes, and valuing the area's changing states as a demonstration of such natural processes. This compromise makes it necessary to define the range of states which are acceptable, and to keep enough control to prevent unacceptable extremes which otherwise natural processes might bring. A changing mosaic of forest, woodland and grassland should be tolerated, even where the relative proportions of the communities alter noticeably; near or complete loss of a vegetation formation would not be acceptable.

With a growing scientific understanding and acceptance of the way herbivore populations wax and wane, and of plant–herbivore interactions, there is less management pressure now to stabilise vegetation states and herbivore populations, or to recreate systems as seen at some previous date. However, there is still extensive debate about related management issues, for example carrying capacity and stocking levels, particularly where both wildlife and domestic stock are involved (Shepherd and Caughley 1987, Homewood and Rodgers 1987). For arid and semi-arid areas the variability of rainfall and of primary production is such that carrying capacity derived from their long-term means is not a useful concept, while their patterns of variability may be fundamental to understanding the system. Linked to the debate over carrying capacity is the persistent idea that reducing a herbivore population must mean a healthier population and a healthier environment. With culling, while surviving individuals may grow larger, suffer lower natural mortality, and show higher fecundity, and the sward carry a higher standing crop, the system tends to lose resilience and becomes more vulnerable (Shepherd and Caughley 1987). It is also no longer 'wild'.

Once the decision has been made that conservation of natural biological processes is among the main objectives, and that change is expected and tolerated, a whole series of potential management interventions become unnecessary and even undesirable. For example, elephant culling has been proposed as a way to reduce woodland damage (see Kaihula 1983 for a recent mention) but rejected by others (e.g. Pellew 1983). Capture and dehorning of rhino was suggested as one way to reduce their attraction to poachers. Other interventions which have recently been considered in NCA are reducing buffalo numbers in the Crater and cropping wildebeest to reduce impacts on the ecosystem. None of these measures has been approved or implemented. Water resources have not been developed specifically for wildlife use, though wildlife benefit from some sources that have been developed for pastoralists. Wildlife control is occasionally carried out when stock (or crops in neighbouring areas) are lost; the raiding animals, or others nearby, are then likely to be shot.

(c) Managing the craters

The exception to the overall policy of tolerating natural change within very broad limits may be the management of Ngorongoro Crater itself (Fig 7.4). Ngorongoro Conservation Area to most people means the Crater. It is the Crater that most visitors want to see, and it is the Crater (together with its buffer forest areas and access to Serengeti), that most conservationists want placed under National Park style management. This 250 km² area is a microcosm of the great variety of habitat types, species and to some extent the processes of the Serengeti/NCA system.

Despite the apparent ease of study of the tame populations in a restricted area, neither individual wildlife species nor community interactions in the Crater have received much research attention. Though existing at very high density, the Crater wildlife populations are small compared to those of the rest of the Ngorongoro/Serengeti system (Tables 2.4, 7.1), and are not self contained. There is evidence that wildebeest, zebra, elephant and buffalo migrate in and out of the Crater (Table 7.2, Estes and Small 1981; Kabigumila

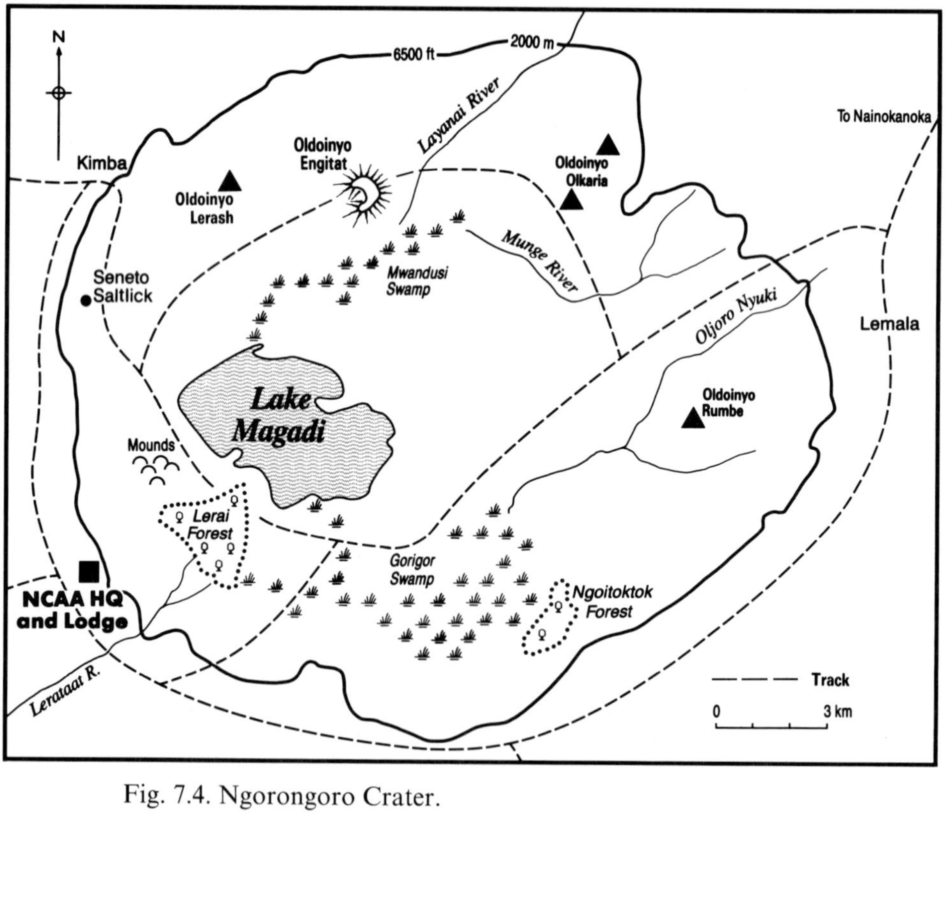

Fig. 7.4. Ngorongoro Crater.

Table 7.2 *The wildlife populations of Ngorongoro Crater (after Estes and Small 1981 and NCAA 1987)*

	Year									
Species	1964 Feb.	1968 March	1970 Dec.	1971 Aug.	1973 Jan.	1976 April	1977 Sept.	1978 Feb.	1986 July	1987 April
Elephant	28	11	20	7	56	42	5	17		
Rhinoceros	27	3	25	8	39	34	16	21	2	9
Zebra	5 038	3 058	2 596	5 523	3 286	5 306	3 312	3 005	4 297	3 127
Hippopotamus	23	2	0	26	0	33	25	47		
Buffalo	11	25	18	0	1 279	1 508	109	1 164	1 455	2 714
Eland	342	355	240	98	499	574	176	177	59	64
Waterbuck	35	26	63	20	49	20	6	29		
Hartebeest	49	19	136	154	150	249	176	129	72	70
Wildebeest	14 922	14 417	17 597	16 797	13 422	10 059	14 451	13 376[c]	11 847	9 011
Grant's gazelle }	2 310	1 376[a]	1 376	1 492	1 833	2 346	1 507	1 361	2 136[b]	
Thomson's gazelle }		4 269[a]	2 860	5 166	2 778	3 407	2 827	3 125	3 312[b]	
Ostrich	37	30	32	13	37	45	23	31	26	23

Notes:

[a] Gazelle data from September 1968.

[b] Gazelle data from strip sample.

[c] Another count in February 1978 gave 9587 wildebeest.

1988; Rose 1975 etc). However, the greater part of the 10 000–20 000 wildebeest are probably resident. Dry season counts 1964–1978 average some 3000 head more than wet season ones, indicating relatively minor seasonal movement (Estes and Small 1981). Some carnivores (cheetah, wild dog) are only present intermittently. The lion population is one of the most dense in Africa (reflecting the exceptionally high year-round preferred prey density) and maintains stable numbers through the emigration of a high proportion of subadults, with no immigration recorded for the last decade. The Crater lion population thus represents an isolated population (van Orsdol 1981, Pusey and Packer 1987). A few species have been studied in some detail in the Crater (wildebeest and gazelle: Estes 1966, 1967, 1969; rhino: Goddard 1967, 1968, Kiwia 1983; elephant: Kabigumila 1988; serval: Geertsema 1985). Many others have been the subject of brief comparative studies by SRI scientists (see Sinclair and Norton-Griffiths 1979). Herbivore populations in the Crater appear to be partly limited by grazing resources (see below), predation (judging by the relatively high numbers of lion and clan-hunting hyaena – Kruuk 1972,1975) and to some extent by net emigration (Estes and Small 1981).

With the possible exception of rhino (and maybe lion) the Crater populations are not of prime importance in biological terms (Table 7.1; section below on rhino). Also, while some components are particularly easily observed (e.g. lions), some of the major processes that make the rest of NCA so interesting are less evident (e.g. the mass migrations) or missing altogether (e.g. the juxtaposition of wildlife and pastoralist stock). The Crater is, however, of enormous importance for wildlife viewing and thus for conservation education and awareness as well as for generating revenue. The variety of habitats, the concentration of large wildlife, and the spectacular scenery make the Crater a tremendously important showcase and symbol: so, while it lasts, does the Crater rhino population. All these factors suggest that the Crater may be a candidate for more intensive and more interventionist management while the rest of the NCA wildlife area is better left to more natural processes.

Fire management is one example. In NCA as a whole, fire management could to a large extent be left to pastoralist range management, with the proviso that forest edges must be protected against grassland encroachment. In the Crater, the exclusion of pastoralists and changing policy has meant that few if any fires took place on the Crater floor from the late 1960s to 1982, while the Crater walls burnt frequently with some loss of forest cover. On the Crater floor the coarse ungrazed and unburnt grass was avoided by wildebeest and gazelle (for example around Korigor Swamp) and tick numbers rose dramatically. Estes and Small (1981) commented that the average total 14 000 wildebeest of the Crater used only the short grass communities limited to under 40% of the area.

In 1982 no-burn policies were relaxed and controlled fires set, and wildebeest and zebra used the post-burn flushes. All burning then stopped again and the NEMP (1989) report the accumulation of rank unpalatable and tick-infested forage, with adverse results for the herbivore populations. A severe wildfire briefly threatening crater rim lodges has led to the reintroduction of fire breaks and of a fire control labour force. NCAA staff say a controlled fire policy must await more research and more equipment. There is also a case for special measures protecting particular habitats (for example encouraging regeneration and maintenance of Lerai Forest) or species (e.g. rhino) and reintroducing or restocking others (again rhino, if the Crater population can be better guarded in future). Intervention is likely to become necessary in the near future controlling tourist access to and impacts on the Crater, as it could rapidly deteriorate under current pressures and with the current lack of regulation (chapter 11).

Similar intensive management might be in order for the spectacular but remote Empakaai Crater (Frame 1982). Olmoti Crater like the other craters has in the past been an important dry season grazing and watering area of Maasai livestock. It is little used by wildlife. Pastoralist livestock were originally banned because of feared *Eleusine* encroachment, but the tussock grass is spreading despite two decades without livestock grazing. Unlike the other craters there is little reason to continue to restrict access to Olmoti and strong reason to open it again to Maasai use.

(d) Subsistence hunting

The Maasai traditionally have little interest in hunting wildlife; they may eat certain species of wild herbivore only in times of severe famine, and value few trophies other than those from rare hand to hand combat with lions, or occasionally elephant and rhino (Makacha, Msingwa and Frame 1982 quoting Fosbrooke 1972). The Dorobo hunters of Maasailand have dwindled to virtual extinction (chapter 3). To the north, west and southwest NCA borders on the Serengeti Park, Loliondo GCA and Maswa GR (Fig. 2.1). The extreme southwest corner faces Sukumaland in the Kakesio/Maswa area. WaSukuma and WaIkoma people are hunters and poaching is common in this zone. Meat poaching is increasingly big business, leading to organised destruction of ranger camps. Some observers suggest that wildebeest snaring in the west may be beginning to affect population size. The eastern and southeast borders of NCA adjoin areas of Mbulu cultivation and settlemen The WaMbulu are separated from the major wildlife concentrations Northern Highlands Forest reserve, but again they are traditionall and meat eaters, and the forest fringes are reportedly heavily h buffalo, bushpig, and bushbuck. There is considerable concern among conservationists over poaching in NCA, mostly on the Maswa GR border, elephant and rhino in the forest, and rhino poaching in the Crater (Makacha, Msingwa and Frame 1982). There are, however, two different classes of poaching, with quite different implications for conservation, which should be managed in completely different ways. These are firstly small-scale hunting (mainly for antelope meat and skins) and secondly commercial poaching (particularly of rhinos but also for elephant).

The original purpose of wildlife protected areas in Maasailand such as the Serengeti Closed Area and the Southern Reserve in Kenya was to control increasing trophy and meat hunting by settlers and tourists. Following excision of NCA from the Serengeti National Park, planned trophy offtake continued. NCA comprised two hunting blocks with set annual quotas (Table 7.3; Rodgers and Nicholson 1973). The full quota was not taken for many species, although there was concern that blackmaned lions were being enticed from the adjacent park by baiting. In 1973 all sport hunting in Tanzania was stopped. It is interesting to compare current estimates of subsistence poaching offtakes in the Endulen Zone of NCA (Makacha, Msingwa and Frame 1982) with the

Table 7.3 *Current poaching offtakes compared to 1970s set quotas for tourist hunting for NCA*

Species	Oct. 1978	Oct. 1979	Tourist quota per block[a]
Buffalo	2	1	20
Dikdik	1	1	?
Duiker	—	—	++
Eland	4	2	5
Giraffe	1	3	—
Grant's gazelle	—	—	24
Hartebeest	—	—	12
Impala	5	—	20
Leopard	—	—	4
Lion	—	—	4
Steenbok	—	—	++
Thomson's gazelle	—	—	36
Warthog, etc.	—	—	++
Zebra	8	5	30
Fischer's lovebirds	—	60	—

[a]*Source:* from Rodgers and Nicholson 1973; ++ = hunting permitted, no set quota; — indicates none found poached (columns 1 and 2), hunting not licensed (column 3).

earlier set quotas for Ngorongoro Hunting Block II (equivalent to the Endulen Zone – Rodgers and Nicholson 1973; table 7.3).

The recent survey was unable to quantify rates of offtake: the carcasses found probably represent individuals taken during a period of several months, and are unlikely to cover all poaching that had taken place. Smaller animals will be particularly under-represented as they may be removed whole leaving no carcass. However, the results indicate that elephant and rhino apart (see below), offtake by local hunters is either similar to or considerably less than those originally designated as sustainable quotas.

The whole issue of subsistence hunting, its conservation implications and its classification as poaching is coming under review in Africa (chapter 11) and in the Serengeti/NCA region (Malpas and Perkin 1986). Conservation in Africa may be viable in the long term only if conservation resources can be used by the local community, which commonly bears many of the costs of wildlife conservation, but few of the benefits (Bell 1987). Subsistence hunting in NCA carries little threat to conservation values, but could represent substantial cultural and economic benefits to several groups. A number of projects elsewhere in Africa are trying to devolve responsibility for controlled wildlife exploitation (as well as for wildlife conservation) to local communities (Martin 1986, Abel and Blaikie 1986). A recent IUCN study of East African antelopes suggest this sort of wildlife exploitation should become more general as part of an integrated and long-term conservation approach (East 1988). Possible future developments along these lines are discussed in greater detail in chapter 11.

(e) Commercial poaching: rhino and elephant

Aerial survey recorded a live elephant:carcass ratio 1.8:1 over the whole Arusha Region in 1979 (i.e. 36% sightings were of dead animals – Ecosystems Ltd 1980:57). Similar surveys in Serengeti suggests 13% dead in 1977 and 38% dead in 1984 (Dublin and Douglas-Hamilton 1987). These figures suggest extremely high poaching mortality (Douglas-Hamilton and Hillman 1981). The inference of high poaching pressures is borne out by the 1970–1980s crash in rhino numbers. While the NCA forest provides some refuge for elephant, it has not been properly monitored or patrolled and it is likely that the NCA population has suffered a proportional decrease similar to that of neighbouring areas.

During their two surveys Makacha, Msingwa and Frame (1982) found a cumulative total of eight elephant and four rhino carcasses in the Endulen zone of NCA. By contrast, during the same surveys 41 rhino and 12 elephant carcasses were found in the adjacent eastern half of the Maswa Game reserve.

The Maswa Reserve has more wildlife that the Endulen zone and poachers are evidently more active there. However, conservationist concern has focused on the future of rhino in NCA where their decline embodies events throughout the species range.

Ngorongoro at one time had two of the densest black rhino populations in Africa: around Olduvai Gorge, and in the Crater itself (Goddard 1967) as well as throughout the forest, higher bushland, Ndutu, Eyasi scarp and the woodland fringes of the plains. Poaching in the 1970s reduced the population of some 300 animals to a present-day total of about 50 in the whole NCA, with none at Olduvai or in the plains. In the Crater where once there were 110 animals there are now 29.

Episodes of rhino poaching are not new in NCA. Conservancy–pastoralist confrontation led to 31 rhinos being speared in NCA in 18 months 1959–60, and a further 12 in 1961. These figures should be compared to the total of 17 taken over the preceding seven years 1952–1959 and (after the 1959–61 outburst) three per year 1962–1967. It was after this that Goddard counted 110 animals using the Crater floor, including some 70 permanent residents and 25 using Lerai Forest alone.

In the 1970s political and economic instability throughout Africa and the search for exportable forms of wealth led to a dramatic rise in the price of ivory and rhino horn. There was an upsurge in ivory and rhino poaching throughout Africa (Bradley-Martin 1979; Hillman 1981, Western and Vigne 1984). Borner (1981), Makacha, Mollel and Rwezaura (1979), and Makacha, Msingwa and Frame (1982) give detailed analyses for Tanzania and the Crater rhino respectively. In 1978 at least 25 were shot by non-Maasai commercial poachers (Borner 1981). Kiwia found five of his animals poached during his study of Crater rhino home range patterns 1980–1982, and estimated 25 rhino left alive in the Crater at the end of his project. The 1987 NEMP aerial survey counted nine in the Crater, suggesting a corrected figure of some 15–20 animals (Borner 1981). In 1989, 16 animals using the Crater floor and walls can be individually recognised.

Fosbrooke (1972) believed the NCA Maasai deterred such poaching by providing information, acting as informal patrols and helping in pursuit. They probably also limited agricultural encroachment and subsistence hunting, although outside the NCA Maasai presence has not deterred the growth of large-scale commercial poaching over the last 30 years (e.g. in Longido and Natron areas – Rodgers 1981a). Within NCA, conservancy – pastoralist tensions have reduced former anti-poaching cooperation. More importantly, the enormous rise in value of ivory and rhino horn means poachers are no longer local and subsistence level. They are now sophisticated operators, armed with modern automatic weapons, working in large gangs across international borders. A number of field staff in northern Tanzania including NCAA have been killed in encounters with such groups. Some Maasai may have become involved with poaching gangs as guides and hunters. There are on the other hand frequent suspicions that NCAA staff may also have been involved. It is certainly the case that what in early times had been a disciplined, highly motivated and well-equipped ranger force in NCA has deteriorated as have infrastructure, roads, patrol posts and the tradition of regular foot patrols. Injections of equipment and funds have done nothing to improve this. Continuous telescopic watch with radio contact from the Crater rim and round-the-clock vehicle patrols on the Crater floor have not prevented rhino being taken in daylight.

The NCA rhino population is still biologically significant. With the adjacent forest there may be a total of around 30 – perhaps the largest single population in northern Tanzania. Rhinos appear to be genetically rather homogeneous: the white rhinos of Southern Africa are all descended from fewer than a dozen individuals at the beginning of this century and NCA still has a potentially viable breeding population. However, the current economic incentive to poach, together with the prevalent ill-will between NCAA and pastoralists, and the inefficient ranger force, makes the future of the rhinos and elephant doubtful at best. New policies of cash reward for poaching convictions may stimulate effective anti-poaching. They may also generate cases of false evidence, and further alienate local people. Currently (in September 1990) poaching is said to be on the decline. Rhino numbers using the Crater are up to 29; elephant are to be seen in the Crater, in the forest and at Lake Ndutu. Anti-poaching inputs must develop alongside education and extension services to be effective. Chapter 11 discusses some of the more radical solutions that have been proposed for systems of conservation elsewhere in Africa – such as licensing of ivory hunting and its regulation by the local community (Bell 1987; chapter 11) – despite international agreements designed to terminate the ivory trade.

Summary and Conclusion

The world-famous wildlife community of the NCA/Serengeti rangelands is dominated by migratory ungulates. Their populations have fluctuated over the last few decades and in some cases shown dramatic increases. In the long term, wildlife population impacts interact with changes in pastoralist herd presence to contribute to natural fluctuations in woody vegetation cover that are as or more important than purely anthropogenic effects.

Management of the wildlife resource is still purely *ad hoc*. There are no wildlife management plans, or overall policies. All inputs are directed towards old-fashioned estate management – borders, roads, water, tourists, anti-poaching. There is still little coordination between organisations. All these issues are addressed by the Ngorongoro Conservation and Development Project.

Conservation management of wildlife populations in NCA can allow for fluctuation within broad limits of vegetation state, and of individual wildlife species. The Ngorongoro and Empakaai Craters require more management control due to their importance and impact as conservation showpieces. Similarly, rhino may require intensive and interventionist management where other individual species populations can be allowed to fluctuate more naturally. Subsistence hunting and ivory poaching have very different impacts on NCA conservation values and should be managed differently. The many separate organisations involved in NCA wildlife management are a source of confusion and duplication and management structure is likely to remain a major problem.

8

Livestock ecology

Erisiore entawuo nabo elukunya olee One heifer is worth a man's head
(With good management any man can become an independent stockowner)
(Maasai proverb: Waller 1979)

Eteyo apa nkerr: 'Ka kirru are' 'The sheep once said: 'Have you ever failed to make me two?'
(Maasai proverb: Waller 1979)

The livestock production aspects of NCA have tended to be overshadowed by wildlife and archaeological conservation issues as far as international perceptions are concerned. This chapter is an attempt to redress the balance. The general nature and significance of pastoralist livestock production in sub-Saharan Africa are outlined and NCA is put in context as a case study in pastoralist ecology and management. We then go on to look at the detail of livestock ecology in NCA. Data on livestock population distribution, dynamics, range utilisation and performance are presented in comparative perspective. With this detail it is then possible to analyse the management debates within NCA on overgrazing, sustainability of land use, interactions with wildlife, and the recurrent controversy over efficiency and productivity of pastoralist performance.

Pastoralist livestock production in sub-Saharan Africa

(a) Importance of pastoralist livestock production

On the basis of 'grain equivalents' (e.g. Jahnke 1982) livestock production accounts for 25% of total food production in sub-Saharan Africa (excluding South Africa). Non-food products such as draught labour, manure as fertiliser and hides are also of considerable significance. The relative importance of livestock is greatest in the most arid countries (Somalia and Mauretania) but Tanzania is well above average in its dependence on livestock, with agriculture contributing 40% gross domestic product (GDP) and livestock production accounting for 25% of this.

The most common pastoralist livestock species in tropical African rangelands are cattle, sheep and goats, with camels restricted to more arid areas and small numbers of equines being found throughout pastoralist areas. Cattle are estimated to make up 75% of all tropical livestock units with sheep and goats contributing 17%, camels 8% and equines 5%. On a regional basis Eastern Africa (including Somalia and Ethiopia) has 56% ruminant livestock for 26% of the land area of tropical Africa. Tanzania has 8% of the continent's livestock on around 4% total land area (Jahnke 1982).

Commercial animal husbandry in Africa has concentrated on land of relatively high potential, and livestock production in the arid and semi-arid zones of tropical Africa is almost by definition subsistence pastoralism (except perhaps in Botswana). Arid and semi-arid areas make up 55% of the land area of tropical Africa and account for over 50% of cattle and over 60% of sheep and goats. Livestock production by subsistence pastoralism thus makes a significant contribution to food production in tropical Africa and is an important component of the livestock production sector.

Tanzania had some ten million cattle, three million sheep and four million goats in 1975 (Raikes 1981) and the cattle population varied from 10 to 14 million between 1965 and 1971 (MacKenzie 1973). Arusha Region, which includes much of Tanzanian Maasailand, had some 20% of these livestock populations. NCA stock densities are intermediate, with 10% of Arusha Region livestock in an area that represents less than 8% of the Region (Ecosystems Ltd 1980). This must be viewed in the context of locally very high wildlife biomasses in NCA and a large area of forest reserve. The wildlife concentrations and the absence of cultivation are special features peculiar to NCA, but the Area is representative not only of the 'purer' pastoralist production but also of the need for, and problems of integration of, this type of production into a wider agricultural economy.

Later, in chapters 10 and 11, we summarise the history and vicissitudes of the livestock economy of Tanzania and their implications for the past and future roles of pastoralist cattle in the national economy. Put briefly, a growing urban demand for meat is currently met primarily by cattle from more easily accessible agropastoralist areas. Urban markets aside, meat is by no means the most important livestock product. Pastoralist cattle populations are traditionally seen by outsiders (Tanzanian and foreign alike) as an underutilised and poorly managed resource (e.g. MacKenzie 1973). Their role in national economies is rarely obvious in official statistics, but traditional pastoralist and subsistence agropastoralist livestock systems are of major importance in the local and regional economies of arid and semi-arid rangelands. A better understanding of pastoralist ecology and economics is gradually emerging.

(b) Pastoralist livestock ecology

Pastoralists and their livestock have been the subject of anthropological interest for decades but the ecology of pastoralist livestock production is a recent field of study. Sub-Saharan livestock production has been a focus of development attempts for some time (Cross 1985). It was the repeated failure of such attempts that eventually stimulated ecological and management-oriented studies of pastoralist livestock. Dahl and Hjort (1976) compiled material from a wide range of sources and stimulated a sudden proliferation of systematic ecological studies (Coughenour *et al.* 1985, Coppock, Ellis and Swift 1986, Coppock, Swift and Ellis 1986, Grandin 1988 and references) together with a number of development-oriented review works (Jahnke 1982, Sandford 1983, Simpson and Evangelou 1984, Niamir 1990).

This section introduces the main characteristics of pastoralist livestock ecology and management as background for subsequent sections detailing the ecology of NCA livestock.

Pastoralist systems are commonly but not exclusively found in arid and semi-arid areas where rainfall and primary production are so unpredictable as to make dry land cropping insecure. The natural vegetation in such a system can support livestock production provided that the animals are able to move so as to exploit areas of new growth. This movement may be a flexible but relatively regular seasonal cycle (transhumance system) based on latitudinal differences as in much of the West African Sahel, or on altitudinal differences (common in East Africa with its diverse topography). In more arid zones, movement may be irregular depending on unpredictable local patches of rainfall and temporary production (semi-nomadic or migratory systems). Arid and semi-arid savannas, with their recurrent but unpredictable disturbance by drought, fire, epidemic, and herbivore population fluctuations have fostered the evolution of a range of risk avoidance strategies in pastoralist groups. In terms of stock management these include

- access to high-potential drought refuges, usually highland or swamp pastures
- maintaining a flexible mixture of stock species with different feeding, ranging, production, disease- and drought-resistance, and reproductive characteristics. Small stock are commonly important in post-disaster herd reconstitution; large stock are the preferred investment once a critical threshold of stock holding is passed (e.g. Mace 1988)
- maintaining herds with a high proportion of females capable of rapid reproductive response in the aftermath of disaster
- maximising stock numbers in the hope of retaining enough to ensure long-term survival of the herd despite heavy periodic drought and disease losses
- splitting the stock holdings into different herding units depending

on species, maturity and reproductive condition and pasturing them in different areas that may be within reach of the same daily base or may be days' journey away
- social systems of stock loan and redistribution among friends and kinsmen that spread risk over a wider geographical area and range of habitat types and thus buffer disaster (chapters 3,10).

Indices of meat, milk and calf production per head are low compared with western commercial concerns but compare well with figures for similarly arid rangelands of Australia and the USA (Sandford 1983). Low animal productivity for such areas is not surprising given the arid climates, low primary production, extensive grazing regime, the mobility of the stock and their exposure to drought and disease stresses. Pastoralist management is aimed not at producing a cash profit but at maintaining everyday human subsistence and long-term herd survival.

There have been numerous productivity comparisons between exotic *Bos taurus* (western cattle), the indigenous *Bos indicus* (zebu) that makes up the main Maasai herds, and more rarely the indigenous *Bos taurus* breeds of West Africa (Rogerson 1970, Rogerson, Ledger and Freeman 1968, Williamson and Payne 1978). The interpretation of these comparisons and of their implications for management has undergone change with time (Cross 1985, Grandin 1988). A more or less universal western perception of *Bos indicus* as low yielding and inefficient, versus the exotic *Bos taurus* breeds as high yielding, efficient and the desired development goal (e.g. MacKenzie 1973, Adams 1988), has given way to a new appreciation of indigenous breeds. It was first observed that in practice pure or even crossbred exotics did not survive pastoralist management stategies. Then experimental observations began to establish the special features of the indigenous breeds: heat tolerance (Finch and Western 1977, Jewell and Nicholson 1989), tolerance of low water availability (Finch and King 1982), trek-hardiness (Western and Finch 1986), adaptability to low food intake (Western and Finch 1986), and disease resistance (Trail and Gregory 1984). It also became obvious that the high production potential of *Bos taurus* breeds could not be sustained in an environmental context combining heat, drought, low quality feed, night enclosure, long treks and lack of veterinary care (Rogerson 1970, Trail and Gregory 1984, Jahnke 1982, Sandford 1983). General reviews of zebu biology (Hacker and Ternouth 1987, Gilchrist and Mackie 1984) confirm that the zebu has evolved through natural selection and breeding manipulation to survive the rigours of a tropical arid to semi-arid environment. It can withstand high ambient temperatures, severe desiccation, a low plane of nutrition and considerable disease and parasite challenge. The disparaging 'average yields are below a litre per day' should be rephrased as 'despite 16 km daily movement, water every second day, an energy and protein deficient diet, high heat load and exposure to tickborne diseases, the zebu raises a calf and provides a milk surplus to its herders!'

During dry periods milk production is insufficient for human subsistence. Storage from times of milk surplus (for example by making clarifed butter and cheese) is practised by some pastoralist groups. Given limited surplus, limited technology and enforced mobility, stored milk product surplus can only go a little way towards making up the dry season deficit. Some groups make use of blood from live male stock during drought. Animals are rarely slaughtered for food purposes alone, and the dry season food deficit is commonly made up by cultivated foods (chapter 10). These are either grown by the pastoralists themselves or acquired by sale or exchange of pastoral produce.

With this background information on general features we now go on to look at the detail of Maasai livestock ecology in NCA.

NCA livestock ecology

The livestock species in NCA are Small East African zebu cattle (*Bos indicus*; females reaching 250 kg and castrates 400 kg at maturity), red Maasai sheep (*Ovis aries*) and small East African goats (*Capra hircus*). Donkeys are important means of transport during transhumance, on grain purchasing expeditions and, in some areas, for carrying water. NCA Maasai are primarily cattle pastoralists (See table 8.1 and chapters 3,10) though small stock have come to make up an increasing proportion of the NCA livestock over the last few decades (Table 8.1).

Our study concentrated on two main aspects of livestock ecology: firstly, measures of utilisation of and impacts on environment, and by inference the resultant effects on wildlife, and secondly, measures of the productivity and efficiency of pastoralist stock as an input to human subsistence.

Data on numbers, distribution and stocking rates in different localities are used to investigate livestock impacts on environment and wildlife populations. Biomass densities and stocking rates in NCA are compared with averages over large areas of Africa that have been estimated to be sustainable on theoretical and empirical grounds. This gives some basis for evaluating allegations of overstocking. A detailed comparison of wildlife and livestock range utilisation patterns allows investigation of the ecological overlap between the two communities (see also chapter 9). The similarities that emerge form the basis for evaluating possible competition between domestic and wild ungulates. Census numbers documenting the performance of wild and domestic ungulate populations over the last 20 years are invoked to complete a discussion of environmental impacts by pastoralist livestock.

Factors affecting overall herd survival, production and performance, are ultimately of interest because of their implications for Maasai subsistence in NCA. Here, activity budgets and ranging patterns are investigated because of their potential importance as production constraints. Livestock population dynamics are described in terms of herd size and composition, fertility, mortality, herd growth and transfers. Performance of cattle is quantified and compared with pastoralist herds elsewhere in terms of condition, calving rates, calf and adult mortality rates, and drought and disease effects, as well as milk production. Patterns of variation with season and area are discussed. Current offtake rates and our observed rates of herd decline are quantified. All this material is fundamental to an evaluation of the efficiency and productivity of Maasai pastoralism in NCA, and to the discussion of development possibilities in later chapters.

(a) NCA livestock population numbers

Perkin (1987) counted 137 398 cattle and 137 389 (*sic*) small stock in his 1987 wet season ground census. The NEMP aerial survey carried out at approximately the same time estimated fewer cattle and considerably more small stock (Table 8.1), though the confidence limits overlap the ground census figures. It is quite possible to get successive livestock counts varying by 100% or

Table 8.1 *Census figures for NCA livestock populations 1960–1987*

Date	Cattle	Small stock
1960	161 034 (161 930)	100 689
1961	(126 760)	(94 160)
1962	142 230	83 120
1963 Dec.	116 870	66 320
1964 Jun.	132 490	82 980
1966 Feb.	94 580	68 590
1968	103 568	71 196
1970	64 766	41 866
1974	123 609	157 568
1977	(110 584)	(244 831)
1978	107 838	186 985
1979	[125 000]	[290 000]
1980	118 358 (126 589)	144 675 (135 501)
1987 Apr.–Jun.	137 398 {113 431 ± 18%}	137 389 {307 832 ± 60%}

Source: NCAA census data as reported in: 1. Ngorongoro Ecological Monitoring Program (NEMP), semi-annual report, Sept 87; 2. ( ) = Arhem 1981[a], where his figures differ from NEMP; 3. [ ] = aerial survey (SRF) Ecosystems Ltd 1980, Table 3D, p. 55; 4. { } = aerial survey (SRF) NEMP 1987.

more over a few months or years (see Sandford, 1983). This is to some extent due to inherent variability of pasture availability and stock movements, and also because of fundamental methodological problems with census techniques (Norton-Griffiths 1978; Caughley 1977). As Arhem (1981a) points out, ground census figures are likely to represent only the domestic stock belonging to settlements inside the Conservation Area because of the gate count method, while aerial surveys will sample all the stock grazing within the boundaries of NCA at a given time. Aerial survey involving sample counts will give unacceptably high variances with highly clumped species. Livestock census figures are thus notoriously unreliable (see also chapters 3, 5 for a discussion of methodological problems).

Figure 8.1 and Table 8.1 summarise census information on numbers of livestock in NCA. The 20 years over which figures are available show considerable fluctuations, with cattle numbers falling from around 161 000 in 1960 to 64 766 in 1970 (early wet season counts) and rising again to 137 398 in 1987. Ground counts of small stock populations showed comparable changes with numbers initially falling from 100 689 animals in 1960 to a low of 41 866 in 1970. There was then a rapid increase to a maximum of 244 831 in 1977, falling again to 137 389 in the 1987 census.

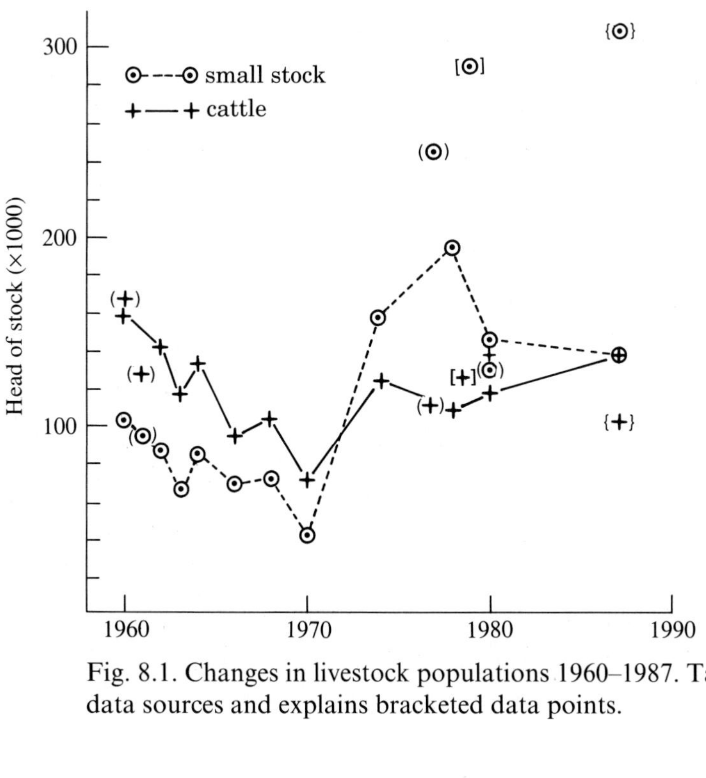

Fig. 8.1. Changes in livestock populations 1960–1987. Table 8.1 gives data sources and explains bracketed data points.

Despite methodological problems much of the observed variation between seasons and years must represent genuine changes in stock numbers within the area surveyed. This is firstly because transhumance movements and semi-nomadic migrations cut across administrative boundaries and shift in an irregular way in response to changing conditions. Secondly, drought and disease can cause rapid stock losses (e.g. Rodgers and Homewood 1986, Homewood and Lewis 1987). Small stock with their short maturation and gestation periods and common twinning can build up more rapidly than depleted cattle herds (Dahl and Hjort 1976). Herd reconstitution by trading small stock for cattle can then bring about locally rapid recovery of cattle herds and equally rapid depletion of small stock (Homewood and Lewis 1987). Such trade takes place over long distances and across regional boundaries.

For example, NCA cattle numbers during the period 1960–1970 may have been influenced in part by the cattle population of the 21 105 km^2 Kajiado District (Fig. 8.2), just over the Kenya border, which fell from 630 000 in 1960 to 200 000 in 1962. This has been interpreted as stock loss through starvation (Meadows and White, 1979; continued as in Grandin 1988) but emigration to areas such as NCA may have accounted for a considerable part of the decline in stock numbers. Dirschl (1966) records that during the years of relatively abundant rainfall following 1962, people and livestock moved back from NCA to 'the areas from which they had come'. Cattle counts in Kajiado showed the District herd had regained its original level of around 600 000 head by 1967 and increased further to 690 000 by 1969. Meanwhile, NCA cattle numbers had fallen by about 100 000 head. Meadows and White (1979) account for the Kajiado changes on the basis of fertility, mortality and official sales, and discount emigration and immigration ('illegal smuggling') of cattle into and out of Kajiado District, thus ignoring a well-documented mechanism for adaptation to local variation in range conditions. For example, Field and Moll (1987) quote NCAA staff as saying that of all NCA stock sold, 70% are taken across the border by unofficial channels and disposed of in Ngong (Fig. 8.2) where prices, taxes and market conditions are more favourable (chapter 10). Whether or not NCA provided a drought refuge for part of the Kajiado herd between 1960 and 1962, or only for cattle from closer lowlying areas, the way in which the dates and reciprocal movements dovetail are suggestive of the sort of movements that could affect NCA cattle populations, and their possible scale.

(b) Livestock biomass densities and distribution

Table 8.2 lists domestic stock biomass (cattle, sheep, goats and donkeys) for each NCA study site and compares these figures with theoretical estimates based on average annual precipitation for each site. There are two

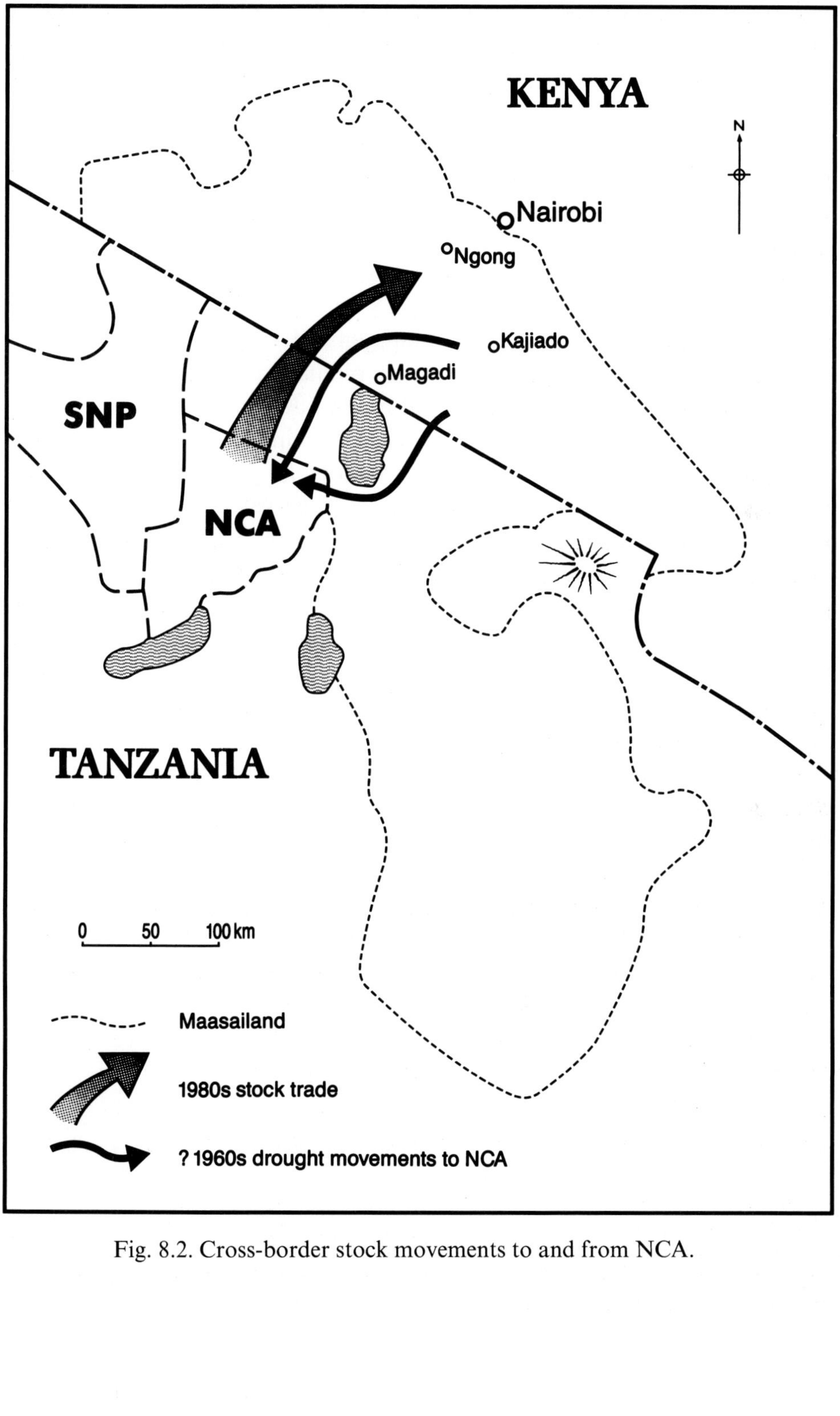

Fig. 8.2. Cross-border stock movements to and from NCA.

sources of theoretical estimates. Coe, Cumming and Phillipson (1976) base theirs on many large ungulate communities from a range of sites across tropical and subtropical Africa (mostly East and Southern), while Le Houerou and Hoste (1977) use data from domestic stock in North African/Mediterranean and Sahelo–Soudanian Africa. East (1984) has shown that similar curves hold individually for the majority of wildlife species for which data are available and that in many cases an even better regression is obtained if the data are divided into nutrient-rich and nutrient-poor savanna soil areas.

Estimates are shown in Table 8.2, together with our empirical values for NCA. Sendui had the highest biomass of 15 100 kg/km² followed by 11 800 kg/km² for Ilmesigio, 7800 kg/km² for Oldumgom and 4500 kg/km² for Nasera. A direct comparison of NCA domestic stock biomass densities with theoretical estimates derived from Le Houérou and Hoste's Sahelian relationship as well as that established by Coe, Cumming and Phillipson (1976) suggests that Oldumgom is higher than average by a factor of 1.59, Sendui by 1.48 and Ilmesigio by 1.3 while Nasera is close to average values for corresponding rainfall regimes. However, figures based on North African/Mediterranean ecosystems (Le Houérou and Hoste 1977) give average biomasses similar to or greater than those found for the NCA sites. These have been shown by Deshmukh (1984) to be much closer to East African grassland production conditions than the Sahelian ecosystems (see chapter 6). The NCA pastures are extremely nutrient-rich and are adapted to high grazing densities, and would be expected to carry higher than cross-continental average biomasses. Before drawing conclusions from this it is necessary to take into account both wildlife biomasses and duration of utilisation by wild and domestic stock respectively.

Table 8.2 *Livestock biomass data for each study site*

			Biomass density (kg/km²)					
				Theoretical				
	Corral	Rain (mm/yr)	Actual	1	2	3	Mean	*A/T*
Gol	Oldumgom	550	7 800	3 568	5 808	5 386	4 921	1.59
	Nasera	500	4 500	3 134	5 318	4 895	4 449	1.01
Ilmesigio	Ngodoo	900	11 800	9 100	9 240	8 815	9 052	1.30
Sendui	Senguyan	1 000	15 100	10 500	10 221	9 717	10 146	1.49

Note:
A/T represents the ratio of actual (observed) livestock biomass density to that expected on the basis of mean rainfall at each site. An *A/T* ratio of > 1.0 indicates potential overstocking.
Source: 1. Coe, Cumming and Phillipson (1976); 2. Le Houérou and Hoste (1977) – linear; 3. Le Houérou and Hoste (1977) – curvilinear.

Nasera plains have a livestock biomass of only 360 kg/km^2, but a wildebeest biomass of many tens of thousands of kg/km^2 use the area for a few days or weeks. Nasera livestock biomass pressure is for a maximum of only six months of the year, Sendui and Ilmesigio for approximately nine to ten months and Oldumgom for a variable period. Thus higher-than-average animal biomass (and therefore apparent overstocking) in several sites will be balanced by its short duration alternating with substantial rest periods. On this criterion none of the sites could be said to be overstocked, and from the point of view of grazing conditions this is probably accurate. Grazing aside, the Landsat evidence of deterioration in the Angata Salei (chapter 6) suggests severe but localised physical damage by trampling. The relative contributions of resident stock and migratory wildlife to this problem are discussed later.

(c) Utilisation patterns

The wide range of environmental conditions within NCA demands a similarly wide range of foraging strategies. The most striking variation in herd management is in the pattern of herd movements in different areas. Herd management relies both on large scale herd movements involving change of base, and also selection of daily grazing areas on a more local scale. Pastoralist decisions influence livestock response to range constraints and opportunities through choice of boma site, pasture and water regime. Sendui and Ilmesigio herds showed regular transhumance from the permanent dry season homesteads to more temporary but frequently re-used boma sites on the rainy season pastures (Fig. 8.3). During the rains some 80% of the Sendui people and herds moved down to Loipukie and used the short grass and ecotone pastures of the Salei Plains until the wildebeest displaced them. The lower wooded ridges of the massif were then used. At Ilmesigio the lower pastures or the adjacent short grass Serengeti plains were used during the rains. In drier months the upper pastures and *Acacia lahai* pastures were used equally.

By contrast, the people of the more arid Gol area show less regular and more opportunist migratory movements. For example, our Gol study boma (which had a permanent homestead in Ndureta: chapter 5) occupied at least 13 different boma sites for periods of over 10 days between August 1981 and August 1983 (Fig 8.4).

Choice of boma locality is determined largely by the availability of water and grazing resources within a few kilometres. Within the locality, actual site choice depends on criteria of slope, soil, drainage and canopy cover (Western and Dunne 1979). Steep slopes are generally avoided for several reasons. The energetic costs of climbing may be critical for animals on submaintenance diets. The downhill stockade must be made extra strong and high or frightened

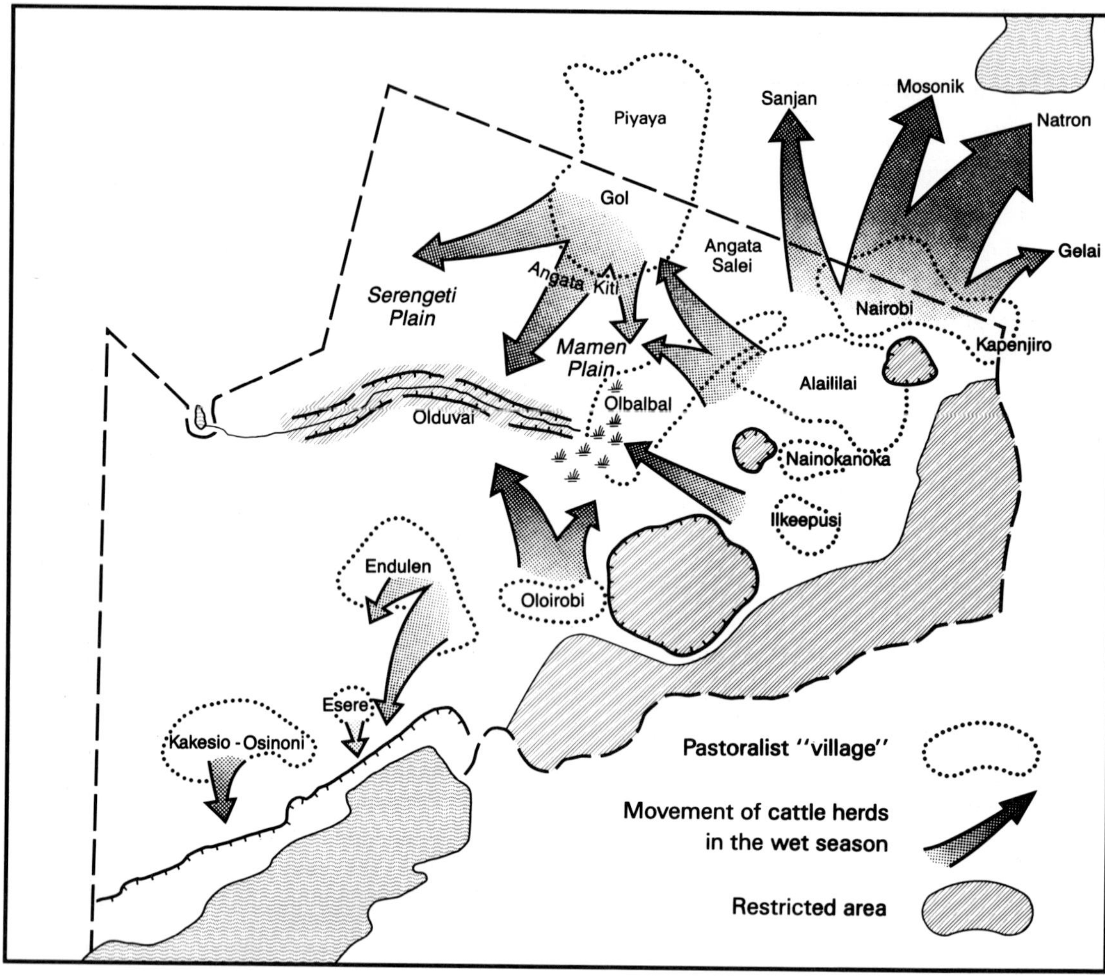

Fig. 8.3. Patterns of Maasai transhumance in NCA (after Arhem 1981a).

Fig. 8.4. (Opposite). Temporary sites occupied by the household and herds of Gol herdowner Andrea Lesian, August 1981–1983. 1. Nasera (July 1981), 2. Kassile (August 1981), 3. Lemuta (September 1981), 4. *en route* to Oitii (September 1981), 5. Oitii, Olduvai (October–January 1982), 6. Oldumgom (January–June 1982); 7. Nasera (July 1982), 8. Lemuta (August 1982), 9. Serengeti edge (August–September 1982), 10. Oldumgom (October 1982–May 1983), 11. Nasera (Lesian, June–August 1983), 12. Oldumgom (Saruni, June–August 1983), 13. Serengeti edge (Godi, June–August 1983). – – – = minor track.

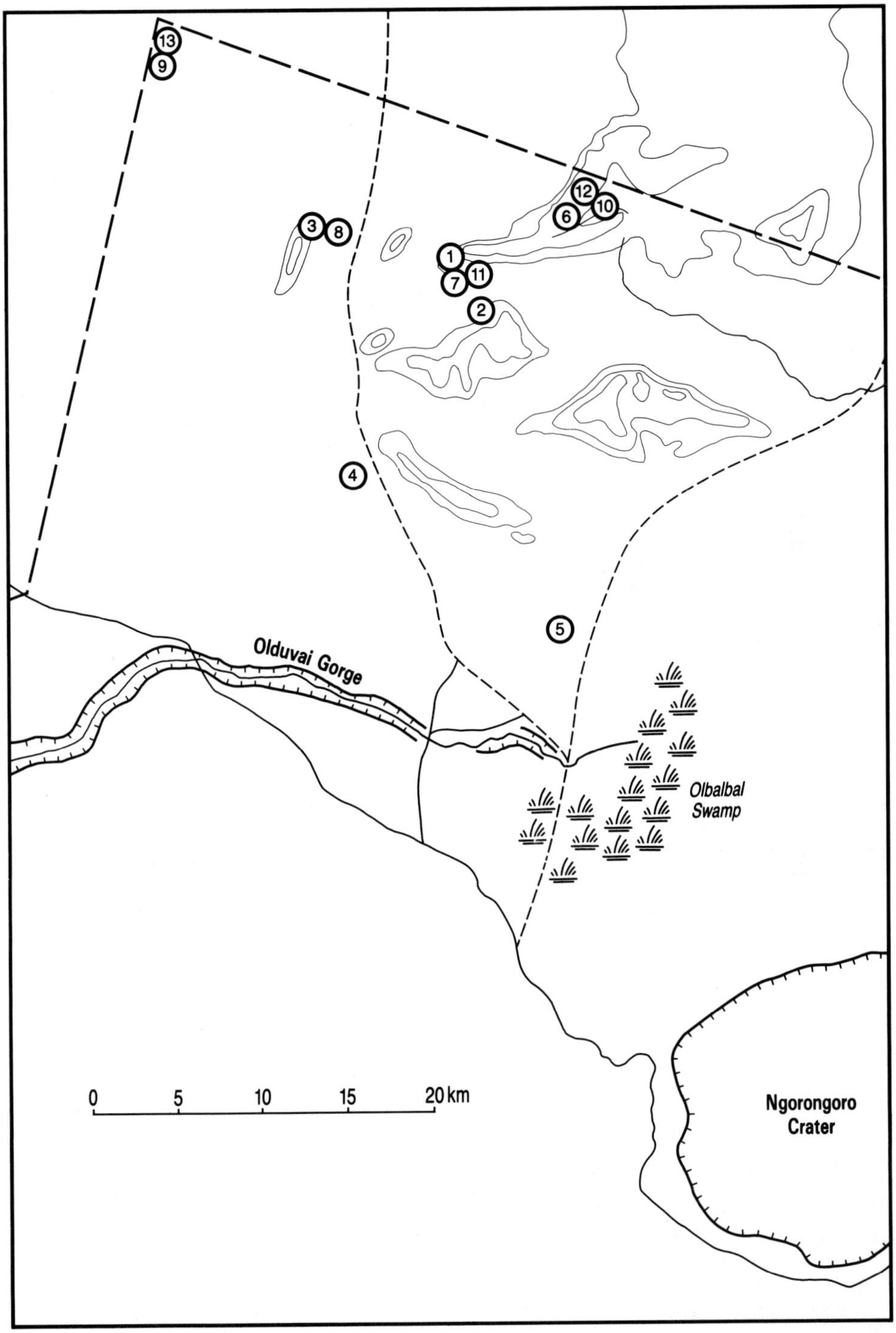
13
9
3
8
12
10
6
1
7
11
2
4
5
Olduvai Gorge
Olbalbal Swamp
Ngorongoro Crater
0
5
10
15
20 km

livestock will breach it, and hut sites on steep slopes require levelling. Hilltops are avoided because the thin soil and rocky substrate make house and stockade construction difficult and the stony ground causes the stock discomfort. Waterlogging at the foot of slopes makes these sites uncomfortable, prone to storm washout and parasite transmission and unsuitable for wet season occupance. In Amboseli dark well drained soils of the upper slopes are favoured; light-coloured soils of basins are avoided, both because of flooding and because of high reflectivity, leading to low night temperatures with adverse implications for the energetics of condition and milk production. Western and Dunne (1979) found their Amboseli boma sites avoided areas of more than 10% canopy cover within a 225 m radius because of the increased danger of predators and large mammals. Casual observation during our study bore out the truth of this last criterion. The Ilmesigio boma had an average of more than 20% canopy cover in an 8 km radius: during our study this boma suffered the loss of over 50 small stock in one day due to predator attack, and had repeated alarms over the proximity of elephants. Western and Dunne (1979) found no settlements in areas devoid of woody vegetation. In our study Ilmesigio and Sendui bomas sustituted timber for the thorn stockades common in Gol. In conversation, NCA Maasai added the aesthetic level of the surroundings as a further criterion. For some areas proximity of resources such as dips, schools and trading routes for grain were of importance. The useful life of a boma in Amboseli is on average seven to eight years, after which subsurface compaction prevents livestock urine percolating through. However, highland bomas in NCA are thought to last 50 years or more (Chamshama, Kerkhof and Singunda 1989). Other factors leading to the abandonment of an established settlement are outbreaks of disease or ectoparasites, death of the senior male occupant, or a poor state of repair. New settlements are built within a few hundred metres of the old, and Western and Dunne (1979) suggest a resettlement cycle of prime sites of the order of 20–25 years for Amboseli. Analysis of aerial photographs from 1955 to 1982 at Nainokanoka shows no reuse of old boma sites despite the rapid population growth in this area over the same period (chapter 10).

Analysis of range occupance is complicated by the fact that cattle may be watered every second day in the dry season and on those days grazing selection and grazing routes may not be those preferred for feeding. On watering days grazing routes were a secondary consideration; movement to water entailed passing through areas which were not necessarily the best grazing available.

In Gol preferred pastures are the wooded hill slopes, the adjacent plains short grass association being largely unavailable due to wildebeest occupance at the time or because of their past grazing impact. In Sendui the *Pennisetum sphacelatum* pastures are preferred by sheep and cattle, the *Eleusine* tussock areas and Embulbul turf being used only *en route* to water. The preferred Sendui wet season pastures are the ridge tops and lower slopes rather than the steep, poorly vegetated ravines, although goats make use of the ravine edges. Ilmesigio has a choice of lower altitude pastures following the rain or *Acacia lahai* pasture during the drier months. Table 8.3 shows the proportions of available pasture for each site and observed utilisation. In Gol the avoidance of short grassland association (SGA) is clear at Nasera but these areas were used at Oldumgom in December 1982 as no wildebeest entered the area. SGA was used more in Sendui in 1981 than in 1983, with a distinct preference for *Pennisetum* on non-watering days. *Acacia lahai* pastures are the preferred type in Ilmesigio. The implications of such habitat preferences for the success of pastoralist livestock production in NCA, for its environmental sustainability and for NCA management decisions, will become clear with a further discussion of food selection, the energetics of travel to grazing and watering sites and the comparison of wildlife and livestock ecology given later.

(d) Food selection

Cattle are coarse bulk grazers, able to crop a grass sward down to 3 cm. They prefer leaf but will take stem. In the dry season they can browse selectively on a variety of forbs and woody species, rarely above 1.5 m in height, but avoid thorny or leathery xerophytes. Sheep are basically grazers, capable of cropping short turf below 2 cm. They will take soft forbs and coarser grass. Goats are largely browsers, taking some greener grass leaf. They can exercise considerable selection due to mobile lips and tongue and can nibble around thorns. Climbing, standing on their hind legs and pushing over shrubs they can

Table 8.3 *Distribution of cattle feeding activity in different habitat types*
(a) Gol

	Nasera		Oldumgom	
	Available habitat (% area)[a]	Observed Aug. 81	Available habitat (% area)[a]	Observed Dec. 82
SGA	78	36[b]	36	37.6
Hill slope	18	60[b]	63	60.4
Hill SGA	4	4	—	—
Ravine	—	—	1	2.1
n		365		386
χ^2		120.6		0.6

Table 8.3 (*cont.*)
(b) Sendui

	Upland					Loipukie	
	Available habitat (% area)[a]	Observed				Available habitat (% area)[a]	Observed
		Aug. 81	Aug. 83 dry	Aug. 83 wet			Dec. 82
SGA	16	29[b]	0[b]	8.0	Ridge	41	88.8[b]
Eleusine	17	22	5.7[b]	22.3	Salei Plain	37	0[b]
Pennisetum	56	49	94.3[b]	69.6			
Ravine	11	0[b]	0[b]	0[b]	Ravine	22	11.1
n		496	326	237			287
χ^2		23.8	61.2	19.3			102

Table 8.3 (*cont.*)
(c) Ilmesigio

		Observed	
	Available habitat (% area)[a]	Aug. 81	Aug. 83 (watering day)
Lowland	35 }	68	9.1[b]
Upland	42 }		32.0
Acacia	17 }	32	58.9[b]
Juniperus	6 }		0
n		382	197
χ^2		3.6	88.2

Notes:
n = number of observations of feeding locality habitat type
[a] habitat type availability within 8 km radius
[b] Significant selection for or avoidance of this habitat type.

browse to an effective height of 1.5 m. Donkeys are coarse grazers, cropping closely but not selectively and taking a diet high in stem.

Table 8.4 compares dry season diet selected by cattle in Sendui (August 1981) with its availability in the habitat (see also Homewood, Rodgers and Arhem 1987). The intertussock turf was made up largely of the nutritious *Pennisetum clandestinum* (with high crude protein and digestible energy – Said 1971) and *Cynodon dactylon*, also a preferred species (see below). Despite small sample sizes and limited area surveillance the data do suggest a clear preference for turf in the lower Sendui pastures. In the upper pasture cattle avoided *Eleusine* and herbs, but turf and smaller tussocks were selected equally. Newbould (1961) and Frame, Frame and Spillett (1975) record that small *Eleusine* tussocks (< 9 inches in height) and the new growth stimulated by firing of old tussocks provide good grazing.

Peterson and McGinnes (1979) list IlKisongo Maasai names of a number of common grasses and their evaluation by Maasai herders in terms of grazing quality. They do not mention *Pennisetum sphacelatum* (= *P. schimperi*). In their survey *Digitaria macroblephara* was considered to be the best grazing throughout Maasailand. The short grass species *Sporobolus ioclados*, *S. helvolus*, *Cenchrus ciliaris* are all good quality grazing, the latter lasting well into the dry season. *Cynodon* spp. are recognised as good grazing, particularly in the dry season, but avoided in the early growth stages when they may contain toxic secondary compounds. Many medium and tall species are seen as reasonable wet season grazing (*Heteropogon contortus*, *Pennisetum stramineum*) or good after a burn and in the wet season (*Pennisetum mezianum*) but worthless in the dry season. Finally, there is a group of species which are considered starvation grazing: cattle reduced to feeding on them are seen by the Maasai as near the point of death (*Dicanthium* spp. high in essential oils, *Microchloa kunthii* which cattle find hard to scrape up, and *Sporobolus spicatus* with a scabrid leaf and high salt content).

Data on browse show a significant time spent feeding on herb and shrub components by cattle in the dry season at Gol and to a much lesser extent at Ilmesigio. At Gol browse made up over one third of feeding time in one sample (35% in Gol, May 1982). Prominent in the diet were browse species *Aspilia mossambicensis*, *Cyathula polycephala*, *Duosperma kilimandscharica*, *Pavonia* sp., *Rhynchosia* sp. and *Withania somnifera*, all soft leaved tall herbs and climbers, although many other species were also taken.

Cattle are primarily grazers but this heavy use of browse is typical for pastoral and agropastoral cattle (Payne and MacFarlane 1963). Homewood

Table 8.4 *Cattle food item selection*

(a) Selection of range components at Sendui (August 1981)

	Frequency in range (% ± SE)	Frequency in diet (% ± SE)
Pennisetum sphacelatum tussock		
Tussock	74.4 ± 8.28	40.2 ± 5.80
Turf	29.6 ± 4.94	59.8 ± 5.68
Eleusine–Pennisetum tussock		
Eleusine	26.0 ± 7.85	10.3
Other tussock	24.1 ± 1.99	41.7
Turf	25.6 ± 2.56	45.5
Herbs	24.9 ± 3.56	2.0
	$n=8$	$n=6$

(b) Selection of browse or graze components

	n	Browse (%)	Graze (%)
Gol – dry	522	24.3	75.7
– wet	303	1.3	98.7
Sendui – dry	451	0.7	99.3
(Loipukie) – wet	461	0.0	100.0
Ilmesigio – dry	185	2.6	97.4

and Hurst (1986) found that browse accounted for up to 40% wet season feeding time in Tugen cattle in Baringo, despite the peak availability of new grass growth at this time. Rees (1974) showed that in a Zambian *chitemene* system in the mid dry season roughly 34% metabolisable energy came from browse. In the late dry season browse alone provided protein and energy sufficient to maintain a 0.75 kg/day weight gain. Browse and forbs totalled 12% of the dietary intake in Galana in Eastern Kenya (Field 1975). By contrast, Ngisonyoka Turkana cattle are recorded as showing little use of browse (Coppock, Ellis and Swift 1986). Browse makes up only 4% of the total annual average intake by cattle in this very much more arid system. Seasonal variation is low but there is a wet season peak with > 15% intake made up by forbs and > 5% by dwarf shrubs. These observations reflect both the avoidance of thorny and leathery xerophytic species in this very arid area and also a herding strategy that maximises separation of cattle feeding areas from those of browsing camels and small stock.

Table 8.5 shows the results of faecal nitrogen analysis and estimates of dietary protein intake for each site for dry and wet seasons. Cattle at Ilmesigio show consistently high protein intakes; Sendui cattle on dry season pastures show much lower levels. In all cases cattle show selection of higher protein than that shown by generally available forage as determined by our clipping (chapter 6), suggesting considerable microsite and plant component selection. Gol dry season values are fairly high, perhaps due to higher browse intake. Crude protein for most East African grasses varies seasonally between 3 and 12% while for shrubs and herbs it ranges from 10 to 20% (McKay 1971). Steers have been found to select plant parts such as to achieve crude protein values in their forage intake that are two or three percentage points higher than what is generally available (McKay 1971).

Table 8.5 *Dietary protein intake*

Season	Place	Mean faecal nitrogen %	*n*	Dietary crude protein %
Dry	Gol	1.43	4	8.0
	Sendui	1.19	4	5.5
	Ilmesigio	1.50	4	8.7
Wet	Gol	1.38	4	7.5
	Sendui	1.25	4	6.1
	Ilmesigio	1.52	3	8.9

Note:
Faecal nitrogen analysis using dietary crude protein = 1.668 faecal protein − 6.93 (Bredon & Marshall, 1962).

(e) Activity, ranging and energetics

Cattle were grazed and watered on alternate days under dry conditions, but this pattern became less rigid in wetter seasons and areas. Only the 'active day' records of the period outside the boma are analysed here, comprising some 400 individual observations on cattle activity for each of a representative grazing day and a representative watering day in each area during the first sample. Follow-up activity and ranging data were taken on subsequent visits.

Results of activity samples are summarised in Table 8.6. The length of the active day for an adult herd ranges from 9.25 hours to 12.1 hours, with time spent walking being as little as 1.7 hours in the wet season or as much as 7.8 hours in drought conditions. On grazing days the time spent walking is more consistent whatever the area or season, averaging 2.5 hours. Time spent grazing varies inversely with time spent walking, from as little as 3.1 hours to as long as 9.1 hours. These figures emphasise the extensive travel and restricted feeding time of NCA zebu, by contrast with the observations made by Harker, Taylor and Rollinson (1954) with an average of 7.7 hours grazing (6.5–10.2). NCA calves and sick cattle have a consistently shorter active day and are herded separately to closer reserved grazing and water. Ilmesigio herds had the shortest trek to both water and grazing. Ndureta and Sendui herds had generally greater distances to travel in both wet and dry seasons, with even calf herds having to move 12–13 km to and from water in dry conditions (cf. < 4 km for Ilmesigio herds).

The main factors affecting the length of the active day and the times spent walking and feeding in NCA were the availability of water and grazing. Clearly, the time spent walking to water constrained the time available for feeding. In August 1981 the Ndureta herd based at Nasera had little time to feed on their watering day as they were forced to spend a considerable time inactive owing to the difficulties of watering in a narrow rocky gorge in the Onianing'arng'ar Hills. In August 1983 Ilmesigio cattle had similar long inactive periods while waiting their turn at small troughs high on Lemagrut. Olmakutian, a subsidiary Ilmesigio boma, had to walk further to water in Sirwa with less time for feeding or resting. Sendui cattle had to walk a long distance to water in the Embulbul Depression (Table 8.6) but were able to feed along the way. All these herds showed a very different grazing day activity pattern, with 60–70% time spent feeding and an appreciable time spent ruminating. By contrast the Olelekando herd in Ilmesigio in August 1981 showed little difference between grazing and water days, though Ilmesigio herdowners stated that when the pastures had burned at the end of the dry season cattle have to walk very much further to graze. Some samples greatly exceed the normally observed maximum day's trek of 16 km (cf. Dahl and Hjort 1976) though similar distances are occasionally recorded by other authors (20 km for Amboseli cattle; Western and Dunne 1979). Cattle on watering days were rarely able to ruminate or rest during the heat of the day. These became dominant activities between 1 and 3 pm on less active days.

The important implications of daily travel become clear with a calculation of the extra energy expenditure involved (see table 8.6 and Western and Dunne 1979 for method). The 15-minute interval location plots on topographic 1:50 000 maps gave minimum estimates for both horizontal distance and vertical displacement. Energetic expenditure on locomotion can then be expressed as a proportion of fasting metabolism (FM) to give an idea of the extra demands of daily travel. These minimum estimates range from 17% FM to over 50% with a mean of 22.4% on grazing days and 37.4% on watering days. By comparison Western and Dunne (1979) describe a typical *maximum* energy expenditure of 35% FM for daily travel in flat Amboseli. Ndureta herds have consistently high mean travel expenditure (around 33%FM) while the Olekando herds in Ilmesigio had the lowest (around 22% FM). This could be expected to have a major impact on both survival and production figures.

These estimates of energy expenditure on travel are likely to be underestimates. Ledger (1977) in a series of experimental observations on Boran, Hereford and crossbred steers kept penned or walked 5, 10 or 15 km/day showed an increase of energy expenditure of 24, 49 and 73% respectively for the 275 kg size class and 34, 69 and 97% respectively for the 450 kg size class. NCA animals were covering greater distances and were negotiating steep difficult hills with considerable vertical displacement while Ledger's were covering laps of a circular beaten track. However, Ledger found unexpected appetite and 'fitness effects' indicating increasing efficiency of food use with physiological adaption to exercise (see below).

Western and Finch (1986) review the adaptability of Maasai zebu to drought conditions, and report the experimental simulation of drought with reduced intake (at 0.7 and 0.5 of maintenance level) and increased walking distance. As did Ledger, they found that FM showed a progressive reduction as the experiment proceeded. A similar effect is commonly observed in fasting humans with declining body weight (and therefore reduced volume of metabolically active tissue). Other postulated mechanisms of physiological adaption may also help improve efficiency of assimilation and utilisation of reduced intake (Davidson *et al* 1979). Their experimental animals showed a loss of body weight with reduced intake but there was no differential loss of weight in experimental groups walking 16 km daily compared with those walking 8 km

Table 8.6 *Daily activity and travel energy expenditure*

			Hours activity			Distance covered (km)			
Date	Season	Day	Walk	Feed	Other	Horizontal	Vertical	Travel energy (kJ)	%FM
Gol									
Aug. 81	dry	w	5.3	3.6	1.9	14.5	1.10	11 134	53
		g	2.8	5.8	1.0	7.5	0.20	3 864	18
Dec. 81	dry	w	—	—	—	16.5	0.24	7 472	36
		g	2.8	7.7	1.5	14.0	0.38	7 245	35
May 82	dry	w	2.8	6.5	2.8	7.5	0.88	7 358	35
		g	3.6	6.0	2.4	5.0	0.70	5 485	26
Dec. 82	rains	w[a]	1.7	7.9	2.3	7.5	0.20	3 864	18
Aug. 83	dry	w[b]	—	—	—	20.6	0.20	8 816	42
		w[c]	7.8	3.4	0.9	23.0	0.24	9 929	47
		g	—	—	—	12.5	0.28	6 166	29
Ilmesigio									
Aug. 81	dry	w[d]	6.9	3.1	0.9	14.0	0.74	9 097	43
		w[e]	2.6	4.8	1.9	5.5	0.20	3 108	15
		g	2.7	4.9	2.2	6.5	0.20	3 486	17
Aug. 83	dry	w	4.3	5.3	2.4	12.3	0.82	8 816	42
		g	—	—	—	4.8	0.36	3 667	18
Sendui									
Aug. 81	dry	w	3.8	5.6	0.9	12.5	0.70	8 324	40
		g	1.6	6.8	2.5	4.0	0.40	3 570	17
May 82	dry	g	2.2	9.1	0.7	4.5	0.35	3 499	17
Dec. 82	rains[f]	w	2.6	8.3	1.1	9.5	0.86	8 013	38
		g	2.1	7.8	2.2	7.5	0.62	6 023	29
Aug. 83	dry	w	4.5	5.1	1.4	12.5	0.70	8 324	40
		g	1.9	6.4	2.3	4.6	0.40	3 797	18

Notes:
[a] Abundant water: no dry days. [b] Osokunwa wells. [c] Olongoyo hill pools. [d] Olmakutian boma. [e] Olelekando (Ngodoo) boma.
[f] Data from Loipukie temporary camp. FM = fasting metabolism. w = watering day. g = grazing day.

daily. Their study bomas divided into two classes of dry season strategy: those that live near water, graze in the swamp, walk only short distances and water daily, and those that live far from water, graze much further afield, and water only every second or third day. Western and Finch (1986) suggest no appreciable difference in performance is incurred through the different travel costs.

Our results suggest a rather different interpretation. It is not possible to dismiss travel as a negligible cost compared to the effect of reduced intake. This is because in real life, though not in Western and Finch's experimental situation, the two are causally linked. More travel means less feeding time and so reduced intake: the two constraints act synergistically. Dry season travel energy costs may also be considerably greater than those covered by Western and Finch's experiments (Table 8.6). In NCA, those herds walking long distances to water were prevented from grazing because of the circumstances of travel – the duration of the walk, the terrain covered, the need to wait their turn to water in areas where no grazing was possible. In NCA, distance covered provides a rough index of all these vicissitudes combined with that of extra energy expenditure on travel, and is better correlated with the consequent reduction in milk yield than is the simple measure of time spent grazing (see section below on milk yields).

(f) Herd size and composition

The herd sizes and compositions discussed here represent management or access herds rather than units of property. In line with the terminology used by Dahl and Hjort (1976), we term the total aggregate of animals for whose daily care a man is ultimately responsible the access herd. The corral herd is the aggregate of access herds which are temporarily being herded and corralled together, while the boma herd represents the aggregate of all corral herds within one boma. Access herds represent the units on which herd management and boma subsistence are based and are therefore the most relevant units of observation for the management plan study. Individual cattle herds for single corrals within each boma were of similar size in all three study areas (157–176: see Table 8.7). The total boma herd sizes varied more (157–331). The 1980 NCA ground counts gave results very close to our 1981 data for Sendui (157 versus 159) and Ndureta (168 versus 151+). The Ilmesigio herd was substantially smaller (176 versus 299 in 1980) as one herdowner was known to be absent with some stock in 1980. Means and medians for NCA counts for all corral herds in the three areas in our study show the study herd sizes to be representative of their localities.

Corral herd composition in Ndureta and Ilmesigio is strikingly similar with

Table 8.7 *Herd counts of cattle in study bomas*

	August 1981						1980 NCAA census						
	Corral herd			Boma herd			Boma herd			Village total			
	Adult	Imm.	Total	Adult	Imm.	Total	Adult	Imm.	Total	Cattle		small stock	*n*
										Mean	Median		
Nasera	126	42	168	178	76	254	109[a]	42[a]	151[a]	167	134	197	31
Ilmesigio	124	52	176	258	73	331	232	67	299	155	180	242	23
Sendui	92	65	157	92	65	157	93	66	159	225	215	450	21

Notes:

[a] These figures are the pooled NCA counts of 1980 corral herds for the two major Ndureta herdowners using the temporary Nasera boma in our 1981 study.

some 40% adult cows, 15% heifers (immature pregnant first time calvers), 5% bulls, 10–15% steers and 25–30% calves under one year (Table 8.8). The Sendui herd had fewer cows but more heifers. The structure of the female herd (calves: adults/heifers) is similar for all study herds.

Herd composition in NCA is typical of pastoralist cattle herds, with a high proportion of female stock and a characteristic age structure. Dahl and Hjort (1976) and, more recently, De Leeuw and Wilson (1987) discuss the reasons for this commonly observed composition. The high proportion of female stock ensures both milk supply and continued calf production. It also represents the potential for rapid herd recovery after stock losses. The retention of animals past their prime growth rate and/or fertility ensures a core of immunologically experienced and physiologically hardy survivors, which may tide the herd over periods of drought or disease stress. Wealthier herds may accumulate a higher proportion of steers as the relative demand for cash per unit animal declines.

Small stock population numbers change very much more rapidly than cattle herds because of high fertility, vulnerability to epidemic disease and day-to-day use of small stock for meat, in exchange for grain, and in stock deals. Their numbers were therefore very much harder to study than cattle. NCA census figures for small stock are given in Table 8.9. Our study bomas had variable numbers of small stock with around 100 goats in each but sheep numbers ranging from very few in Ilmesigio to around 300 in Gol (Table 8.9). On average there were some 270 small stock per boma, close to the overall mean figure of 210 for NCA (NCAA 1980). The 1987 ground census indicated an overall mean of 128 small stock per boma (Perkin 1987) but the aerial survey carried out at the same time suggested 271 small stock per boma.

Table 8.8 *Herd composition of cattle in study bomas*

	% Total herd					Total *n*	% Female herd		
	Cow	Heifer	Bull	Steer	Calf < 1y		Cow	Heifer	Calf < 1y
Nasera	42.9	15.5	5.4	11.3	25.0	168	60.5	21.8	17.7
Ilmesigio[a]	40.6	14.5	2.1	14.6	28.1	96	60.0	21.5	18.5
[b]	38.1	12.5	4.6	15.3	29.6	176	58.6	19.3	21.9
Sendui	29.4	26.8	4.6	13.0	26.1	157	42.5	38.7	18.7
Dahl and Hjort 1976 theoretical average composition							60.0	25.0	15.0

Notes:
[a] Ngodoo.
[b] Ngodoo + Olelekando.

(g) Performance of cattle herds: condition, fertility and mortality

Cattle performance during the study was assessed in terms of condition, fertility, and mortality and milk production. Milk production, the index whereby Maasai monitor performance, is discussed in the following section.

Adult females were in very much poorer condition than other age–sex classes due to pregnancy and lactational loads. There was an overall decline from initial good condition in August 1981, and an eventual general recovery was noted during the 1983 rains at Gol and Sendui (Table 8.10; Pullan 1978 for condition criteria). Observations on condition bear out later identification of those critical factors that are major constraints on production and therefore on Maasai subsistence in NCA.

Data on births and deaths for each individually identified cow and her offspring were used to calculate calving rates, and calf and cow mortalities. Additional data on deaths, sales, gifts and exchanges for all cattle age–sex classes allowed the estimation of the exponential rate of increase (or decrease)

Table 8.9 *Small stock holdings*

	Study bomas			
	Gol	Ilmesigio	Sendui	NCA mean
Study boma				
Sheep	285	26	131	
Goats	144	119	115	
Total	429	145	246	273
Mean small stock/boma[a]				
1.	197	242	450	210
(*n*)	(31)	(23)	(21)	(960)
2.	182	187	272	212
(*n*)	(12)	(13)	(13)	(38)
3. sheep	105	34	105	
goats	112	73	118	
total	217	107	223	128
(*n*)	(10)	(32)	(51)	(1070)

Note:
[a] Averaged across all bomas in the 'village' to which study boma belongs. The number in brackets represent the number of individual bomas whose small stock holdings were counted in each 'village'.
Source: 1. NCAA 1980; 2. Arhem, Homewood and Rodgers 1981; 3. Perkin 1987.

'*r*' for each herd (Caughley 1977), as well as of the finite rates over the whole study period. There was little difference between sites for most rates and ratios (Table 8.11). All sites showed a decline in herd size during the study period. Cow mortality was lower in Ilmesigio than elsewhere but calf mortality was similar. The lower frequency of site visits in Ilmesigio may have led to an artificially low value for calving rate (0.53 per annum, cf. 0.61 and 0.69 for Gol and Sendui respectively). However, observations at a second Oloirobi boma situated on the Crater rim rather higher than Ilmesigio found a similarly low average calving rate of 0.51 during 1981–1983 (Rodgers and Homewood 1986). Offtake rates for sale or slaughter were consistent at around 8% per annum, giving rates of population decrease '*r*' equivalent to the cattle population halving in 9 years for Gol, 11 for Ilmesigio and 12 for Sendui. With the great variability in the NCA environment, two-year means do not necessarily reflect long-term patterns or indeed consistent differences between sites. The single Oloirobi boma studied in 1981–83 in addition to our three main study sites showed their herd halving over the 25-month study period (Rodgers and Homewood 1986).

Calf mortality shows different patterns in the different areas. In Ilmesigio, calves less than three months old showed high mortality, compared with calves in Gol and Sendui which showed low or intermediate values (Table 8.11). Gol calf mortality peaks around the age of weaning (around 12 months) whereas for Ilmesigio it is linked to high disease incidence from a few months of age (Rodgers and O'Rourke 1987), and possibly to a higher offtake of milk from the mothers for human use (see below and chapter 10). In the separate Oloirobi boma studied in parallel to the three main study sites, over 50% of calf deaths

Table 8.10 *Cattle condition scores*

Site	Date	Bull	Steer	Heifer	Cow	Sample size
Gol	Aug. 81	4.8	4.9	4.8	4.3	354
	Dec. 81	4.0	4.0	4.2	2.9	206
	May 82	4.0	4.1	3.5	3.2	286
	Dec. 82	4.5	4.6	4.6	3.5	138
Sendui	Aug. 81	5.0	5.0	5.0	3.8	718
	Dec. 82	4.6	4.4	4.3	3.4	150
	May 82	4.0	4.1	3.4	3.0	128
	Aug. 83	4.1	3.7	3.7	2.7	246
Ilmesigio	Aug. 83	4.3	3.6	3.3	2.7	246

Note:
Data are mean values for each class on a scale from 1 = very poor to 5 = very good condition (Pullan 1978 for criteria).

took place during three months of the 25-month study period, underlining the importance of drought and forced weaning as well as a high background level of disease in this case (Rodgers and Homewood 1986). Births were concentrated in the rains, over half being born in the period January to March, with fewer than 3% being born in the dry July–September period. In the higher altitude Oloirobi boma a significantly higher proportion (over two-thirds of births) took place in the November–April period. Sex ratio at birth was not significantly different from unity in our sample and there was no differential mortality with sex (Homewood, Rodgers and Arhem 1987) despite suggestions in the literature that female calves may be left more milk and shown more care.

Calving rates and mortality levels may be compared with those for a range of pastoral situations during normal and drought periods (Dahl and Hjort 1976, Wilson, Diallo and Wagenaar 1985, de Leeuw and Wilson 1987, Homewood and Hurst 1986, Homewood and Lewis 1987; Table 8.12). Dahl and Hjort (1976) give a range of calving rates from 50–80% with a median of around 70% as 'normal' and suggest that this drops to zero in drought years. Wilson, Diallo and Wagenaar (1985) suggest that Sahelian cattle have a mean calving rate of 60%. De Leeuw and Wilson (1987) record values ranging from 48–59% for East and West African pastoral and agropastoral systems. Their two Kenyan Maasai groups averaged calving rates of 56–57% during a 'relatively favourable period' in 1981–1983. De Leeuw, Bekure and Grandin (1988) suggest calving rates of 62–63% for these same Kenyan Maasai. Baringo calving rates ranged from 69–100% for cattle in the favourable conditions of an inter-drought period. This fell to a mean of 69% in the first year of drought and eventually to zero as the drought progressed (Homewood and Lewis 1987). NCA calving rates (53–69%) are all within the 'normal' range.

Table 8.11 *Cattle population data 1981–1983*

	Gol	Ilmesigio	Sendui
No. adult cows monitored	70	38	45
Calving rate	0.61	0.53	0.69
Cow mortality	0.129	0.079	0.133
Calf mortality	0.231	0.279	0.276
Offtake	8.3%	8.8%	7.3%
r	−0.074	−0.065	−0.057
Time to halve (years)	9.3	10.7	12.2
% calves dying at <3 months	6	41	18
3–12 months	86	35	46
>12 months	8	24	36

Dahl and Hjort (1976) suggest that calf mortality ranges from 10–40% (median around 20%) in a 'normal' year, rising to around 90% in a drought or epidemic. Wilson, Diallo and Wagenaar (1975) estimate calf mortalities of 20–35% for Sahelian herds, rising to over 50% in drought years. De Leeuw and Wilson (1987) record 'extremely low' calf mortalities of 5–13% for their Kenya Maasai but higher rates of 17–36% for the West African groups. Baringo data suggested annual pre-drought calf mortality rates of 0–19% rising to 57–100% in the first year of drought (Homewood and Lewis 1987). Calf mortality rates for our three main study sites average 26%, within the 'normal' range (if on the high side). The Oloirobi boma showed an exceptionally high calf mortality rate of 71% during the same period, reflecting a combination of drought and

Table 8.12 *Comparative performance of NCA cattle 1980–83 (figures are expressed as % per annum)*
(a) Normal conditions

	NCA mean for study	Baringo[1]	sub-Saharan Africa[2]	Sahel[3]	Kenya Maasai[5,6]
Calving rate (births/adult females)	61(53–69)	83(69–100)	70	60	60(56–63)
Adult cow mortality	11(8–13)	4–17	10(5–15)	5	6(2–10)
Calf < 1yr mortality	26	ca.11(0–19)	20(10–40)	25(20–35)	9(5–13)
Offtake from total herd (sale/slaughter/gift)	8	25	8		

(b) Drought and/or epidemic

	Oloirobi epidemic[4]	drought[1]	drought[2]	drought[3]
Calving rate	51	69[a]	0	
Adult cow mortality	43	45	(50–80)	
Calf < 1yr mortality	71	79(57–100)	90	50+

Note:
Range given in brackets. [a] Mean value for first year of drought: thereafter, calving rates declined to zero until after the drought.
Source: [1] Homewood and Lewis 1987, Homewood and Hurst 1986; [2] Dahl and Hjort 1976; [3] Wilson, Diallo and Wagenaar 1985; [4] Rodgers and Homewood 1986; [5] de Leeuw and Wilson 1987; [6] de Leeuw, Bekure and Grandin 1988.

disease in this densely stocked and crowded area around NCA HQ (Rodgers and Homewood 1986).

Cow mortality has been reported as around 5–15% for a range of 'normal' situations compared to 50–80% in drought or disaster years (Table 8.12). Our main study site values (8–13%) are well within the 'normal' bracket. The Oloirobi boma studied in parallel showed a mortality rate of 43% among adult cattle due to drought and disease during the study period.

For the three main study sites, though not the Oloirobi boma, the study clearly revealed 'normal' rather than disaster conditions of fertility and mortality. Despite local, seasonal and age/sex class variation, our scoring showed the generally fair or medium condition of most of the stock for most of the study. It is notable therefore that once offtake is taken into account, all study herds showed a net decline over the study period. This becomes of interest in our summing up of the efficiency and productivity of Maasai pastoralist livestock.

Disease (a major cause of mortality) was investigated in some detail in the August 1983 mid dry season and reviewed in 1987 as part of the Ngorongoro Conservation and Development project (Table 8.13, Field and Moll 1987, Rodgers and O'Rourke 1987), Field and Moll (1987) list a series of factors acting synergistically to bring about high disease incidence:

- nutritional deficiencies (pasture/water/mineral availability)
- boma conditions favouring transmission of infectious disease
- travel stress in drought
- disease vector and reservoir populations

Field and Moll (1987) report most NCA stock as being in fair to poor condition in their early wet season visit (see last section, this chapter). In our study, heifers and male animals were generally in better condition than adult females with a lactation/pregnancy burden (Table 8.10). Such animals were anaemic and frequently presented signs of stress and disease – lung murmurs, jugular pulse, swollen lymph nodes, etc. (Rodgers and O'Rourke 1987). Calves were in very poor condition showing serious anaemia and dehydration. Herds at Gol had less overt sign of cattle disease and blood parasites were not seen. *Anaplasma* and *Theileria* were seen in both Sendui and Ilmesigio, and trypanosomes at Ilmesigio. Intestinal parasites were not frequent in cattle. Pneumonia and respiratory infections were common in small stock, and hydatid cysts were seen in all animals examined. Cattle dipping to reduce tick burdens (and hence tickborne disease, especially East Coast Fever, Theileriosis) was commonly practiced in Oloirobi village (including Ilmesigio) in the past (e.g. Fosbrooke 1972) but acaricide shortages have led to infrequent dipping and a great increase in cattle deaths (Machange 1987). The importance of cattle disease in

Table 8.13 *Diseases reported in NCA cattle*

Disease[1]	Confirmed[2]	Notes
(Rinderpest)[a]		1982 wildlife outbreak
Foot and Mouth		
Trypanosomiasis	+	Lowland endemic
Malignant Catarrhal Fever		Seasonal endemic: major cause of range limitation
Tickborne		
East Coast Fever (ECF)	+	Major calf mortality
Anaplasmosis	+	
Babesiosis		
Heartwater		
'Olmilo' = ? cerebral theileriosis		
Anthrax		
Blackquarter		
Pneumonia		
TB		
Actinomycosis		
White scour		
Internal parasites		
Liverfluke		
Neoascaris		
Monieza		
Strongyloides		
Cysticercus	+	
Echinococcus		
Coccidia		
Leeches		
Ectoparasites		
Tick load (6spp. including major disease vectors)	+	
Ringworm		
Lice		
Fleas		
Infertility diseases		
Brucellosis		
Trichomoniasis		
Mineral deficiency		
Blood conditions		
Anaemia	+	
Dehydration	+	

Note:
[a] Cattle are vaccinated against rinderpest (Scott 1985, Machange 1987, 1988). This largely controls its impact among livestock.
Source: [1] Field and Moll 1987; [2] Rodgers and O'Rourke 1987.

Ilmesigio was emphasised by the amount of energy and effort herdowners put into traditional medicine as well as husbandry and management methods designed to reduce infection in this area. The synergism of drought and disease is illustrated by the heavy mortality among cattle in the separate Oloirobi boma.

(h) Performance of cattle herds: milk production

For the Maasai, milk production is perhaps the most immediate index of cattle performance. We analysed over 1000 individual milk yields from over 100 different cattle for a representative range of study herds, areas and seasons (Table 8.14). Our figures measure the milk taken for human consumption and exclude milk taken by calves. Grandin (1988) analyses the variation in proportional offtake for human rather than calf use with such factors as wealth and number of milking cows. Milk offtake per person was fairly constant among the different study families on Olkarkar ranch in Kenya, and richer households with more milking cows milked individual cattle less thoroughly or only once a day, leaving more for the calf and achieving better calf growth rates and survival. As our study herds and households were all of intermediate wealth and stock holdings, and different zones showed both different milk yields per cow and correspondingly different milk intakes per person (chapter

Table 8.14 *Milk production*

Date	Area	Boma	Mean milk yield (g)	SD	No. cows
Aug. 81	Gol	Nasera	300	97	38
	Ilmesigio	Ngodoo	597	148	24
	Sendui	Senguyan	340	113	33
Dec. 81	Gol	Olduvai	352	151	21
	Sendui	Senguyan	243	136	13
May 82	Gol	Ndureta	339	122	25
	Sendui	Senguyan	511	120	9
	Sendui	Lera	656	174	13
Dec. 82	Gol	Ndureta	547	169	21
	Sendui	Loipukie	487	145	14
Aug. 83	Gol	Nasera	227	102	18
	Ilmesigio	Ngodoo	424	247	11
	Sendui	Senguyan	379	187	13

Notes:
NCAA herd total yields average 4.7 l/day (8 cows × 8 weeks; SD = 0.51), NCAA cows fed supplement: 5.3 l/day (8 cows × 10 weeks; SD = 0.62), full yields: Ilmesigio, Aug. 81: mean 635 g/milking (1270 g/day), range 240–1000 g/milking, $n = 4$.

10) the variations in yield that we observed are attributed primarily to ecological effects rather than differential milking. However, it is possible that heavy offtake for human use in Ilmesigio was a factor in the poor survival rates of calves from this site (chapter 10, also Grandin 1988).

The lowest mean yields were in Ndureta (227 g – August 1983; 300 g – August 1981) and Sendui (243 g – November 1981). The highest were in Ilmesigio (596 g – August 1981) and at a temporary intermediate altitude boma at Lera (Naibor Ajijik) belonging to a Sendui herdowner (656 g – May 1982). Yields recorded from the improved breed x zebu dairy herd kept by the NCAA and herded on the Crater rim are not directly comparable as they represent full yields (calves are bucket fed). The dairy herd also had some supplementary feeding, daily watering and no travel stress. These cattle averaged some five litres/day, an order of magnitude more milk than do pastoralist cattle in NCA. Field and Moll (1987) use our 1981 figures to extoll 'the Ngorongoro dairy herd as an example of what can be achieved under intensive management'; they then (apparently without intended irony) document the subsequent progressive demise of this dairy herd since the time of our study. All are now dead. Chapters 10 and 11 investigate the severe constraints on the potential for dairy herd development.

A detailed analysis of milk yields is given in Homewood, Rodgers and Arhem (1987). Our main interest was firstly to identify main ecological constraints affecting milk production, secondly to document the input from milk to the pastoralist diet (see chapter 10), and thirdly to investigate the possibility of increasing milk production in NCA for more reliable subsistence or even commercial production (see chapters 10, 11). Milk yield data were analysed for correlation with energy demands of travel, season, feeding time, time of day, grazing versus watering day effects, age and sex of calf, maternal age and parity, and coat colour (which affects heat balance and therefore physiological performance – Finch and Western 1977).

Despite the differing energy costs and feeding times there was no clear pattern of grazing versus watering day yields, nor morning versus evening yields. There was a general trend towards grazing day yields being greater than watering day yields, and morning greater than evening production, but in most areas and seasons this was not significant (Homewood, Rodgers and Arhem 1987). A breakdown of yield by calf age showed no obvious pattern in individual areas. There is no clear pattern of yield with parity. The few animals which were pregnant, lactating and still being milked showed a fall-off in yield with advancing pregnancy as expected. Other than this there was no consistent pattern for the yields of individual cows at different stages of lactation. De Leeuw, Bekure and Grandin (1988) calculate that the Kenya Maasai in their study took 24% of the total milk production, as their growth rates show that calves take 76% (they assume that 1 kg calf equals 9 kg milk).

Finch and Western (1977) suggest that heat stress is a major selective pressure on East African pastoralist cattle and that light coloured cattle which reflect incident radiation perform better at low altitudes. By contrast, night cold stress becomes increasingly important with altitude. Dark coloured cattle which rewarm more rapidly could thus be at a selective advantage in high altitude herds. Robertshaw and Katongole (1969) found cold thermogenesis at temperatures below 20 °C in zebu cattle. Boran cattle show cold thermogenesis at temperatures below 24 °C and the energetic costs may be considerable for animals on submaintenance diets. This would be the case, for example, in Sendui where night temperatures may drop below 5 °C. Our results on coat colour distribution for herds at different study sites suggest that temperature is indeed a significant selective pressure acting on NCA herds and may be expected to have differential impact on survival and production in the different areas (Homewood, Rodgers and Arhem 1987). An influence of coat colour on energy balance and hence milk production cannot be ruled out. If any such influence exists in our milk yield data, however, it is masked by the impact of food availability and energy expenditure on travel, the estimation of which is detailed in the earlier section on activity and range utilisation.

The overriding constraints on milk production were those of daily travel distances, (in themselves a function of range conditions), and resultant limitations on time spent feeding. All the other variables mentioned may be expected to play a part in determining milk yields but in practice they are overridden by the tradeoff between food intake and energy expenditure. Milk production showed a close inverse correlation with the energy demands of travel (see Tables 8.6, 8.15, Fig 8.5, Homewood, Rodgers and Arhem 1987, Homewood and Hurst 1986) as well as a significant inverse correlation with time spent walking. Conversely, there is a significant correlation between milk yield and time spent feeding (though cattle may compensate for restricted feeding time by increasing intake rate – Lewis 1978). Our study cannot distinguish between restriction of feeding time and food quality, and extension of walking time as prime determinants of the reduction in milk production. Western and Finch (1986), using weight loss as an indication, suggest that reduced intake is the most important factor and conclude that extended walking time has negligible energetic costs for zebu. However, this was established under experimental conditions where time spent walking did not limit time spent feeding. Homewood, Rodgers and Arhem (1987) and Homewood and Hurst (1986) found a close inverse correlation between milk yield and estimated energy expenditure on travel over a wide range of grazing and

Table 8.15 *Main factors affecting milk production*

Date	Area	Mean milk yield (g/milking)	Mean estimated travel energy (%FM)	Time spent walking (h)	Time spent feeding (h)
Aug. 81	Gol	300	35.5	4.1	4.7
	Ilmesigio	597	16	2.7	4.9
	Sendui	340	28.5	2.7	6.2
Dec. 81	Gol	352	35.5	2.8	7.7
May 82	Gol	339	30.5	3.2	6.3
	Lera	656	17	2.2	9.1
Dec. 82	Gol	547	18	1.7	7.9
	Sendui	487	33.5	2.4	8.1
Aug. 83	Gol	227	36.8	7.8	3.4
	Ilmesigio	424	30	4.3	5.3
	Sendui	379	29	3.2	5.8

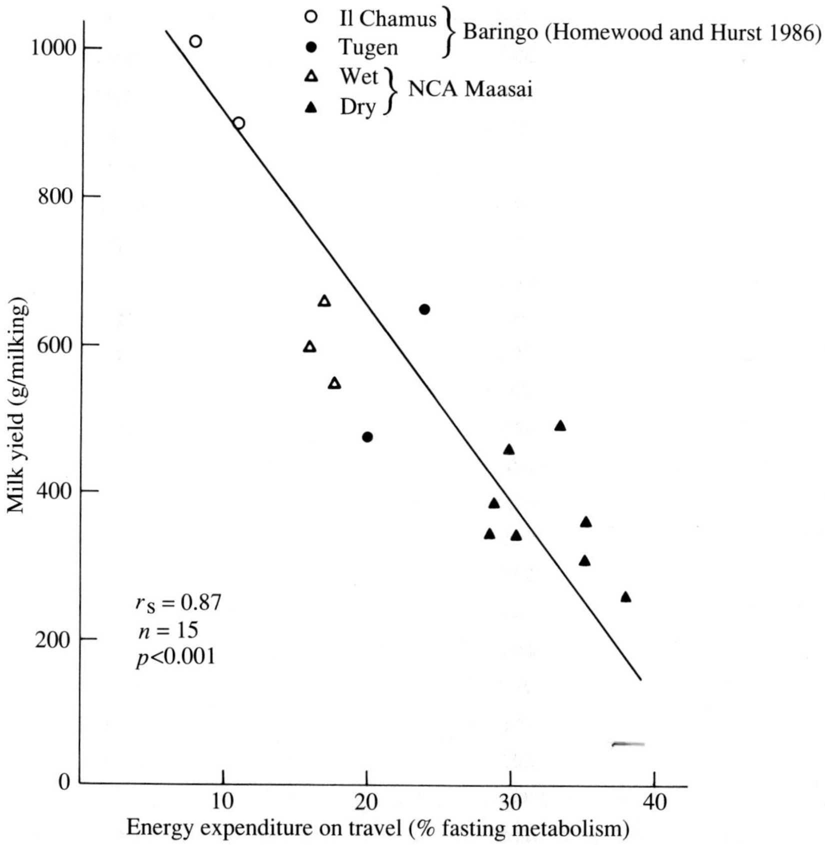

Fig. 8.5. Relationship of milk yield to energy expenditure on travel.

travel conditions. The important point is that given the remote nature of dry season water in NCA and its difficulty of access, the combined effects of increased travel time and of the consequent restriction of time spent grazing limit livestock performance, milk production and presumably survival. These relationships account for the significant differences in milk production between areas and seasons, with high yields where adequate grazing and water are freely available (e.g. only Ilmesigio in August 1981; all three main study sites in December 1982). Again, this finding is of interest when we come to evaluate Maasai livestock efficiency, and potential for future production.

One final aspect of production was that of milk composition (Table 8.16). This was analysed to give a basis both for comparison with other studies and for calculating the pastoralist energy balance. Protein values for Ndureta and Ilmesigio corresponded with most other data. Sendui figures are significantly higher than other NCA values but come within the range of Maasai zebu dry season figures given by Little (1980). Zebu cattle show a marked decline in milk quality with a single long dry season (Table 8.16: see Sahelian zebu fat content figures, and energy values given by Little for Maasai zebu wet and dry season milk). However, Nestel (1986) found significantly higher fat (and therefore energy) content in dry versus wet season milk from Maasai zebu in Kenya (Table 8.16). These contradictory findings may be due to the relative severity of conditions, with fat content initially rising but eventually declining as drought conditions and nutritional status deteriorate. Where NCA experienced a very long dry period (e.g. from before August 1981 until after May 1982) a parallel continued decline may be implied. Sendui cattle could be expected to do better at high altitude with heavy dew and low evapotranspiration: our range study established availability of at least some green grass at all times in this area (chapter 6: see also Homewood, Rodgers and Arhem 1987). Protein values for the NCA model dairy herd correspond to those for Ndureta and Sendui. There is no significant difference between areas in lactose values. These correspond with the expected Maasai zebu average quoted in Dahl and Hjort (1976), though they are higher than the values given by Little (1980). Phosphate and calcium levels vary significantly between areas. Ilmesigio shows the lowest and Sendui the highest concentrations. Maasai maintain that the Ndureta pastures are richest in necessary minerals. This applies chiefly to the short grassland association which was not productive during our dry season visit. By contrast, Ilmesigio pasture is held to be the least rich in minerals. This pattern would be expected on the basis of the soil, rainfall and vegetation patterns described in chapter 6. This low mineral status of Ilmesigio and Oloirobi cattle is of course the reason why people from the village are so anxious to maintain periodic access to the salt licks of Ngorongoro Crater. This relatively minor requirement

Table 8.16 *Milk composition*

Study	Fat %	Protein %	Lactose %	Ash %	Energy (J/100 ml)
This study Aug. 81 Gol	—	3.78	4.61	—	—
Ilmesigio	—	3.61	5.01	—	—
Sendui	—	4.72	4.88	—	—
NCA herd	—	3.68	4.69	—	—
1. Maasai zebu	5.5	3.7	4.9	0.73	349
2. Tropical cattle	4.8	3.3	4.8	0.47–0.68	—
3. Sahelian zebu – dry	2.5–2.8	3.0			
– wet	5–7	3.8			
4. Maasai zebu – dry	3.5–4.3[a]	4.4–6.5[a]	3.4–4.3[a]	—	273–378
– wet	3.5–4.1[a]	6.0–10.0[a]	3.4–4.0[a]	—	336–525
5. Maasai zebu – dry	6.0 ± 0.2	—	—	—	323 ± 8.4
– wet	4.4 ± 0.4	—	—	—	244 ± 12.6
6. Europe	3.5	3.5	5.0		277
7. USA	3.2[a]	3.8[a]	4.9[a]		273

Notes:

[a] g/100 ml

Source: 2,3,6 Dahl & Hjort 1976: 154, 5 Nestel 1986, 1,4,6 Little 1980: 484.

has been treated as a major issue of conflict by the NCAA and by some visiting conservationists (e.g. Struhsaker *et al.* 1989). Currently Maasai are allowed as a concession to use the Crater at Seneto for dry season water and mineral licks.

Summary and conclusion

1. Livestock production by subsistence pastoralists is an important component of food production in sub-Saharan Africa generally and in East Africa and Tanzania in particular.

2. The breeds commonly used in subsistence pastoralism differ from those of commercial livestock production and have in many cases been shown to be ecologically better suited to the special conditions of arid and semi-arid Africa.

3. NCA livestock numbers have fluctuated but show no overall increase over the last 25 years. Livestock biomasses do not suggest unsustainable stocking rates.

4. NCA herds have the preponderance of female animals typical of pastoralist herds. Fertility and calf and adult mortality at the time of the study were average for subsistence pastoralism.

5. Offtake rates at the time of study were low compared with commercial herds but unsustainably high given NCA livestock fertility and mortality rates.

6. Performance in terms of milk production was inversely correlated with energy expenditure on travel to water and grazing, and positively correlated with time spent feeding.

7. Pastoralism is a productive and viable form of land use in NCA, as in many other parts of sub-Saharan Africa. Our detailed data on livestock ecology gives a baseline from which it is possible to go on to consider comparisons and interactions between livestock and wildlife and livestock performance in a joint land use area (chapter 9), important factors in Maasai subsistence ecology (chapter 10), the future of joint pastoralism and wildlife land use in NCA (chapter 11), and the potential for technical and other interventions (chapter 12).

9

Livestock and wildlife

Ejo ilparakuo eji mikirisio kirisio	The pastoralists say we are not equal, yet we are
Teneru olkiteng' neru munkarro toldoinyo	When the ox lows, the buffalo echoes on the hill
Neru osikiria neru olosira marae	The donkey brays, so does the striped-backed zebra,
Needo olkiteng' nebaiki enkai, olmeut!	Tall is the ox, it touches the sky, giraffe!
Kumok intare neibor kidong'o, nkoiliin!	The goats are many and have white tails, gazelles!

(part of Maasai song teasing Dorobo: Kipury 1983)

One common theme that runs through management debates in NCA is the comparison of livestock with wildlife. There is an overt assumption throughout most of the sub-Saharan rangelands in general and in NCA in particular that wildlife do not overgraze while stock do (Lamprey 1983); that wildlife land use is sustainable while livestock land use is not (Lamprey 1983); that wildlife should have the freedom to range throughout the Ngorongoro (and Serengeti) system while stock should not (Dirschl 1966, Ole Kuwai 1980); that stock present a competitive challenge detrimental to wildlife (Pearsall 1957; Ole Kuwai 1980; see also chapter 7) and that stock are generally less efficient and productive than wildlife (Simpson 1984a). The dynamics of the wildlife community were examined in chapter 7, and chapter 8 set out the detail of livestock ecology. The present chapter considers comparisons between the wild and domestic herbivore communities (particularly between wildebeest and cattle, their respective dominant species) in terms of feeding, ranging, and environmental impacts, and looks at their population interactions. The relative dominance of the wild versus the domestic herbivore communities is analysed for a range of wildlife/pastoralist joint land use systems. Finally, the efficiency and productivity of Maasai pastoral livestock in the Ngorongoro joint land use system is evaluated.

Comparison with wildlife

(a) Feeding

The literature on African land use has many examples of grazing communities where wild and domestic herbivores are, or were, closely integrated. The Lake Rukwa grasslands in southwest Tanzania (Rodgers 1982a); the Nile floodplain grasslands in Sudan (Howell, Lock and Cobb 1988), the Simanjiro Plains in Tanzania (Kahurananga 1981) and the Amboseli Lake Basin in Kenya's Maasailand (Western 1975) are cases in point. The wild and domestic grazing communities are usually perceived to coexist in harmony when resources are abundant. However, both pastoralists and conservationists in these systems complain of competition when resources are in short supply whether due to herbivore population increase or to drought, fire, lake or flood level changes. With human help in providing water and locating available fodder, domestic stock become extreme generalists and can dominate the system (e.g. Kahurananga 1981; Western 1971; see also section on joint land use systems below).

The wild and domestic large mammal herbivores of the Serengeti/Ngorongoro region overlap in their feeding and habitat requirements. Wildlife feeding habits are reviewed by Jarman and Sinclair (1979) and by McNaughton (1983, 1985) for the grazers. Niche separation is achieved by differential selection for coarse and fine scale habitats, plant species and plant part as well as by sequential movement patterns. Casebeer and Koss (1970) in a comparative study of the feeding habits of wildebeest, zebra, hartebeest and cattle on the Athi Plains of Kenya found that all four herbivore species preferred *Themeda triandra* to *Pennisetum mezianum*, which in turn was preferred to *Digitaria macroblephara*. Zebra were the least selective, their diet showing the closest similarity to the composition of the sward, and hartebeest the most selective. Cattle and zebra were the most similar in their diets. All four ungulates had a wide range of species in the diet, the range being widest in the dry season. However, cattle diets were the most consistent across seasons, more so than the varying combination of grasses available at different times of year.

(b) Ranging

The continuum of range use patterns by livestock corresponds closely with that shown by the large ungulate wildlife. Both pastoralist stock and individual wildlife species show seasonal transhumance in the Crater Highlands, but semi-nomadic or migratory movements in the Gol/Serengeti area, (e.g. wildebeest – Maddock 1979, Sinclair 1983a). The close parallel between ranging strategies of pastoralist herds and of wildlife is dictated by their common dependence on critical grazing, mineral and water resources. Response to these main factors is modified according to additional pressures such as administrative bans, grazing competition and disease interactions.

Foraging patterns may have evolved to maximise Darwinian fitness (e.g. Pike in Swingland and Greenwood 1983). Maasai pastoralist management strategies for optimising the production of calves and milk are equally bound to affect the relative success of different individuals and herds in succeeding generations of livestock. Sinclair (1983a,b) stresses that with a fluctuating environment species that remain resident are regulated by resources available at the worst time of year (unless they are able to store). Such residents tend to be much less numerous than migrants, the small resident Serengeti population of 50 000 hartebeest contrasting with the peak population of 1.5 million migrant wildebeest of similar size and ecology. Most wild ungulates and their associated predators have adopted the strategy of migration by following good resources. As resources become increasingly unpredictable so more nomadic strategies are adopted. Until pastoralists have means to store surplus produce their best stategies remain those migratory and nomadic movements that have evolved by natural selection in these fluctuating environments. Land use managers must provide for these strategies.

Pennycuick (1979) presents a general scheme for estimating limiting foraging radius of a large mammal herbivore dependent on a fixed water source and needing to travel an ever-increasing distance out from this source to find grazing under dry season conditions. Substituting a mean body weight of 180 kg for zebu into his equation gives a predicted day's travel of 16 km if the animal is to maintain condition. This is the observed maximum day's trek often quoted for pastoralist cattle (Williamson and Payne 1978, Dahl and Hjort 1976). It may of course be exceeded (this study: Gol, December 1981; Western and Dunne 1979) but only at severe cost in terms of condition, production and ultimately survival. Adult wildebeest have a similar mean body weight and mean foraging radius. Pennycuick (1979) quotes Kreulen's estimate that wildebeest travel an average 10 km/day throughout the year. Taking our adult cattle figures, firstly for all study areas pooled and secondly for Gol alone, we estimate the mean day range to be between 9.4 and 10.4 km/day – remarkably close to the comparable wildlife value.

The standard pastoralist practice of splitting the herd into different units (e.g. calves versus adults) exploits the relationship between body weight and foraging radius. Adult pastoralist cattle may weigh 250 kg (or even 350 kg for steers) and could achieve an equilibrium day's journey of 18–20 km, but herding labour constraints limit the number of possible units into which the herd may be split.

(c) Impact of range restrictions on livestock

Since the Maasai and their livestock were first evicted from the Serengeti pastures and water sources they have been subjected to various range restrictions. Various grazing policies have been planned though none has been put into formal practice. Dirschl (1966) divided NCA into 17 land zones of which Ngorongoro Crater, the Highlands Forest Reserve, the Olmoti Forest and Empakaai Crater were given full protection with no grazing access, for aesthetic or watershed conservation reasons. A number of zones were suggested for controlled grazing, for example Loshonyokie (a more or less treeless section of Forest Reserve which was put forward for experimental pasture schemes); and Lemagrut mountain where exclusion of stock and burning from forest and forest edge areas was suggested. In other areas Dirschl suggested restricting season of use or banning permanent settlement. His proposals were never implemented but later range management plans tend to repeat variations on the theme Dirschl outlined.

Currently substantial areas of the 8292 km² NCA are banned to livestock – 40 km² around Olduvai; the 250 km² Ngorongoro Crater, with its mineral rich grazing, salt licks and permanent water; 650 km² of forest reserve, the higher portions of which have reserved dry season grazing in forest glades; and the Olmoti and Empakaai Craters. All of these areas are used occasionally, whether illegally or by special permit, and are a perennial cause of tension between the pastoralists and the NCAA. Perhaps the most serious restriction is one that is not directly imposed by the administration but that is most rigorously observed: the short grass plains must be avoided during the rains because of the wildebeest and the MCF they carry.

Given the availability of other types of productive pasture during the rains, is the exclusion of livestock from the short grass plains of any importance to pastoralists? The answer is yes, for a number closely related reasons: loss of a prime resource during its peak production; loss of potential dry season reserve grazing which must be used in the rains as well; and exclusion from a range of mineral-rich pasture types which may be critical in reestablishing condition at the end of the dry season (Swift 1983, McNaughton 1988, 1990) and thus determining survival and fertility in the ensuing year. Finally, the disease transmission implications of range restrictions have a serious effect.

Swift argues that the beginning of the rains may be more critical for stock than the long period of deteriorating condition as the dry season progresses. Change from low quality forage to nutrient-rich new growth; high secondary toxic compound concentrations in some herb new growth; rapid travel over long distances to exploit this new growth; and the fact that all these influences are acting upon already debilitated animals, combine to make the transition season difficult. Skilled management decisions (about how much to graze, where, on what selection of potential pasture types and with what access to minerals) make for a herd which rapidly gains condition and which subsequently shows a high general performance. Such management is felt to be more important at this time than, for example, during the dry season when the availability and management of herding labour may be more critical. If choice of grazing patterns and timing of mineral supply are crucial, then restrictions imposed by administrative bans and wildlife–livestock disease interactions may have a greater impact than is immediately apparent. The Serengeti and NCA plains are mineral rich. Kreulen (1975) suggested that wildebeest select areas of short grassland with high calcium and protein content as soon as the rains begin. This is thought to influence successful lactation, and survival both of newborn calves and of adult females which otherwise deplete their own bone calcium. McNaughton (1988, 1990) has demonstrated overwhelmingly significant selection for a range of minerals. Similar requirements probably influence cattle performance and carry long-term repercussions even during seasons previously considered problem-free by outside observers. In a village survey, Arhem (1981a,b) identified restrictions on land use as the most serious perceived problem for both Olbalbal (including Gol) and Alaililai (including Sendui). The resulting dry season hardship is obvious; the transition season implications may be equally serious in the long run.

(d) Disease interactions

Disease interactions of wildlife and domestic stock have strongly influenced land use and conservation practice in East Africa (Ford 1971; Kjekshus 1977; McCracken 1987). The drastic effects of the 1890s rinderpest epidemic were described in chapter 3. Vaccination of stock against the disease may have triggered the wildebeest eruption and buffalo population increases 1960–1980 (see below and Sinclair 1979). The subsequent reappearance of rinderpest in the Serengeti ungulates may have been a factor in the levelling off of that increase, for buffalo at least.

Most disease interactions, however, affect cattle to a much greater extent than wildlife. In NCA the viral disease of MCF is particularly important to the interaction between wildlife and domestic stock. MCF is endemic to wildebeest and it is common for newborn wildebeest calves to suffer a mild version for a short time. The disease is highly contagious and fatal to cattle. Cattle coming into contact with calving wildebeest are likely to succumb and the Maasai believe that it is the wildebeest placentas which harbour and transmit the infection. They therefore keep their stock well away from wildebeest during the calving period until no placentas remain on the pasture (Rossiter, Jessett and Karstad 1983). MCF is really transmitted in saliva and mucous from the nasal passages, and is possibly airborne over short distances, but the practical outcome of the Maasai theory is that cattle are protected during the critical period when transmission is likely. The corollary is that cattle are excluded from the high quality short grass pastures during their period of peak production, with far reaching implications for cattle performance (see below).

Trypanosomiasis, transmitted by tsetse fly, limits the distribution of stock throughout much of Tanzania allowing extensive wildlife populations to survive by default in bush and woodland areas (see for example the Tanzania Atlas distribution maps). Ford (1971) suggested that the decimation of wildlife, stock and human populations in the rinderpest epizootic of the 1890s allowed the spread of both woody vegetation and tsetse vectors through extensive areas of Africa previously maintained as grassland by grazing and burning. The result is that many areas once used by farmers and pastoralists remain disease foci excluding livestock in the present day. Much of Maasailand is affected, particularly the Southern Steppe. Tsetse eradication in the 1940s allowed the expansion of cultivation in the Mbulu areas, cutting the forest corridor between Manyara and NCA (see section on forest wildlife below). However, tsetse levels in NCA today do not seem to deter stock movements to any area despite some morbidity and mortality (Rodgers and O'Rourke 1987). There is no record of recent changes in fly induced wildlife or cattle distribution. During drought periods tsetse populations and disease transmission decline and areas that are normally infested and avoided become relatively more attractive because of the availability of pasture and water (e.g. Homewood and Lewis 1987).

Rinderpest has had an enormous impact on the ecology of Tanzania, both historically (chapter 3), and more recently when its eradication allowed the eruption of the Serengeti wildebeest (Sinclair 1979) with all the attendant implications for NCA cattle. It is still a major disease hazard, which could potentially decimate both wild and domestic herbivores. Vaccination programmes prevent major epidemics in cattle, though recent outbreaks have been of low virulence (Machange 1987, 1988, Scott 1985).

Many other diseases are transmitted between wildlife and domestic stock by tick vectors (Machange 1988, Field, Moll and Sonkoi 1988). Tick control is a major management issue. The NCAA has in the past held responsibility for acaricide dips for livestock: these are largely non-functional. Traditional Maasai management of tickborne disease relied on an effective combination of transhumant movements, burning, and grazing sequence as well as hand grooming of cattle to remove individual ticks:

> One of the best examples of a stable, integrated program of pasture rotation to achieve the dual objectives of continuous feed supplies and the control of parasites was that of the Masai of Tanzania (Branagan 1974). Cattle grazed the Ngorongoro crater in the dry

> season and moved down onto the surrounding arid plains at the onset of the wet season. Movements were carefully timed to avoid contact between cattle and both ticks and helminths in the crater, which are aroused from dormancy by rain. The wild herbivores were temporarily deterred from grazing the plains by fencing-off of the waterholes with thorn bushes prior to reintroduction of the cattle. The return to the crater also involved a complex ritual to reduce the numbers of the tick *Rhipicephalus appendiculatus*, the vector of *Theileria parva* which causes the severe disease known as East Coast Fever. The *Themeda triandra* pastures in the Crater were first burnt to kill ticks. Grazing by sheep and goats then harvested any surviving ticks. The burning killed ticks and also controlled the unpalatable grass *Pennisetum schimperi* which formed dense tussocks favourable for tick survival. Only after these procedures were completed were cattle reintroduced. Unfortunately, that elegant system was lost when game conservation measures were imposed and the disruption of the system led to devastating outbreaks of East Coast Fever.
>
> Sutherst 1987

Theileriosis or East Coast Fever is common in NCA (Rodgers and O'Rourke 1987) and a new form has recently surfaced in Oloirobi (Machange 1988). ECF can be fatal to cattle and causes huge production losses, but has little recorded effect on wildlife. Ticks in very high numbers cause anaemia and predispose wild animals to disease (Sinclair 1977), but tall-grass tick-infested pasture is avoided by wildlife on nutritional grounds, leading to parasite avoidance (Estes and Small 1981). Surveys of perceived problems showed many Maasai villages listing disease (usually ECF) as one of their major problems (Arhem 1981a,b; Chamshama, Kerkhof and Singunda 1989:14). The breakdown of tick control seems to have been a major factor in the high mortality and overall decline of the cattle population 1980–84 in NCA (Rodgers and Homewood 1986; chapter 8). Animals on a low plane of nutrition due to drought were faced with huge tick challenges. Tick populations can be controlled by keeping grass short, whether by heavy grazing (cf. Sukumaland – Birley 1982), or by burning. However in the late 1970s and early 1980s not only were dips not maintained or operated, but also traditional range burning was banned. Full transhumant movement patterns were (and still are) severely limited by a combination of administrative restrictions and MCF interaction with the growing wildebeest population. Machange (1988) gives a detailed discussion of the failure of the NCAA dipping programme, due to lack of acaricide and water. Both are symptomatic of the lack of input into pastoral resources by the NCAA.

Perhaps most serious is the express prohibition of traditional disease control by burning, in an area which was zoned for cattle development and had major disease problems. As one small example of the outcome, the deaths of half the Oloirobi boma herd over a two-year period were largely attributable to this loss of control over disease transmission (Rodgers and Homewood 1986).

(e) Environmental impacts

The main impacts of herbivores on the environment are defoliation, trampling and nutrient cycling (Chapter 6). Selective defoliation may bring about a shift in cover values and species composition, in some cases stimulating and in others adversely affecting growth and spread of particular species. Reduced cover values and increasing bare ground may exacerbate erosion together with physical changes in the soil adverse to plant growth. Trampling in some cases assists tillering and vegetative spread but also leads to soil compaction and adverse structural changes, especially at higher intensities. Chapter 6 set out the main debates on livestock impacts on short, medium and tall grasslands of NCA together with the evidence for the effects on particular grass species. The main grassland formations of NCA have coevolved with, and are sustained by, heavy grazing. Accusations of overgrazing have typically been poorly defined, unsubstantiated, and based on spot judgements which themselves relate to standards of range condition inappropriate to semi-arid rangelands and pastoralist/wildlife land use (Homewood and Rodgers 1987; Caughley 1983; Sandford 1983; Warren and Agnew 1988). Our figures on stock and wildlife densities and durations of use for different NCA rangelands suggest that biomasses (both of stock and of stock and wildlife combined) are within those found to be average and suggested to be sustainable for East African grasslands (see chapter 6).

In general, NCA shows low rates of erosion compared to surrounding agropastoralist areas (Ecosystems Ltd 1980; see also chapter 6) despite its geomorphological and climatic predisposition to erosion. However, Landsat evidence (King 1980, 1982; and more recently by the NCDP in prep.) suggests severe trampling in Angata Salei and progressive spread of bare areas in the unconsolidated volcanic dust soils of this migration corridor. Evidence reviewed in this and previous chapters on the relative numbers and durations of use by wildlife and livestock allows an evaluation of wildebeest versus cattle trampling effects. Ssemakula (1983) has made a comparative study of hoof pressures of wild and domestic ungulates, estimating hoof pressure by dividing body weight by the sum of hoof area. He showed a generally linear relation of hoof pressure with body weight. Light-bodied sheep and goats exert rather low hoof pressures; large bodied eland and cattle have correspondingly heavier hoof pressures. There is no evidence for the persistent notion that domestic stock exert more severe hoof pressures than do wildlife of comparable size: in fact cattle showed a rather lower hoof pressure than expected versus eland. Wildebeest hoof pressures have not been measured, but given their similarity of body size, Maasai zebu and wildebeest probably exert rather similar hoof pressure. Trampling effects are as much a function of animal density as of hoof pressure. Wildebeest use of the Angata Salei increased considerably in the 1970s (Chapter 7). Short grass plains species resistant to (or even dependent on) moderate trampling are killed by repeated simulated hoof pressure (Belsky 1986c). While close herding and frequent passage over the same area might exacerbate domestic stock trampling effects, one must conclude that the concentrated passage of tens of thousands of kg/km^2 of wildebeest during a period of a few weeks is more likely to be responsible for the damage to the Angata Salei migration corridor than the rather low biomass of resident livestock.

Elsewhere, there are restricted areas of tracking converging round corral and boma sites. However, this appears to be of negligible importance. Examination of aerial photographs of old settlement sites at Nainokanoka shows no discernible increase in gulley erosion from 1955–1982 despite the rapid increase in human and livestock populations in this area during the period covered. Maasai, for obvious logistic reasons avoid siting their settlements on erosion-prone slopes, drainage systems and soils (Western and Dunne 1979). Cobb (1989) suggests an erosion problem may have developed between 1973 and 1978 in the valleys of the northwest slopes that are used as trek routes for cattle from the highlands down to the Angata Salei plain, but there are no data on more recent changes in this area.

Herbivores assist nutrient recycling through removal of plant matter and deposition of dung: the latter is often concentrated in particular areas (McNaughton 1985 for wildebeest; Stelfox 1986 for cattle). A number of workers have expressed concern about livestock feeding over – and stripping nutrients from – a wide area, then concentrating those nutrients in the form of dung or urine in particular sites (within boma corrals or at water points) thus effectively removing those nutrients from circulation. Tolsma, Ernst and Verwey (1987) studying nutrient levels in soil and vegetation around artificial waterpoints in Eastern Botswana suggest the cattle strip the savanna of specific nutrients to the point of generating phosphate deficiency both in the vegetation and ultimately in their own diet. The deposition of phosphates and other nutrients over a small radius round waterpoints in an inevitably heavily trampled area precludes the growth of anything other than trampling-resistant woody species and a few toxic herbs. By contrast, Stelfox (1986) presents a rather positive view of the nutrient concentration effect of Maasai cattle in the Athi Plains. He found high levels of N, P, SO_3, Ca salts and organic material in and around boma sites compared with control sites 250 m from the boma. Grasses immediately around the boma were higher in crude protein (CP) and lower in crude fibre (CF), and the high quality grass *Cynodon* predominated, as is common on disturbed and fertile soils. By contrast, the control sites were all *Pennisetum–Themeda* grassland. Stelfox (1986) pointed out the advantage of *Cynodon* growth immediately around the boma which is used to pasture calves while older animals can be taken further and pastured on lower quality forage (see earlier discussions of the implications of body size for foraging radius and of the energetics of travel). Stelfox (1986) goes on to point out that frequent relocation of bomas, as well as transhumance, encourages redistribution of nutrients concentrated in dung and urine throughout the range. Western and Dunne (1979) estimate the mean life of a boma site as seven to eight years with a maximum of ten years. The formerly common (currently forbidden) practice of cultivating small fields of maize and tobacco on past boma scars exploited this temporary concentration of nutrients for human use and hastened their recycling.

Comments on range degradation by livestock in NCA have tended to be based on supposition, and not infrequently motivated by ulterior political designs rather than objective ecological criteria (e.g. Pearsall 1957). The failure to maintain long-term grazing experiments or (until recently) to undertake ecological monitoring in NCA makes direct evaluation of livestock impacts well-nigh impossible, but the other lines of evidence reviewed here are suggestive. In addition to this evidence on herbivore effects in comparable and related ecosystems, there are historical and archaeological points. Pastoralists, livestock and wildlife have coexisted in the area for over 2000 years (Mehlman pers. commun, Phillipson 1977, Collett 1987). Pastoralist grazing and burning activities have helped to shape the present highly valued landscape. Traditional pastoralist management and utilisation is not thought to have undergone any sudden recent quantitative or qualitative change (eg Waller 1988, Spencer 1988, de Leeuw, Bekure and Grandin 1988).

(f) Population interactions: effect of cattle on wildebeest

Watson, Graham and Parker (1969) suggested that cattle replace wildebeest as the dominant large herbivore in the Loliondo GCA, while total large herbivore biomass (and presumably production) remain the same there as in the adjacent Serengeti. Similarly, McLaughlin (1970) reported a 5% increase in wildebeest numbers one year after cattle were excluded from Nairobi National Park. In the view of the NCAA, the pastoralists and their livestock are increasing to the point of becoming an environmental threat incompatible with other land use interests in NCA (chapter 4; Serengeti Committee of Enquiry 1957, Ole Kuwai 1980, Chausi 1985, Malpas and Perkin 1986). They are thought to be causing habitat damage ultimately prejudicial to wildlife condition and population size.

Historical and archaeological research suggest pre-1890s population and stock densities comparable to those of today (Collett 1987). Livestock numbers monitored over the last 20 years have fluctuated but show no overall trend of increase. Individual areas (e.g. Ilmesigio) have shown major local increases, for example the doubling of the Oloirobi cattle population 1960–1980 due to compression from Barabaig raiding to the southwest, wildebeest encroachment to the west and expulsion from the Ngorongoro Crater. Current high densities may affect incidence of infectious disease but are not seen as overtaxing grazing and water resources. In general, our livestock biomass figures compared with theoretical estimates show no indication of overstocking when seasonal movements are taken into account. By contrast, over the last thirty years wild herbivore populations have undergone a dramatic increase (Sinclair and Norton-Griffiths 1979, Malpas and Perkin 1986). It is specifically the wildebeest, closest to cattle in body size, ecological requirements and strategies, that show the most striking increase. This makes the idea of adverse competitive impact of NCA livestock on wildebeest and other wildlife dubious if not untenable. Disease interactions between wildlife and livestock favour the former whether in the case of MCF, tickborne diseases or the current rinderpest outbreak (Scott 1985). Administrative bans exclude livestock from the use of selected areas, while wildlife are free to move anywhere in NCA. Livestock in NCA are subject to major constraints, both natural and imposed. The performance of the relatively favoured wildlife populations, documented over three decades, is the best proof that range utilisation by livestock is both sustainable and compatible with conservation values.

(g) Population interactions: effect of wildebeest on cattle

McNaughton (1979) concluded that wildebeest dominate the large mammal community of Serengeti/NCA because their combination of nomadic habit and potentially dense herding behaviour allows them successfully to exploit their patchy habitat, where new growth on the short grass plains occurs in unpredictable, widely spaced patches associated with random isolated showers, rather than uniformly across the region. The success of pastoralist cattle in NCA would depend on their maintaining a precisely comparable strategy. The NCA short grass plains show high primary production of nutritious mineral-rich fodder in the rains which is stimulated by grazing.

Where possible, NCA cattle herds make use of the lowland plains during such periods: however, their utilisation is drastically limited by serious disease interactions with the massively increased wildebeest population. In the Gol (Nasera) area, wildebeest disease and grazing pressure are enough to preclude livestock using this resource. The same is true to a lesser extent for all areas adjacent to the plains: livestock are effectively locked up in traditionally dry season pastures where wet season disease transmission and lower quality grazing take their toll. This immobilisation of the pastoral herds is in part the reason for the eruption of the wild ungulates.

(h) Ungulate eruption and decline: the roles of grazing, disease and Maasai livestock

Sinclair (1979) has described the dramatic increase of the wildebeest and the lesser but still impressive increases in buffalo and gazelle. Malpas and Perkin (1986) bring this material up to date. Sinclair saw these population eruptions as initially triggered by the elimination of rinderpest when the 1960s cattle vaccination campaign throughout East Africa created a protective ring around the Serengeti/NCA system, and fuelled further by the mid-1970s increase in dry season rainfall and dry season food availability.

Neither cattle nor domestic stock as a whole show any trend of increase for the same period (chapter 8), though cattle are ecologically very similar to wildebeest, and were subject to the same release from rinderpest and the same improved dry season rainfall. They have been assumed by some authors to be undergoing the same process of eruption (see for example Makacha and Frame 1986) but this is unfounded (chapter 8). It is well established that wildebeest exclude cattle from wet season pastures because of the susceptibility of cattle to MCF. This exclusion from preferred wet season grazing is in the first place likely to have helped trigger the wildebeest eruption and secondly is probably enough to limit the cattle population.

It seems likely that the expulsion of the Maasai and their livestock from the Serengeti may also have contributed to the wildebeest eruption. An eyewitness recalls huge herds of sheep and lower numbers of cattle coexisting with extensive wildlife populations as he travelled through Seronera in the 1930s (Read and Chapman 1982). Pearsall (1957) describes the Maasai excluding wildlife from waterholes around the Moru kopjes as well as from the grazing in Angata Kiti, by building thorn fences. Grant (1954) describes as standard practice the construction of thorn barriers completely preventing the wildebeest migration from entering the Angata Kiti. When the Maasai were expelled from the Serengeti they were no longer able to use such fencing to protect chosen areas of wet season grazing. Fosbrooke (1972:79) records the failure of a later attempt by Newbould and Orr to repeat the technique in Angata Kiti.

To assess the possible effect of the past removal of Maasai cattle on the wildebeest population, and of the subsequent wildebeest eruption on later cattle population dynamics, it is necessary to look at the importance of limiting food quantity and quality. Classic ecological theory suggests that plant foods are superabundant, that herbivore populations use only a small proportion of them, and that herbivores tend therefore not to be food-limited. Sinclair (1975) challenges this idea for the SEU by demonstrating a severe dry season deficit. Sinclair, Dublin and Borner (1985) conclude that dry season shortage and associated mortality ultimately regulate the wild ungulate populations. We would take this further. We suggest that nuances of quality and availability within generally abundant and nutritious wet season forage affect later susceptibility or tolerance to ensuing dry season conditions. Early wet season access to the best grazing may be critical in reestablishing condition, and therefore in determining survival and fertility in the ensuing year. Dry season food availability influences mortality, but achieving good condition in the prior wet season is also essential for survival.

The migratory herds can exploit and physiologically store the pulse of high-energy, high-protein, high-mineral production in the short grasslands and then disperse to mark time on lower quality dry season grazing until the next rains. We suggest that the expulsion of the Maasai and some 50 000 cattle together with their small stock, and especially the restriction of their practice of fencing of grazing and water resources, turned key areas of high quality short grass plains over to wildebeest use and helped foster their increase. By contrast, the cattle are now excluded from the prime resource during its peak production because of disease interactions with the massively increased wildebeest population. Wildebeest remove most green leaf, and after they move on the short grasslands produce nothing in the dry season. Cattle are restricted year round to lower quality resources that produce at a lower if less variable level from year to year and over a longer season. They are also forced to remain in disease-infested pastures where previously migration and burning would have interrupted parasites' life cycles. For livestock, release from rinderpest coincided with a severe curtailment of grazing resources. Their population has had no opportunity to undergo an eruption analogous to that of the wild ungulates.

This analysis of competition between cattle and wildlife is of importance not only historically but also for future joint land use management. Put simply, when people occupy key resources such as water at Moru in such a way as to exclude wild herbivores, wildlife numbers are drastically limited. When resources are shared (but separated in time) then both wild and domestic herbivore numbers can remain high. It is the exception rather than the rule for

East African pastoralists to exclude wildlife from key resources. Most commonly their settlements are sited so that stock travel to and away from point resources of water or minerals, and thus allow time-sharing. This is one of the key factors determining the relatively high wildlife:cattle ratios in most East African pastoralist areas (see below). It contrasts sharply with the effect of pastoralist or agropastoralist land use that excludes wildlife access to water (see Rodgers 1988 for a comparison of Indian and East African pastoralist effects). Patterns of regional migration in Tanzania, driven by local high population densities in Sukuma, Chagga and Meru areas for example, have led to these agropastoralists settling around permanent water in what were once Maasai areas. Wildlife numbers in such areas are declining as a result.

Joint wildlife/pastoralist land use areas

The healthy state of the NCA wildlife populations is confirmed by a comparative look at other joint land use systems (Fig 9.1). Joint wildlife/pastoralist grazing systems can be broadly categorised as those which are predominantly managed for wildlife (like the core area of Amboseli), those where wildlife and pastoralist interests have more or less equal weight (as in Loliondo and Simanjiro Game Controlled Areas), and open areas where wildlife is secondary to human interests (such as Kajiado group ranches in Kenya Maasailand).

(a) Amboseli

The history of Amboseli is set out in detail by Western (1982, 1984), Lindsay (1987) and Drijver (1990) among others. Amboseli is located in Kenya Maasailand, and its central swamp has provided dry season grazing and watering for livestock and wildlife for millennia. By 1948 the dry season concentrations of wildlife and of competing livestock in the Amboseli swamp were felt to warrant Reserve status for an area of 3260 km^2. In 1974 the central area was upgraded to form the 488 km^2 Amboseli National Park, centred on the swamp grazing and watering. From the start Park management, itself largely Maasai, has sought to work collaboratively with the Maasai users of the surrounding buffer and open areas. The problems of joint resource use, of conservation impacts on wildlife, and of devolving benefits to the Maasai to compensate for their forfeit of the Park grazing and water are all instructive for the case of NCA. Both wildlife and pastoralist communities are essentially the same as for NCA. Similar problems of wildlife conservation and pastoralist development are central to the management of both areas. The wet season dispersal area necessary for the long term survival of the Amboseli wildlife populations comprises some 5000 km^2 round the swamp. This land is owned by

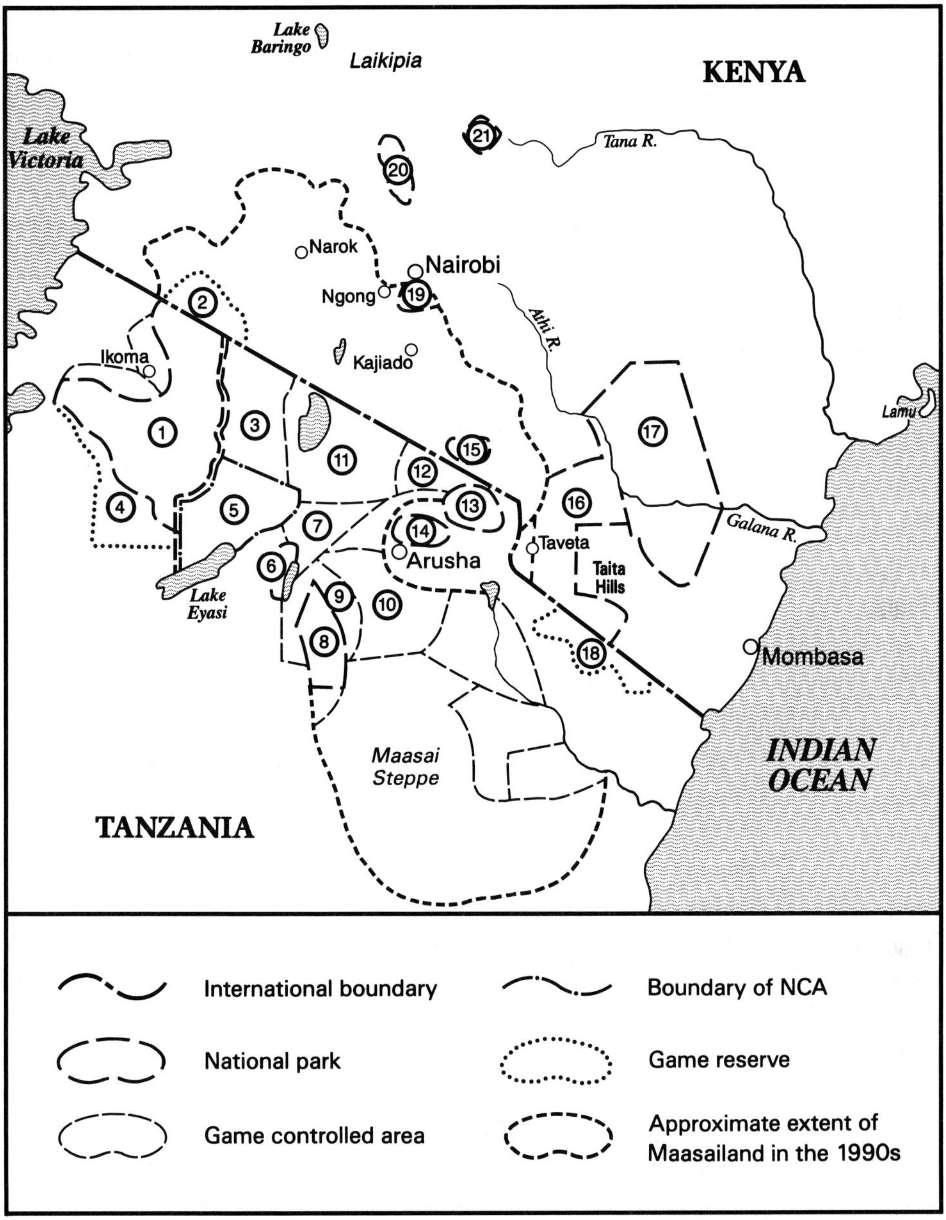

Fig. 9.1. Conservation and joint land use areas in and around Maasailand. 1. Serengeti National Park, 2. Maasai Mara Game Reserve, 3. Loliondo GCA, 4. Maswa Game Reserve, 5. NCA, 6. Lake Manyara National Park, 7. Mto Wa Mbu GCA, 8. Tarangire National Park, 9. Lolkisale GCA, 10. Simanjiro GCA, 11. Lake Natron GCA, 12. Longido GCA, 13. Mt. Kilimanjaro National Park, 14. Arusha National Park, 15. Amboseli National Park, 16. Tsavo West National Park, 17. Tsavo East National Park, 18. Mkomazi Game Reserve, 19. Nairobi National Park, 20. Aberdare National Park, 21. Mt. Kenya National Park.

the Maasai as group ranches. Continued access for the wildlife throughout the wet season dispersal area depends on the goodwill of the Maasai and on their abstention from fencing or extensive cultivation.

In NCA the Crater Highlands provide a dry season refuge (as does Amboseli swamp) but the Highlands are dominated by livestock, not wildlife. By contrast with Amboseli, the wet season short grass pastures represent NCA's most important conservation grazing resource. These are dominated by migratory wildlife to the exclusion of domestic stock. In some ways the Amboseli situation is perhaps more directly comparable to that of the Ngorongoro Crater. The Crater forms a concentrated wildlife spectacle which is the major viewing attraction, as does the Amboseli swamp. The Crater, like Amboseli, is a dry season refuge with permanent water and swamp grazing. The Crater wildlife populations, like those of Amboseli, are probably only viable in the long run if able to move across the surrounding buffer zone of livestock-dominated rangelands. Crater wildlife movements largely consist of a limited wet season dispersal/dry season concentration pattern as do those of the Amboseli wildlife, while the main biomass of wildlife associated with NCA is migratory, moving over a much larger 25 000 km^2 SEU.

Amboseli is a highly rated conservation area, and it is of some interest to look at the relative biomass composition of the two ecosystems as some measure of NCA's relative conservation status and the relative wellbeing of its wildlife community. Biomass contribution is only one measure of conservation interest, and does not take into account diversity, rarity or endemism, but is nonetheless a useful indicator of conservation potential. Table 9.1 compares the relative contributions of stock and wildlife biomass to the Amboseli ecosystem (defined as the dispersal area of the migratory wildlife – Western 1971) with our estimates for NCA.

There are complications in comparing a migratory wildlife biomass present only for part of the year with one showing more limited wet/dry season dispersal movements, and in using 1971 figures for Amboseli versus 1980 figures for NCA. It is a crude approximation to use an Ngorongoro District average to represent NCA, and to pool all NCA zones. Nonetheless, this simple comparison makes fairly clear the point that the wildlife to livestock biomass ratio in NCA is around five times that for the Amboseli grazing system. The proportionally lower wildlife presence does not detract from the conservation or viewing importance of the Amboseli National Park, and in many ways the surrounding Maasai land use adds to the Park's attraction. Contrary to alarmist views on the expanding Maasai and livestock population of NCA (Makacha and Frame 1986), NCA is not only extremely well off in wildlife terms but has considerable margin for experiment or error with the wildlife:

Table 9.1 *Proportional biomass contributions of livestock, wildlife and pastoralists to NCA and Amboseli grazing systems*

	Grazers		Secondary consumers		All large mammals	
	Wild	Domestic	Wild predators	Maasai	Wild mammals	Maasai plus stock
Amboseli (Western 1971)	?	?	15%	85%	35%	65%
NCA 1980[a]	74%	26%	87%	13%	74%	26%

Notes:

[a] Using the following figures from the Ecosystems 1980 report, averaged across all Ngorongoro District zones and therefore biasing towards an underestimate for NCA wildlife and an overestimate for NCA stock. Estimates for local stock biomass in chapter 8 are two to five times higher because they refer to local concentrations not overall averages.

Mean wildlife biomass (all large mammals) = 11 200 kg/km^2, mean biomass wild grazers = 10 416 kg/km^2, mean predator biomass = 784 kg/km^2, mean domestic stock biomass = 3739 kg/km^2, Maasai (mean weight 50 kg at 2.5/km^2) = 125 kg/km^2.

livestock balance (see chapters 8 and 10 for trends in livestock and human populations respectively). The Crater provides the concentrated viewing attraction, as does the Amboseli National Park swamp; the surrounding rangelands in both cases provide the biologically essential continuity and wider grazing system, where wildlife are free to move and graze, but are not found in dramatic concentrations. Both in Amboseli and to an even greater extent in NCA, the dry season grazing areas provide an essential drought refuge for domestic stock from a wide area of Maasailand. Over and above this NCA has the spectacular migratory invasion as well as the important biological role of providing the wet season short grass pastures that sustain the migratory biomass.

The Maasai around Amboseli receive financial compensation for wet season use of their land by wildlife, but they also need alternative dry season grazing and watering facilities. These were meant to be provided as a condition of the Maasai ceding their use of the National Park area. Periodically, when water developments or grazing fail there are incursions into the Park, with consequent friction between Park management and herdowners. On the part of the Maasai such friction can take the form of infringing access regulations, or of direct attack on the wildlife values of the Park. Lindsay (1987) discusses periods of park/pastoralist strife in Amboseli, and shows how the *murran* have several times retaliated to conservation impositions that are perceived by them as unacceptable by spearing species of particular conservation value (rhino, cheetah) as a gesture of political dissent.

For political reasons and because of the development history of Kenya Maasailand, the Amboseli Maasai have better security of land tenure than their NCA counterparts (chapter 10). Subsistence conditions, including returns from the Park, transport and communications, cultivation possibilities, and livestock/grain exchange terms are all more favourable than for NCA. Nonetheless, Amboseli experience of problems and compromises will be useful in designing future participatory and consultative management structures and development in NCA (e.g. Drijver 1990:11–12). As well as issues involving the Maasai, Amboseli has interesting parallels and contrasts to NCA in terms of wildlife viewing tourism and in its commercial, rather than political, ivory poaching problem.

(b) Tanzania Maasailand: Simanjiro/Tarangire and Loliondo/Serengeti

The conditions in other areas of Tanzania Maasailand suggest that the pattern of coexistence of wildlife alongside Maasai livestock that is found in Amboseli as well as NCA is the norm rather than the exception for Maasai rangelands. In particular, the common dependence of conservation areas on adjacent Maasai rangeland as dispersal areas for wildlife is clear. The 570 km^2 Simanjiro plain in Northern Tanzania is a part of the wet season dispersal area for the 2600 km^2 Tarangire National Park. Tarangire is the dry season wildlife concentration area for all of Eastern Maasailand (Borner 1985, Lamprey 1964). The Tarangire/Simanjiro areas make up a grazing system comparable to that of Amboseli. Maasai domestic stock make up 60% of the 8500 kg/km^2 grazer biomass in Simanjiro in the wet season; the remainder is composed of seasonally mobile wildebeest and zebra (Kahurananga 1981, using 1970–1972 census figures). Borner (1985) sees the progressive conversion of Maasai rangelands to large-scale farming, permanent subsistence cultivation and settlement, as well as formal ranching, as rapidly blocking off vital wet season dispersal and migration routes. He suggest a management system like that of NCA as the only hope of survival for the Tarangire wildlife. However, the population density today is so great that compensation payment for resettling those within 10 km of Park borders alone would cost millions of shillings. It is probably too late to save more than a fraction of the past migration.

In the 6734 km^2 Loliondo Game Controlled Area, domestic stock made up 47% of an overall grazer biomass equivalent to that of the Serengeti Ecological Unit (Watson *et al.* 1969). Migratory herds of the SEU move into Loliondo on a seasonal basis. The main difference in grazer species composition between Loliondo and the SEU was that in the former the short grass feeders included cattle.

Conditions in Amboseli, Loliondo and Simanjiro bear out the suggestion that savanna conservation areas in East Africa are dependent on the sort of buffer zone provided by Maasai pastoralist rangeland, which allows long-term coexistence of wildlife and livestock. The presence of unfenced, unimproved and uncultivated joint land use rangelands effectively increases the total area and range of resources available to wildlife in associated conservation areas. This enhances their long-term survival as predicted on the basis of island biogeography theory (Western and Ssemakula 1981). Of all the joint land use areas in East Africa NCA retains the highest wildlife:livestock ratios. In any of these areas, agropastoralist immigration could be a death blow to major wildlife concentrations.

(c) Kenya rangelands

Regular aerial censuses of wildlife and livestock carried out by the Kenya Rangeland Ecological Monitoring Unit (KREMU) estimate wildlife: livestock ratios for different parts of Kenya (Peden 1987). The results are given by administrative district irrespective of land use, though intensive agriculture clearly produces some very low wildlife counts (e.g. Nakuru) just as the presence of large National Parks accounts for high wildlife presence elsewhere (e.g. Taita-Taveta, with Tsavo East and West National Parks). Despite this, and other confounding factors such as aridity limiting both wild and domestic ungulates, it is clear that Maasai areas retain outstandingly high wildlife populations and show correspondingly high wildlife:livestock ratios. Thus Narok District (which contains the Maasai Mara, but also extensive Maasai rangelands and agricultural schemes) had the highest wildlife density for the whole of Kenya, well over twice that of any of the runners-up: Lamu, Laikipia (a former Maasai area–chapter 10) and Taita-Taveta. Kajiado is a district of more arid Maasai rangelands, adjacent to Narok, containing the small Amboseli National Park but otherwise given over to Maasai group ranches. It had fifth highest wildlife density with around one-quarter of the wildlife biomass density of Narok and an overall wildlife:livestock biomass ratio of 1:4. The traditionally Maasai areas of Narok, Laikipia and Kajiado all have both high stocking levels for their ecoclimatic conditions and high wildlife:livestock ratios. Not surprisingly, all the areas with 1:1 or better wildlife:livestock biomass ratios are primarily pastoralist (mainly Maasai) or National Park areas.

(d) Marsabit/Kulal in North Kenya (IPAL area)

The Integrated Project for Arid Lands (IPAL:Lusigi 1981) study area provides a striking contrast to Maasai rangelands. It comprised some 22 500 km² in North Kenya adjoining the Ethiopian border. At the time of the main United Nations Environment Program (UNEP) study it was being used by 38 000 Gabra, Rendille, Boran, and Samburu pastoralists. The study was designed on the assumption of desertification and environmental misuse and located in an area which has seen considerable disruption amounting to civil war (see for example Dahl 1979, Hogg 1985, 1987). The people of the IPAL area were more seriously destitute than any Maasai group since the turn of the century. The IPAL study concentrated on making an inventory of vegetation, stock and wildlife. Stock land use, feeding patterns and productivity were studied. The area was estimated to carry 400 000 ungulates of which 97% were domestic; biomass proportions were equivalent or even more heavily weighted towards livestock. KREMU figures (Peden 1987) indicate marginally higher wildlife:livestock biomass ratios for this area. Wildlife is sparse and little is known of its utilisation, but IPAL recommendations include establishing reserves and national parks.

(e) Jonglei Canal Impact Area, Sudan

The Jonglei Environmental Project surveyed an area of 67 500km² (Howell, Locke and Cobb 1988). The Nuer, Shilluk and Dinka livestock biomass averaged 1400 kg/km² in the mid wet season, falling to 1189 kg/km² in the early dry season and rising to 2231 kg/km² in the late dry season. The corresponding wildlife biomasses were 372 570 and 1093 kg/km². Thus wildlife to livestock biomass ratios varied from 1:3.8 in the mid wet season to 1:2 in the dry. Pastoralists and wildlife are seen as making up a grazing succession whereby people burn the tall coarse grass rangelands to make pasture for cattle which are then followed by ever more selective species – tiang, reedbuck and finally gazelles. Subsistence hunting of wildlife contributes an important 25% of the annual meat intake. Because the wildlife is migratory the wild meat is fattened elsewhere but harvested in Jonglei. Earlier studies (Payne 1976 quoted in Mefit-Babtie 1983) saw the wet season pastures as limiting cattle production. The 1983 report sees dry season grazing (on the drawdown area of floodplain) as being both the most nutritive grazing and also the critical resource limiting cattle production. This had implications for the impact of construction of the Jonglei canal, but as it now seems unlikely that the canal will ever be finished perhaps this is not important. There are strong points of similarity with the Ngorongoro system: the migratory wildlife, the scale of migration, the major transhumance system of the pastoralists. The wildlife:stock ratios are more as for Amboseli than for NCA, but are still considerable. There are also points of contrast. The grazing system centres on dry season grazing, with the most nutritive species found in dry season swamp/drawdown areas. This contrasts with the NCA situation where the wet season short grass association is the most fertile, nutritive and productive grazing. The Jonglei impact area is assumed to operate for human rather than conservation priority, but local pastoralist interests are seen by some goverment agencies as subordinate to other development possibilities. However, the isolation of the Jonglei area precludes much wildlife viewing tourism or indeed wildlife conservation control, and also severely limits the potential of commercial livestock development.

(f) NCA in perspective:joint wildlife/pastoralist livestock grazing systems

Conditions in the Amboseli ecosystem, the Simanjiro Plains and Loliondo Game Controlled Areas, as well as KREMU figures, all suggest that wildlife conservation areas throughout Maasailand are dependent on Massai pastoralist rangelands as buffer zones for the survival of migratory or seasonally dispersing wildlife populations. Nationwide censuses in Kenya show Maasai areas as retaining high wildlife:livestock ratios (as well as overall high animal populations through their limiting of large-scale cultivation). More remote joint grazing systems are also of interest. The IPAL area in northern Kenya illustrates conditions in a pastoralist area with severe political and attendant ecological problems. The Jonglei impact area in Sudan gives points of comparison with a wildlife/pastoralist system that has seen little development intervention, and is sufficiently remote and far ranging for the troubled political circumstances of the southern Sudan to have little effect on the ecology.

The important points that emerge from this review of joint wildlife/pastoralist livestock areas in East Africa are that:

1. NCA has a far higher proportion of its animal biomass made up of large mammal wildlife than do other joint land use systems. The trends continue to favour wildlife.

2. Other systems (especially Amboseli) are still of outstanding conservation value despite wildlife:livestock biomass ratios markedly lower than for NCA. This suggests that NCA has scope for weathering considerable fluctuations in wildlife populations before its conservation values are jeopardised.

3. Maasai rangelands that operate as open areas, with human interests predominating and without formal protection of wildlife populations, retain comparatively high wildlife:livestock biomass ratios, showing the compatibility of wildlife conservation and pastoralism. Similar conditions apply in the Jonglei area with the pastoralists of the Sudd region. Many East African conservation areas are continuous with Maasai rangelands (and were excised from them). Such rangelands are invaluable buffer zones. By comparison, creeping agropastoralism with its concentration on water sources rapidly eliminates wildlife.

4. In northern Kenya the IPAL area used by Gabra, Rendille, Boran and Samburu pastoralists has an ungulate population overwhelmingly dominated by livestock. The paucity of wildlife may be due to greater aridity, the consequently higher competition, and the greater ability of livestock with the assistance of human herders to exploit water and grazing or browse in marginal areas. Other possible contributory factors might be the history of civil disruption, dispossession, and destitution in the IPAL area, or cultural attitudes less favourable to wildlife survival than those of the Maasai.

Pastoralist cattle production in a joint land use system

Wildlife appear to do well under the current regime in NCA. How well do the pastoral livestock do? The prevailing view of the NCAA is that Maasai pastoralism is a primitive and inefficient form of land use (e.g. Malpas and Perkin 1986, Kitomari 1986). This attitude colours official perceptions of the way that NCA should be managed in the future. In particular, it affects the extent to which pastoralism is seen by the NCAA as a legitimate and viable form of land use relative to wildlife conservation and tourism. This section looks at the efficiency and productivity of Maasai pastoralism in the context of the available resources and prevailing constraints.

The comparison of migratory wildlife and pastoralist stock emphasises the divergence between subsistence pastoralism and commercial livestock production which is often seen as the natural development alternative (Barnes 1979, Ole Saibull 1978, Simpson 1984b). Subsistence pastoralists at the arid end of the spectrum pursue an opportunistic strategy suited to an unpredictable, fluctuating environment (Dyson-Hudson 1980) in which a combination of risk avoidance and tracking resources may be more productive for pastoralists than any attempt to maintain stable production levels (Sandford 1982). Pastoralist herds show maximum potential for rapid increase, high mobility, efficient colonisation of temporarily utilisable areas followed by resource exhaustion and renewed dispersal. Bottleneck periods of unfavourable conditions lead to massive emigration (where still possible) and sometimes heavy mortality among livestock (e.g. Meadows and White 1979, Kjekshus 1977, Homewood and Lewis 1987). By contrast, the development alternative is of controlled production and high proportional offtake of high quality individuals from smaller populations, limited mobility and flexibility, with numbers (both human and stock) kept to a steady level corresponding with maximum reliable financial yield. It is not clear how sustainable and efficient this strategy may be in unpredictable arid and semi-arid areas (Sandford 1982) but there is no doubt that in such areas it would require major technological investment and social change not currently feasible in many sub-Saharan rangelands. Caughley, Shepherd and Short (1987) dismiss the possibility of sustained yield cropping in the highly variable arid and semi-arid systems of Australia and stress the need for flexible management.

Chapter 8 gave a review of the changing interpretation of productivity comparisons between pastoralist and commercial livestock breeds. Comparable insights extend to the level of herd management. Early comparisons of subsistence and commercial herds concentrated solely on yield of milk or meat or cash per animal as a criterion (e.g. Barnes 1979), and found subsistence systems inefficient by contrast. More recently, comparisons have begun to take into account the fact that management aims of subsistence pastoralism are very different from those of commercial concerns using improved breeds and sophisticated technology. It is increasingly clear that pastoralist systems are more productive than western-style ranches in similar environments in terms of energy, protein or money equivalent per unit of land (e.g. Grandin 1988). The commercial beef rancher sees as optimal stocking rates, cattle breeds and management strategies that give a maximum growth rate per individual animal over the first few years of life. This allows rapid sale and turnover of a particular culturally desirable quality of meat fetching a high price. Similarly, a commercial dairy farmer's management must maximise milk yield per individual cow over a limited production life. By contrast the subsistence pastoralist is more likely to seek stocking rates, animal breeds and management strategies which allow the maintenance of larger numbers of animals in any given area, with lower rates of milk or meat production per animal but an optimal balance in the output of a much wider range of products necessary for long-term subsistence of a larger number of people. One of the prime criteria for the subsistence pastoralist must be the long-term survival of his herd. This entails optimising herd composition for continued calf production, as well as retaining animals which may be past their prime in terms of milk, meat or calf production, but may by virtue of prior exposure be more likely to survive drought and disease. When all the inputs and outputs are taken into account, rather than the yield of a single product per animal, the subsistence systems are as or more efficient than commercial ranches (Behnke 1985; Cossins 1985; Grandin 1988; Sandford 1983:123–127; Jahnke 1982). De Leeuw, Bekure and Grandin (1988) compare production indices per unit of land for a Maasai pastoralist system and a commercial beef ranch (Table 9.2). Livestock outputs and gross cash incomes were similar but the Maasai system had costs less than one-tenth those of the commercial system. Household labour was the main input. Subsistence pastoralism, as a system that generates employment and uses few capital goods, has clear advantages with obvious significance for developing countries. The main component of the high productivity of pastoral systems is milk, along with other dairy products. Some of the other potential outputs do not apply in NCA. For example, ox labour for ploughing or transport of produce, and dung

Table 9.2 *Comparative production of Maasai cattle pastoralism and commercial cattle ranching*

	Maasai herd	Athi River Plains commercial ranch
Livestock output (kg/ha/yr)	29	25
Gross income ($/ha/yr)	13.5	13.6
Costs of production ($/ha/yr)	0.9	11.0
Planned stocking density (ha/head)	c.3	3.6

Source: From de Leeuw, Bekure and Grandin 1988.

(as fertiliser) are both livestock outputs of enormous value in agropastoralist systems elsewhere (Jahnke 1982; McCown Haaland and de Haan 1979) but of limited scope in NCA under present circumstances.

Our study of livestock performance suggested that for the three main study areas mortality and fertility rates, as well as milk production, were average for subsistence pastoralism. This and other studies of Maasai pastoralism show calving rates, survivorship and milk production similar to those of West African pastoral systems (chapter 8; Table 8.12). Within NCA the separate Oloirobi boma illustrates the sort of mortality crisis which can afflict any pastoralist subsistence unit at any time, halving the subsistence herd and decimating the number of lactating cows and the day to day availability of milk over a couple of years' run of drought and epidemic disease. Although the Oloirobi boma illustrates in microcosm the problems that concern any pastoralist settlement, the three main study areas are perhaps more representative of conditions in NCA.

Measuring meat production and offtake from NCA is not straightforward. Both cattle and small stock are sold at NCAA-sponsored auctions, which are irregular, poorly publicised and poorly attended (Arhem 1981a) and which are now little better than clothing marts (Machange 1987). Arhem (1981a) shows the official commercial offtake from NCA livestock sales declined from 3–4% of the NCA herd in the early 1960s, when market conditions were more favourable to the pastoralist, to 0.4–0.7% in the late 1970s. However, there has been a corresponding increase in unofficial livestock sales. Recent estimates suggested that 70% of NCA stock sales were cross-border trade (Field and Moll 1987, Field, Moll and Ole Sonkoi 1988). Arhem (1981a but using 1978 figures) estimated a commercial offtake (sales through both legal and unofficial channels) of around 3–4% of the NCA herd per annum in the late 1970s as necessary to pay for Maasai grain needs. This consistent 3–4% offtake estimate contrasts with the 8–15% gross offtake rates of group or privately owned ranches quoted by Sandford (1983:125). Our own observations of transfers and transactions, involving identified stock monitored throughout the study, gave estimated total offtakes of around 8% for all areas, double the offtake estimated by Arhem (1981a). This is because our study recorded not just commercial offtake for cash sales but also animals leaving the herd for internal slaughter and consumption, or as stock payments to other herdowners. Gross or total offtakes, counting non-commercial transactions, are thus within the range given by Sandford for commercial enterprises.

Summary and conclusion

1. Pastoralist livestock use resources and strategies similar to those on which wildlife depend. Despite this overlap and the potential competition it

entails, there are so far no serious environmental impacts of pastoralism in NCA, nor have the wildlife populations suffered from joint land use.

2. Expulsion of the Maasai from the Serengeti together with the curtailment of isolated instances of fencing off grazing and waterholes may have contributed towards triggering the eruption of the wildebeest population. Conversely, the massive increase of the wildebeest has severely limited the growth and performance of the cattle herds by excluding them from critical wet season grazing through disease interactions that act to exacerbate the effect of administrative bans.

3. Livestock are not causing problematic overgrazing, soil erosion or soil nutrient stripping problems in NCA. Local erosion in Angata Salei is due as much or more to trampling by wildebeest than by livestock.

4. Broad comparisons with other joint wildlife/livestock grazing systems in East Africa confirm that NCA has quite outstanding wildlife biomass, and that Maasai areas whether under conservation regimes or operating primarily for pastoralist livestock retain a high wildlife component.

5. Maasai pastoralist rangelands are essential buffer zones for the wildlife of a number of East African Parks. Other (non-Maasai) pastoralist areas of East Africa have sparse wildlife, but the Sudd grazing system shows wildlife:livestock biomass ratios comparable to Amboseli and the Simanjiro Plains.

6. The aims and techniques of subsistence pastoralism differ from those of commercial livestock production and have in many cases been shown to be ecologically better suited to the special conditions of arid and semi-arid Africa.

7. The high mortality, low milk yields and low fertility of NCA Maasai livestock (compared to commercial systems with exotic breeds) are average for subsistence pastoralism. Performance reflects ecological and economic constraints rather than poor management or an inefficient system. In NCA, characteristic pastoralist livestock yields are further prejudiced by range restrictions imposed by the NCAA and by the realities of disease interactions with wildlife, again exacerbated by administrative bans on traditional disease control techniques.

10

Maasai ecology: development, demography and subsistence

Milo ang'ata miata olesuama.
Do not go to the wilderness without someone to remove the dust from your eye.
(ie. Never go alone – Maasai saying: Mol 1978)

This chapter covers three main aspects of Maasai ecology. Firstly, the ways in which past livestock management and development policies have interacted with Maasai social and ecological organisation are discussed. Secondly, demographic trends are analysed and their implications for living standards and future management considered. Lastly, our food survey is described. This completes the link between our information on livestock production and current conditions of human subsistence. The milk, grain and meat components of the diet are quantified, food adequacy estimated and grain requirements estimated for different areas and seasons. Our results underline the special problems of Maasai subsistence in NCA.

Past interventions in Maasailand

Kinter olemodai pee kintoki oleng'eno
We begin by being foolish and become wise by experience (Maasai saying)

The history of the last hundred years reveals the fundamental importance of Maasai social structures in determining the outcome of different attempts by outsiders to manipulate their pastoralist system. It therefore has implications for management in NCA. There is a vast literature on past interventions from the early colonial days to post-Independence times (Waller 1976, Anderson & Grove 1987, Raikes 1981, Evangelou 1984, Moris 1981, Hoben 1976, Jacobs 1978, Galaty 1980). Development attempts in sub-Saharan Africa have generally been motivated by Western perceptions on the one hand of pastoralists overstocking, overgrazing and degrading rangelands, and on the other hand of pastoralist inefficiency and low livestock productivity. These twin (and frequently contradictory) assumptions have been used in Maasailand as elsewhere to justify progressive expropriation of large areas of land (e.g. Sindiga 1984), and to impose destocking (Homewood and Rodgers 1987). They have also been used as the theoretical basis for designing systems of land tenure and access to veterinary and livestock husbandry facilities which aim to reduce stocking rates and thereby to improve livestock health and productivity, and intensify commercial offtake. These perceptions were almost certainly due to the tendency to see the drastically reduced pastoralist and livestock populations of the early colonial period as a baseline.

(a) 1890–1960

The Maasai and their herds were drastically reduced by the 1890s epidemics and went through a time of sweeping social, political and ecological change brought on by disaster (chapter 3; Waller 1988). Normal processes of negotiation between sections disintegrated during this period of terrible stock deaths, famines and epidemics remembered as *Emutai* (chapter 3). In Kenya, the 'Maasai Moves' of the early colonial period (up till 1911) saw eviction of the Maasai from Laikipia and their removal to what is now Kajiado District, with the consequent bringing together and compression of in some cases mutually hostile sections (Galaty 1980). In the 1920s, official concern on overgrazing began to be expressed. This intensified through the droughts of the 1930s and has been used ever since as a major 'ecological' argument to justify specific livestock development interventions, particularly destocking attempts (Anderson 1984, Homewood and Rodgers 1987). Between the 1920s and 1960s progressively greater areas of land were alienated from the Maasai in both Kenya and Tanzania (Sindiga 1984), with areas being set aside for expatriate settlers (e.g. Laikipia); for cultivators from politically more dominant tribes (e.g. around Kilimanjaro and Loitokitok); for wildlife (e.g. the National Parks of Amboseli, Serengeti and Tarangire) and since Independence for major agribusiness or parastatal plantation schemes (wheat; beans in Lolkisale–Borner 1985). This resulted in ever greater compression of Maasai within the ecologically less desirable remaining rangeland areas.

Throughout this time the colonial administrations provided little incentive to develop pastoralist production. Despite the strong evidence now available to the contrary, Maasai were seen as reluctant to sell stock, unresponsive to market incentives and economically and socially isolated from the rest of the nation (Raikes 1981, Berntsen 1976). One surprisingly poorly publicised exception to this was the stock gift scheme (Fosbrooke 1980) whereby the Ngorongoro Maasai contributed considerable numbers of stock to the British Government throughout World War II. For the rest, Maasai initiatives largely took the form of evasion of imposed restrictions. For example, during the 1950s, when Kenyan African producers were excluded from lucrative markets to protect settler production, Kenya Maasai moved considerable numbers of cattle across the border to sell in Tanzania (Raikes 1981:209). When later the relative price situation was reversed, Tanzanian Maasai began to move their stock to Ngong to sell (and still do so – chapter 8).

(b) 1960–1980

The 1960s saw the emergence of parallel pastoralist development schemes in Kenya and Tanzania Maasailand. In Kenya, group ranches (as well as individual ranches) were registered throughout Kajiado and Narok Districts (Evangelou 1984, Galaty 1980), while in Tanzania the US Agency for International Development (USAID) initiated a livestock and range development scheme that was to involve ranching associations with a structure in some ways comparable to the group ranch (Moris 1981). Group ranches were intended to be ecologically self-sufficient, demarcated areas, the title to the land in each case being effectively held by a group of pastoralist families who would jointly take out loans, plan and commission water developments and other improvements, control stocking rates and cooperate on day to day grazing and marketing management. Tanzanian ranching associations were meant to establish communally owned and intensively managed group herds as a nucleus of commercial production alongside more extensively managed and individually controlled subsistence herds (Raikes 1981:163–4). The Kenyan group ranches and Tanzanian ranching associations were intended to maintain lower stocking rates and to show a higher offtake for commercial meat than the traditional sector, thereby bringing about an economic and ecological transition among the Maasai and their closer integration into the national economy in each case.

The course of events did not entirely follow the expected pattern. In Kenya, the Maasai showed themselves eager to register for membership of group ranches. In retrospect the main incentive was to get title to the land. In this sense the group ranches have improved Maasai subsistence security in Kenya (Galaty 1980). The guidelines on which the group ranches were set up failed to allow for differences between stock numbers of different herdowners. Registration fees were the same for all members irrespective of herd size. Graham (1988) shows how since adjudication traditional wealth redistribution mechanisms have deteriorated; rich members have become richer and the poor, poorer. Rather few group ranches took advantage of the loan facilities and where these were taken up few have been repaid. The ranch boundaries conformed to some extent to traditional divisions of land rights, but where these were contravened there have been murderous disputes (Galaty 1980:167–8). During the droughts of the 1980s it became apparent that the group ranches were anything but ecologically self-sufficient units. Many did not include drought refuge grazing. However, the membership of group ranches made it easy for herds to be moved between group ranches because of section, clan and ageset ties (e.g. Sandford 1983, Grandin and Lembuya 1987). This was precisely the effect that the planners had intended to preclude. Where the effect of group ranches on the material wellbeing of the Maasai has been studied (e.g. Grandin 1988) it becomes apparant that any extra milk potentially available to richer households is consumed directly by calves (boosting calf survival, calf growth rates and presumably herd growth) rather than being taken to feed people. Similarly money derived from an increased participation in commercial livestock sales tends to be translated into more stock holdings, (Nestel 1985, 1986) again confounding the intentions of the planners.

In Tanzanian Maasailand the Maasai Livestock and Range Development Project (MLRDP) took shape in the late 1960s. This cooperative project involved $23 000 000 of USAID money, expatriate technical personnel and the logistic and administrative support of the Tanzanian Goverment. The project set out with admirable guidelines of working towards an understanding of traditional Maasai aims and methods before encouraging development along already planned lines. In practice there was little or no consultation with Maasai on geographical boundaries or specific aspects of development. Maasai attempts to participate through fund raising and other initiatives were ignored (Hoben 1976, Jacobs 1978, Moris 1981). Several ranching associations were set up but the only one to achieve the promised title to the land was Monduli (home of a prominent Maasai politician). The project managed a fairly rapid installation of technical services like dips, dams and other water development measures. This brought about massive uncontrolled immigration both of pastoralists and even more so of cultivators, to ranching associations thus favoured (Ndagala 1990a and below). Marketing arrangements originally set up under the control of ranching associations were brought back under parastatal control. With the reimposition of highly taxed, poorly organised and low-price official stock sales, stock marketing began to circumvent legal channels (Moris 1981). It became apparent by 1978 that the establishment of ranching associations was seen by the government as conflicting with the Tanzanian policy of establishing *ujamaa* villages. After Monduli, no further ranching associations were granted occupancy rights (Moris 1981).

The Maasai saw the failure of the MLRDP land tenure plans as just one more case in a long history of broken promises. They complied with subsequent government resettlement programmes in the hope of getting some political reward and security. Their mobility and lack of possessions other than stock made the process easier than for cultivators (Ndagala 1982). Much of *Ujamaa* resettlement in the pastoralist case was a purely cosmetic (if time-consuming) exercise rearranging traditional settlements, though Jacobs (1978) cited changes of boma size and composition, with the emergence of single-family bomas, as a result of 'villagisation' (chapter 3). The dams, dips and other USAID technical inputs deteriorated rapidly. By 1979 the MLRDP was defunct. Numerous post-mortem reports evaluated the reasons for its failure (Jacobs 1978, Moris 1981). Far from any lasting development having resulted, Jacobs (1978) saw Maasailand infrastructure as having dwindled since the 1950s. He cited the declining availability of even the simplest introduced technologies (hand churns and hand grinders for maize), and the revival of traditional beaded leather clothes for women, as evidence for the failure of the 1970s development initiatives.

(c) NCA

One of the four Tanzanian ranching associations to be established on paper, but never given land occupancy rights, was NCA (Moris 1981:105–6). It is typical of the history of livestock development for Tanzania that this information is not generally known outside USAID. It is not available in NCAA files, nor in current NCA range/stock development documents (see e.g. Cobb 1989) though some USAID range management reports from this period are quoted (Clebowski 1979) describing Oloirobi range resources. The 1970s also saw the establishment of a World Bank-supported beef ranch at Endulen, within the boundaries of NCA, based on stock 'volunteered' by NCA pastoralists and set up in conjunction with the USAID programme. Little information is available on the way this ranch operates ecologically or commercially (but see Machange 1987).

(d) Lessons for development

Considerable sums were spent on pastoralist development in Tanzania in the decade 1970–1980 in an attempt to bring about the standard livestock development aims of reduced stocking rates, raising productivity, raising commercial offtakes and encouraging integration into the national economy. All the now-familiar problems common to such projects can be recognised (Wyckoff 1985; Ellis and Swift 1988). These were for example:

- failure to consult the pastoralists (for example on hydrological matters before installing what proved to be non-viable water developments),
- failure to identify pastoralist priorities (particularly the relative importance and viability of subsistence milk production versus commercial beef production; and the need to invest returns in breeding stock rather than in an unreliable and inaccessible banking system),

- over-emphasis on perceived range degradation (in retrospect acknowledged not to be a problem – Peterson 1978, Jacobs 1978)
- Maasai, Tanzanian government and project personnel all working towards different aims.

Together with political circumstances peculiar to Tanzania, these all combined to make the MLRDP another classic example of failed pastoralist development programmes. By 1979 the MLRDP had collapsed as had many of its technical inputs and the project had been superseded by villagisation. The main lasting impact in NCA seems to have been the rapid immigration of the 1970s, when the population trebled in seven years (see below).

The main effect of the villagisation programme in NCA has been in terms of Maasai political and administrative infrastructure. This has involved the replacement of the traditional elders' council with a village council which comprises the elders but also admits younger age-set members and Western-educated individuals (Sandford 1983:241). Representing as they do a working compromise between traditional and national structures for communicating information and making decisions these committees could be both effective and reasonably equitable. Unfortunately, within the NCA their position is anomalous and they do not have the same degree of responsibility and self-determination open to villages in the rest of Maasailand (chapters 3,4). No political structures have been developed above the village level in NCA.

Demography of NCA Maasai

Meirag te entim olotoishe He who has children does not lie down in the forest
(i.e. He who has children will always have a home: Maasai saying)

As with livestock numbers (chapter 8), pastoralist population numbers are both inherently difficult to census and also liable to major fluctuations (chapters 3, 5; also Hill 1985). Seasonal migrations and transhumance take place both between different zones of NCA, and across NCA boundaries into adjacent areas. With NCA less than 100 km from Kenya it is possible that some NCA Maasai and livestock cross national boundaries in the course of their seasonal peregrinations (possible without crossing sectional frontiers). This is quite apart from the well-established cross-border stock trade which undoubtedly exists. Conversely it is possible that Maasai and stock from Kenya do on occasion use NCA. These movements make population estimation and analysis problematic and generate major management and legislation complexities.

A recent count puts the pastoralist population of NCA at 22 637 Maasai (Perkin 1987). The 1979 census counted 17 932 NCA Maasai (Table 10.1) representing 22.5% of a total population of 80–90 000 Tanzanian Maasai (Arhem 1981a). Kurji (1981) suggests that NCA has for decades been progressively acquiring a steadily greater proportion of Tanzanian Maasai. However, current estimates put Tanzanian Maasai at around 100 000 implying that NCA in 1987 (as in 1979) provided a base for some 22.5% of Tanzanian Maasai. This is considerable given that the 8292 km^2 NCA covers only some 13% of Tanzanian Maasailand (which Jacobs 1975 estimated as 24 000 sq. miles).

(a) Population fluctuations

The scale of population fluctuation from year to year is indicated by census figures from 1954 to 1987, presented in Table 10.1 and Fig. 10.1 Different counts were undertaken in different seasons. However, Arhem (1981a) shows that the recorded trends cannot be accounted for simply on the basis of wet season dispersal patterns in the Highlands or dry season dispersal in Gol, and he concludes they represent genuine population changes. Cobb (1989) is sceptical of the accuracy of all these counts and suggests they may underestimate by 50%. According to the counts, NCA Maasai numbers have fluctuated from around 10 600 in 1954 (Grant 1954), to 7400 in 1966 (Dirschl 1966) and to an all-time low of around 5400 in 1970, rising again to around 22 600 in 1987 (Perkin 1987). Dirschl (1966) reports substantial immigration into NCA during 1960–61 from drought-affected adjacent lowland areas. An influx of thousands of head of cattle means a corresponding influx of herders caring for these stock,

Table 10.1 *NCA Maasai census figures*

Date of count	NCA Maasai[a]	Source
1954	10 633	Grant 1954
1966	7 387	Dirschl 1966
1970	5 435	Arhem 1981a
1974	12 655	NCAA/Kurji 1981
1977	16 705	Makacha *et al*/Kurji 1981
1978	17 982	National census/Kurji 1981
1980	14 645	Ecosystems Ltd/Arhem 1981a
1987	22 637	Perkin 1987

Notes:

[a] omits non Maasai NCAA staff.

All these figures are open to question because of methodological and other problems (see text). To the extent that they can be trusted, note the overall decline 1954–1970, the rapid increase 1970–1978, and the fact that post-1978 growth slows to under one-fifth of the 1970–1978 rate:

1970–1977: $r = 0.150$, doubling time 4.6 y, 1978–1987: $r = 0.026$, doubling time 27 y, using $r = (\ln N_t - \ln N_0)/t$, where r = exponential rate of increase, N_0 = Number at start, N_t = Number at time t, t = time in years.

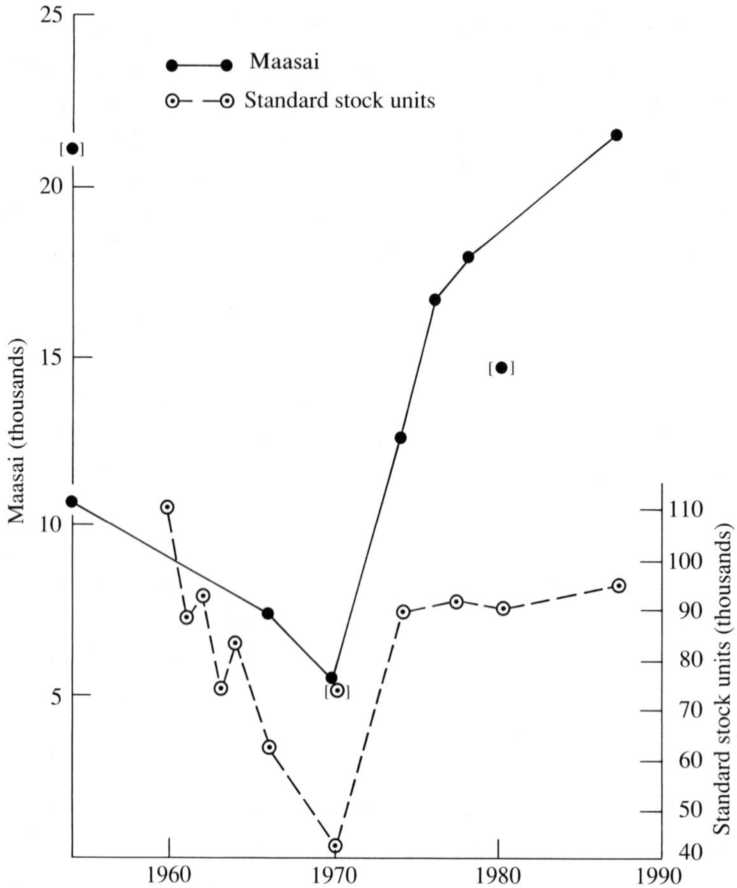

Fig. 10.1. Maasai and stock population trends in NCA. See Table 10.1 for Maasai data. Standard stock unit (SSU) data ⊙ are taken from the Ngorongoro Conservation and Development Project Report 1987. They use the following conversion:

1 mature bovine = 2/3 SSU
1 immature bovine = 1/3 SSU
1 sheep or goat = 1/10 SSU
1 donkey = 2/5 SSU

The 1970 SSU figure [⊙] given by NCDP is suspect (see Table 8.1, Figure 8.1). We have recalculated the data point ⊕ from the 1970 census figures in Table 8.1, assuming 25% bovines immature (NCDP assumption for 1974 and 1978); also assuming same ratio of donkeys to other stock as in 1987, giving 1000 donkeys = 400 SSU. The 1954 and 1980 data points [●] for the Maasai population are suspect (Pearsall's 1954 figure probably originated in a misreading of Grant; the 1980 data point is based on a nonstandard method – see text). The 1954 data point we use is derived from Grant 1954 and the 1980 figure is ignored in our interpolated curve.

and of families dependent on their produce. Dirschl (1966) suggests that in the following years of relatively good rainfall these people and their livestock dispersed back to their home settlements bringing about a decline in numbers of people and stock in NCA by 1966–1970. At the same time the insecurity of Maasai tenure in the Serengeti/Ngorongoro region may have prompted emigration. According to Arhem (1981a) another severe drought in the mid-seventies was responsible for a major part of the increase in NCA human and livestock populations at that time (Grandin 1988 cites a major 1970s drought in Kenya Kajiado). The now-defunct ranching association programme together with villagisation was without doubt as or more important in the rapid growth of the NCA population in 1970–77. NCA was one of the first-established and highest-potential Ranching Associations (RA) planned by the MLRDP 1969–78 (Moris 1981). As for all the other RAs except Monduli, the promised land rights were never confirmed, but NCA along with four other RAs did receive dips and dams in the initial phase of the project. Maasai were quick to respond to these benefits and there was rapid immigration into NCA (Moris 1981:106). Project staff saw this immigration could be controlled by granting the RAs occupancy rights and the ability to enforce against trespassers. However, the MLRDP gave way to the Tanzanian *ujamaa* programme, which outside NCA led to increased cultivator pressure in Maasailand as farm villages were rapidly established in higher-potential areas of former grazing land (Moris 1981). Many Maasai, particularly poorer households, were driven to move to the NCA which does not allow for settlement of cultivators.

Figures for 1966 and 1970 show a decline to around half the 1954 values. The period 1970–1977 saw a trebling of the population. Other fluctuations have been less dramatic and some may owe more to methodological problems than to population changes. Arhem (1981a) estimated a population of around 14 600, a 20% drop from the 1978 census figure. As Arhem's figure was calculated from an aerial count of settlements it is as much (or more) open to doubt than a total ground count of households and their inhabitants. Pearsall (1957:11) reported 21 000 Masai (equivalent to the 1987 population count) in an area equivalent to the present day NCA. This would indeed have been dramatic confirmation of the thesis that NCA population has fluctuated rather than showing overall growth for the last few decades. However, a rigorous search suggests that Pearsall's figure is based on a misreading of Grant (1954: with summation of subtotals and totals giving a final total twice the correct value of 10 500). Such slips aside, the year to year fluctuations are as important to the management ecology of NCA as is the overall growth trend. They stress the importance of immigration and emigration in NCA human ecology. The immigration rate may have reached around twice the rate of natural increase given by the excess of births over deaths for the period 1966–1978 (Kurji 1981 and below) and could potentially be greater still. Immigration and emigration allow for rapid response to environmental, political and economic change. The scale of migration highlights the imminent management problem of defining a user population and legislating user rights (see chapter 12).

(b) Population growth

Arhem (1985a) interprets the fluctuations in human and livestock populations as being due to large-scale population movements in response to temporary range and water conditions. To some extent this is probably the case, but the dislocation between stock and human population trends (Fig. 10.2) suggests that NCA has not remained entirely isolated from the national processes of population growth and subsistence decline. Numbers of people have increased to more than double their 1960 level by 1987, whereas cattle numbers have not yet regained their 1960s levels, and small stock numbers currently show an increase of about a third over the 1960s figures. The exponential rate of increase of the NCA Maasai population 1954–1987 can be calculated from initial and final figures as $r = 0.023$. This is approximately equivalent to a population growth rate of 2.3% per annum and a doubling time of 30 years. An alternative method, using the mean exponential rate of increase over the 33-year period (Caughley 1977:109) irons out irregularities and short term trends. This gives a value of 3.2% per annum (doubling time of 22 years).

Henin and Egero (1972) estimated a rate of 'natural increase' (births–deaths) of 2.3% for the then Masai District. However, this 'natural increase' measure omitted immigration, which accounted for a further 2.2% per year, giving a total 4.5% growth rate per year for the Masai District in 1957–78. Kurji (1981) estimates an even higher immigration rate for NCA of some 3.9% per year, and arrives at an aggregate estimate of 6.2% total growth per annum in NCA in 1966–1978. Kurji's high growth rate figures result from choosing the low 1966 figure as the baseline rather than 1954. Kurji's figures span the period during which NCA was scheduled for Ranching Association development, and during which the population trebled in seven years. This was presumably largely a result of immigration in response to the lure of development and the possibility of land tenure (Moris 1981). Since 1979 when the MLRDP programme collapsed the rate of growth of the NCA population has slowed considerably (Table 10.1,Fig.10.1).

These estimates can be compared with the average figure of 2.2% per annum growth estimated for pastoralist populations in Kenya (Evangelou 1985:30). This is slow compared to cultivator population growth in Kenya (around 4%

per annum). Sindiga (1987) argues that traditional social controls on fertility (particularly the fact that elders control livestock deals and hence bridewealth, effectively monopolising women of childbearing age), prolonged breastfeeding and postpartum sexual abstention, high secondary sterility, seasonal food shortages, spousal separation, and general environmental health hazards causing high infant mortality all conspire to make Maasai population growth as low as or less than the 2.2% attributed to Kenya pastoralists overall.

The NCA figures suggest there has been a population growth rate of 2–3% per year for the last few decades, resulting from both natural increase and immigration, and that on top of this, temporary in- and out-migration have created peaks and toughs in the overall trends of increase. Human population increase in NCA is lower than for other areas in Arusha Region (e.g. Arumeru District). Nonetheless, it has clear implications for development, which must deal with the problems of a growing population and dwindling per capita resources. Development planning must take into account that demographic trends in NCA show both population growth and also major fluctuations.

(c) Distribution of Maasai within NCA

The Maasai population is by no means evenly distributed throughout NCA, and distribution patterns change with seasonal migration and transhumance as well as longer-term shifts. Arhem (1981a:39) analysed the population trends within different parts of the NCA over 1974–1980. He found a general increase in the population of most villages and zones in 1974–80, except in the Nairobi/Kapenjiro zone. Here there was a decline up to 1977 and then an increase in 1977–78. Arhem (1981a:39) suggests the decline was due to the newly enforced prohibition on cultivation, followed by emigration of the agricultural and agropastoral Waarusha originally concentrated in Nairobi/Kapenjiro. Arhem, apparently unaware of the existence and impact of the USAID programme, attributed the general population increase in other zones during the same period firstly to immigration from outside NCA during drought years (especially 1973 and 1976), and secondly the 1974 villagisation which tended to concentrate people round trading centres. The rapid influx in 1970–1976 which occurred in all the high-potential MLRDP-sponsored RAs (Moris 1981:106) is also likely to have concentrated people around those centres where dips and dams and other facilities were built. The 1974–78 period shows the most marked divergence between human and cattle population trends, possibly as a result of an influx of poor pastoralists with villagisation and the Ranching Association programme as well as the effect of drought losses. Arhem (1981a:39) interprets the 1980 boma count as showing an overall decline in human population numbers in all zones except Melenda and Olbalbal. Kakesio–Endulen zone showed a particularly marked decrease. Arhem (1981a) concluded that besides net emigration from NCA in 1978–1980 there had been net movement within NCA to the centres of Endulen and Oloirobi. These trading centres are the best equipped in terms of shops, transport, communications and public services. Cattle raiding has resulted in a net shift away from Kakesio and Osinoni villages which border on Sukumaland (Ecosystems Ltd 1980, Arhem 1981a, Cobb 1989). For example, the villages of Kakesio and Osinoni claimed to have lost over 3000 head of cattle to Sukuma raiders during 1980 (Arhem 1981a) of which only 111 were recovered.

The most recent figures (Perkin 1987) show that the overall 26% increase of the NCA Maasai population from 1978 to 1987 occurred mainly in Nainokanoka (> 50% increase over 1978 figures) and Ngorongoro administrative zone (35% increase) while Endulen/Kakesio has shown a net 8% decline. The increase in human population during the period 1978–1987 was accompanied by a trivial increase of cattle and small stock; livestock populations have actually declined in Nainokanoka despite the influx of people in this zone. The limitation of livestock populations is largely attributable to livestock–wildlife interactions (chapter 9) and to some extent to administrative bans (chapter 8).

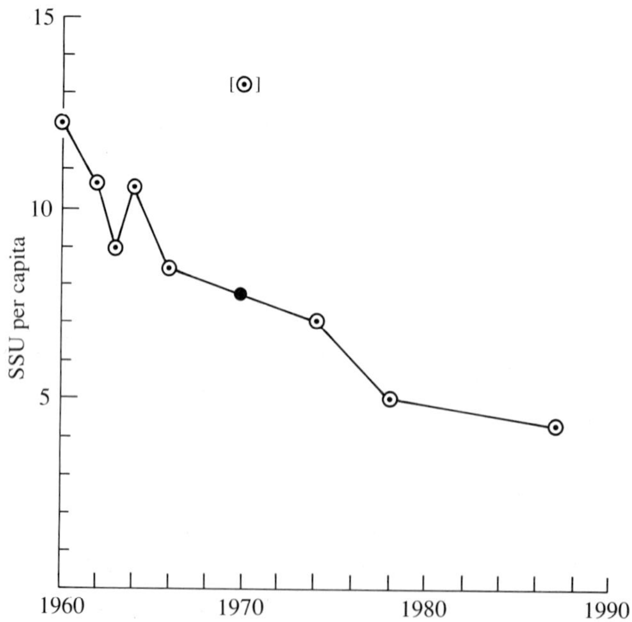

Fig. 10.2. Standard stock units (large and small stock) per capita for NCA Maasai. See Fig. 10.1 for data and conversion rates from livestock numbers to SSU figures as well as for calculation of 1970 data point (●).

(d) Boma size, composition and stock holdings

Bomas vary in size but on average at the time of our study consisted of some seven houses with a total of around 20–30 inhabitants (Table 10.2). Individual herdowners often have two or sometimes three wives (66% were polygynous). Arhem estimated there to be around four people per house (Table 10.2) as did Perkin (1987:3.8). This compares closely with earlier estimates for NCA (six to eight per 'household' i.e. per aggregate of houses associated with one married man – Kurji 1981) and current estimates for elsewhere in Maasailand (four to five per 'subhousehold' (i.e. per house) – Nestel 1986).

The majority of herdowners (70%) controlled between 10 and 100 cattle as well as small stock; around 15% had fewer than 10 and 15% more than 100 cattle. Fewer than 5% of the herdowners had more than 300 cattle, and very few had over 500 head. Arhem (1981a) noted a roughly inverse relationship between cattle and small stock numbers. On average there were some 150 cattle and 200 small stock per boma. All these figures show marked changes from those reported by Jacobs (1975) who quoted a typical 'camp' in the 1960s as having 50–80 persons and 1200–1500 livestock units. Jacobs (1978) saw this decline in boma size as a direct outcome of the villagisation programme. Over the last twenty years NCA stock holdings have declined from about ten cattle and ten small stock per head of population in 1960 to around six cattle and six small stock in 1987 (Perkin 1987).

(e) Discussion: Trends in NCA demography and living standards

NCA Maasai population has fluctuated over the last 35 years but overall has shown 2–3% per annum growth similar to that for pastoralist areas elsewhere in East Africa. Immigration and emigration are extremely important: the Ranching Association programme was associated with a trebling of the population in seven years. Meanwhile, stock populations have also fluctuated but show no overall trend of increase, and average holdings per capita have declined (Fig.10.2).

It is of interest to compare NCA stock holdings wth conditions in Kenya Maasailand. Nestel (1986) gives estimates of Kajiado group ranch Maasai stock holdings as 14 cattle and 8 small stock per person. She also presents the results in terms of the ILCA system of Livestock Equivalents (LE) per Active Adult Male Equivalent (AAME). It is not possible to calculate the figures for NCA in precisely the same way, as the detailed age structures are not available either for stock or for human populations, but a close approximation can be obtained by averaging LE values across all classes of immature stock and similarly averaging AAME values across all age classes of under-fifteens. This

Table 10.2 *Composition of Maasai households*

Area	Locality	Number of houses	Adult (> 15 yr)		Child	Total	Source
			M	F	(< 15 yr)		
NCA	Gol	7	8	7	11	26	1
	Ilmesigio a	7	6	9	13	28	
	b	6	4	6	11	21	
	Sendui	8	6	9	26	41	
Mean boma composition		7	6.0	7.8	15.3	39	
Mean house composition			0.9	1.1	2.2	4.2	1
NCA (mean of total count)			0.8	1.0	2.0	3.8	2
Kajiado (a) High potential zone			0.4	1.4	3.0	4.8	3
(b) Low potential zone			0.3	1.0	3.1	4.4	

Source: 1. Arhem, Homewood and Rodgers 1981; 2. Perkin 1987; 3. Nestel 1986.

gives an approximate average value of 6.1 LE/AAME for NCA Maasai, which puts them into the lower level of Nestel's intermediate wealth stratum (0–4.99 is poor; 5–12.99 intermediate; 13+ wealthy). This is borne out by comparison with Evangelou (1985) whose averages for the whole of Maasailand are around four cattle and five small stock per person. The conclusion is that NCA Maasai on average are less wealthy in stock terms than the inhabitants of the Kajiado group ranches but are not worse off than average for Kenya Maasailand. Stock numbers are of course only a rough indication of prosperity, firstly because productivity levels differ with habitat (see chapter 8 and section on the food system below), secondly because the livestock–grain exchange terms differ and thirdly because many inhabitants of Kenya Maasailand also have some cultivation.

A number of authors (Brown 1971, Dahl and Hjort 1976, Jewell 1980) have attempted to estimate the numbers of stock required to maintain pastoralist families subsisting on a largely milk diet. However, it is more realistic to calculate the minimum subsistence herd in what Kjaerby (1979) defines as exchange-oriented pastoralism – a system where stock and/or pastoral produce is exchanged for grain. Kjaerby (1979) gives estimates for the Barabaig (adjacent to NCA). Using an exchange value of 450 kg of maize meal per cow (realistic for NCA in the late 1970s/early 1980s – Arhem 1981a) Kjaerby estimates approximately five cattle per capita are required to meet minimum subsistence. According to this estimate, NCA Maasai after 20 years of declining living standards are now near the minimum standard of living within the constraints of exchange pastoralism imposed on them by the cultivation ban. Given the distribution of livestock, at least 15% families are below this minimum. Maasai perceptions as reported in a village survey (Arhem 1981a) were that food, both milk and maize meal, had become progressively more scarce over the last 20 years.

Food survey

En-jipati olalae. A little food to keep life going

(Maasai expression; Mol 1978)

Our NCA survey investigated patterns of food consumption in different areas and age/sex classes for the August 1981 dry season (Homewood, Rodgers and Arhem 1987). Later study visits allowed extrapolation of baseline milk availability and consumption to different seasons.

(a) Diet items and food sharing

The main items of food were milk (fresh, sour or cooked with maize meal); meat and animal fat (roast or boiled as soup); grain (maize flour cooked

Table 10.4 *Energy requirements and food composition values*

(a) Energy requirements

Child y	Estimated energy requirements (kJ/day)	AE	Male y	Estimated energy requirements (kJ/day)	AE	Female y	Estimated energy requirements (kJ/day)	AE
<2	3 570	0.3						
2–4	4 620	0.4	15–34	11 760	1.0	15+	10 920	0.9
5–9	7 140	0.6	35+	10 500	0.9	pregnant	12 180	1.0
10–14	10 500	0.9				lactating	13 020	1.1

(b) Food composition values

Food	Energy (kJ/100 g)	Reference protein (g/100 g)
Milk – dry season	315	2.8
Milk – wet season	420	2.8
Maize flour	1508	4.4
Meat	1155	20.0
Honey beer	210	—
Honey	1218	0.4
Sugar	1638	—

Note:
Reference protein requirements taken as 34 g/day for an adult male, 27 g/day for an adult female. See text for sources.

Table 10.5 *Composition and protein/energy adequacy of NCA and Kajiado Maasai diets*

		Contribution to dietary energy (%)				Dietary energy (% RDI/AE)		Dietary protein (% RDI/AE)		
Area	Site	milk	meat	grain	other					Source
NCA	Gol	38	+	56	+	73	91[a]	156	190[a]	1
	Ilmesigio	73	+	23	+	59	71[a]	149	150[a]	
	Sendui	27	+	71	+	69	67[a]	123	126[a]	
	mean	34	+	53	+	67		127		
	mean[a]	30[a]	+	64[a]	+		75[a]		160[a]	
Kajiado (a)		64	4	16	16	70		233		2
(b)		31	12	34	22	66		187		

Notes:
[a] adjusted for reciprocal sharing.
Source: 1. Arhem, Homewood and Rodgers 1981, 2. Nestel 1986.

season energy intake would have been less than 75% of estimated requirements. This points to a marked energy deficit, at least on a seasonal basis (see below). During the August 1981 survey pastoral foods (milk and meat) provided 73% and grain 23% of the dietary energy in Ilmesigio. In Nasera the corresponding figures were 38% and 56% while in Sendui pastoral foods provided only 27% dietary energy and grain 71%. Grain thus accounts for an overall average of 53% dietary energy (and milk 34%) during the August 1981 dry season sample. Allowing for additional food possibly received through reciprocation of sharing, these figures become 64% dietary energy from grain and 30% dietary energy from milk. Pastoral foods provided the bulk of the protein intake in all three areas (average 72%, ranging from 88% in Ilmesigio to 58% in Sendui).

(e) Dietary composition and adequacy: 1981–1983

It is possible to combine our August 1981 dry season survey with milk availability measures made during later visits to give a picture of the NCA Maasai diet through the two years of our study. Milk availability for each sample visit was calculated from average milk yields and numbers of cows in milk (Table 10.6). The seasonality of calving together with that of high calf mortality means that more young calves and hence lactating cows are present at the end than at the beginning of the rains. Thus both December samples had 55% cows in milk and May–August almost 65%. A long intercalf interval (18–23 months) and seasonality in calving mean that different numbers of cows come into milk each year. This difference was most pronounced for Gol (Table 10.6). A large proportion of Gol cows were reported pregnant during the August 1983 sample so they would have a higher proportion of milking cows in the following year.

Table 10.6 incorporates data on numbers of adult equivalents (AE) and estimates availability of milk per person during each period on this basis. This gives rather different estimates from those made on the basis of house food intake. Dividing the total milk yield by the number of adult equivalents will produce an overestimate as it does not allow for sharing with occasional visitors from outside the boma (Table 10.3, cf. Table 10.6)

Milk availability varied seasonally, falling drastically low, for example in August 1983, and necessitating major stock sales for grain purchase. Sendui had the most severe shortfall in milk availability, with an average over the two years monitored by our study visits of only 580 g/AE/day, compared with 1080 g/AE/day in Gol. Our figures for Ilmesigio may be artificially low as there was no wet season sample. Although we have no information on grain intake during the later samples, it is striking that the two-year averages for milk availability are lower than, or approximately equal to, Arhem's estimated mean dry season potential milk intake of 1065 g/AE/day for August 1981. Bearing in mind that the follow-up figures are likely to be overestimates, and the August 1981 house survey intakes underestimates, our results indicate a run of bad years in terms of subsistence. This is despite the fact that cattle performance was, if anything, average in all bomas other than Oloirobi, which was hit by severe epidemic. The high offtake levels we observed, resulting in overall herd decline despite 'normal' to low fertility and 'normal' mortality, can now be seen in the context of reliance on imported grain to make up the dietary deficit. Declining standards of living are not so much a product of poor herd performance as of an increasing human population together with the practical difficulties and growing expense of obtaining grain in exchange for pastoral products.

(f) Nutritional status: July 1989

The most recent survey of Maasai nutrition in NCA was a brief but intensive anthropometric study carried out as part of the NCDP programme (McCabe, Schofield and Pederson 1989). Over 55% children sampled were found to be malnourished (see Table 10.7: 19% below 80% weight for height) or undernourished (38% between 80–90% weight for height). Over 35% of adults were undernourished or malnourished, with 12% males and 15% females having body mass index (BMI) values indicating malnutrition. There are problems in applying standards derived from one country to define cut-off points indicating 'malnutrition' or 'undernutrition' in another (see e.g. Pacey

Table 10.6 *Variation in milk availability*

Area	Date	Mean milk yield/ cow (g/day)	No. milk cows	Total yield (kg/day)	Adult equivts	Milk kg/AE	Mean
Gol	Jul. 81	600	38	22.8	20.3	1.12	
	Dec. 81	704	29	20.4	20.3	1.00	
	May 82	678	41	27.8	20.3	1.36	
	Dec. 82	1094	27	29.6	19.5	1.52	
	Aug. 83	454	17	7.8	19.8	0.40	1.08
Ilmesigio	Aug. 81	1194	23	27.4	21.1	1.30	
	Aug. 83	848	19	16.2	22.9	0.70	1.00
Sendui	Aug. 81	680	34	23.2	29.6	0.78	
	Nov. 81	486	21	10.2	29.6	0.34	
	May 82	1022	23	23.6	30.8	0.76	
	Dec. 82	974	16	15.6	28.8	0.54	
	Aug. 83	758	20	15.2	30.0	0.50	0.58

and Payne 1985:92) particularly where, as with the Maasai, the associated patterns of morbidity and mortality risks are not known. There are also problems in using a brief single-round survey to assess nutritional status in a population whose diet is strongly affected by seasonal conditions. Nonetheless, the results have been taken to indicate an unacceptably high level of malnutrition, whether they represent the low point or worse still the peak of the annual cycle of food availability and nutritional status (McCabe, Schofield and Pederson 1989). Comparison of these results with anthropometric studies on Kajiado Maasai and Turkana pastoralists suggest an alternative interpretation (see Table 10.7 and below).

McCabe, Schofield and Pederson (1989) found low weight-for-height children in both prosperous and poor bomas, with few significant differences in child anthropometric status between wealth strata. This is a common finding in rural Africa, where households include not only full lineage descendants of the household head but also considerable numbers of 'poor relations' and other dependents (see Wilson 1990 for review). If anything, the wealthier the household, the larger will the number of such attached individuals be. The variance within households may thus be as great as the variance between households. In Maasai bomas two further mechanisms even out potential differences in nutritional status between bomas: food sharing with visitors, and the reluctance of wealthy families to strip more milk from the cows than is absolutely necessary, because of the effect on the calf (see below, and Grandin 1988). However, Nestel found significant differences between the populations of 'high-potential' and 'low-potential' ranches in Kajiado (Table 10.7)

Table 10.7 *Anthropometric status of NCA and Kajiado Maasai children*

		Kajiado 1982–83	
	NCA 1989	Semi-arid High-potential	Arid Low Potential
% thin (<90% weight for height)	55	57	72
% very thin (<80% weight for height)	19	20	33
% short (<95% height for age)		42	51
% very short (<90% height for age)		15	19

Source: NCA data from McCabe, Schofield and Pederson 1989, Kajiado data from Nestel 1986, see also Nestel 1985.

(g) NCA Maasai diet compared to Maasai elewhere

As Nestel (1985, 1986) points out, despite the numerous recent studies of Maasai diet the only other research to have attempted quantification of their dietary intake was that of Orr and Gilks (1931), who give no details of their methods. Our own study used methods closely comparable to those of Nestel (although we cannot begin to match her sample size, and her observations relate only to women and children while ours also included men).

The main point of comparison between Kajiado and NCA Maasai diets is in terms of the amount of milk available per AAME per day. Nestel (1986) gives these in terms of litres. In our study one litre of milk was found to weigh 1.1 kg and this conversion factor is used here to adjust Nestel's observations. The higher adult equivalent values we apply for pregnant and lactating women (Table 10.4) cancel out to some extent against the lower values for children and the results are roughly comparable. Nestel calculated that in the higher-potential group ranches there was an average of 2.63 l/AAME/day equivalent to 2.89 kg/AAME/day. In the low-potential area there was an average of 1.18 l/AAME/day equivalent to 1.30 kg milk/AAME/day. Grandin (1988) found that on the higher-potential (Olkarkar) ranch wealthy households showed the same milk consumption per person as poor families with one-fifth the number of cattle.

These values are almost exactly double the intakes measured for NCA in August 1981, with Ilmesigio having 1.47 kg/AE/day and Gol 0.65 kg/AE/day. On an annual basis Nestel's figure indicates mean monthly values ranging from 2.09–3.96 kg/AAME/day in the higher-potential area and 0.88–2.09 kg/AAME/day in the lower-potential area. Annual figures for NCA estimated from our knowledge of milk production, number of cows in milk and number of adult equivalents present in any one month suggest the August 1981 values for NCA milk availability are representative of or higher than availability over the ensuing two years (see below).

Table 10.5 also shows estimates for the dietary energy and protein adequacy of the Kajiado group ranch Maasai (Nestel 1986). Our 1981 dietary energy adequacy values were remarkably similar to figures for the Kajiado Maasai. Both Maasai populations were apparently existing at an average of around 70% dietary energy adequacy, and this perhaps calls into question the validity of the RDI figures used in the calculation. Pastoral Turkana also seem to exist on an average of 70% recommended energy intake for their weight (Little, Galvin and Leslie 1987). The more important point is that in terms of adequacy of intake the NCA Maasai population seems to have been rather closely comparable to Kajiado figures at the time of our study, though in terms of dietary composition the Kajiado Maasai could clearly rely on a greater milk intake and correspondingly lower contribution of grain to the diet. Most importantly, the recent survey data suggest that NCA Maasai currently maintain a nutritional status comparable with or better than that of the Kajiado Maasai studied by Nestel (1986, 1989, see also Table 10.7. Kajiado figures quoted in the report by McCabe, Schofield and Pederson 1989 do not correspond with Nestel's published data).

(h) Pastoralist diets and grain dependence

The formerly widely held idea that Maasai populations traditionally lived on pastoral produce alone (e.g. Jacobs 1975, Brown, 1971) has been gradually modified, though it persists (e.g. Nestel 1986:4, Grandin 1988:6). While the warrior age-set may traditionally eat only milk, meat or blood, the Maasai as a group make extensive use of grain and it is now thought by historians and anthropologists that they have probably always done so (Waller 1979). The interdependence of pastoralist and cultivator is now better understood (e.g. Berntsen 1979, Toulmin 1983). It may entail the exchange of products and facilities between different groups, between different individuals within groups, or, with sedentary agropastoralists, individual families may maintain both cultivated fields and resident or migratory herds. Over the generations one family may pass through a cycle of subsistence from pastoralist to cultivator and back without losing sight of their Maasai identity. The main point is that a purely pastoralist diet is the exception rather than the rule, and this has probably always been the case. Annual averages for a range of pastoralist groups suggest that the single largest source of calories in many pastoral diets is from grain, with milk products giving most of the protein (as well as an important contribution to energy requirements) and meat playing rather little part in the diet.

Nestel's Kajiado group-ranch Maasai obtain 43–68% of their dietary energy from pastoral products: one-third to over a half came from cultivated foods, and a considerable proportion of the families were cultivating to supplement their diet (Nestel 1985, 1986). In all the Kajiado study sites around 50% of non-stock expenditure was on cultivated food (and 40% of this was on maize). At the other end of the scale perhaps the highest values for milk and meat dependence in East African pastoralists have been recorded for the Ngisonyoka Turkana where 76% dietary energy came directly from pastoral produce, 16% from cultivated foods obtained by sale or exchange of pastoral produce and 8% from wild plants and animals (Coughenour *et al.* 1985). Such a heavy dependence on pastoral produce is rare. Little (1983) found that for every litre of milk consumed, grain consumption in Il Chamus families declined by approximately 0.6 kg, but that some grain was consumed even during periods of peak milk availability. A similar (though non-significant) inverse relationship between milk and grain consumption exists in our data. Further evidence for the dependence of herder on farmer comes from the recent studies of nutrition in Sahelian pastoralists, as well as more sedentary agropastoralists (Hill, 1985, Benefice, Chevassu-Agnes and Barral 1984). Even the most livestock-oriented groups (Wodaabe, Fulani, Tamasheq), with an almost entirely milk-based rainy season diet, show major or even total dependence on grain during the dry season. The overall 12-month median contributions to daily energy intake among pastoral Tamasheq were 28% from milk, 62% from grain and 9% from meat (calculated from Wagenaar-Brouwer 1985:235).

On the basis of the August 1981 dry season survey and the assumption that wet season milk yields would double we estimated comparable annual energy intake figures for the NCA Maasai of 56% from milk, 39% from grain and 5% from meat (Arhem, Homewood and Rodgers 1981). However, the follow up data on milk availability in 1981–83 presented above and in Table 10.6 suggest that these estimates of average annual milk intake may have been overoptimistic. The August 1981 figures of 30–34% dietary energy from milk, 6% from meat and 53–64% from grain may have been a better indication of the average requirements for non-pastoral supplementary food sources over the period 1981–1983 in NCA.

The NCA Maasai are thus dependent on imports of grain in exchange for stock sales. Most grain imports take place through private traders (Arhem 1981a), and in most cases livestock sales are through unofficial or even illegal channels (chapter 8; Field and Moll 1987), becoming invisible as far as national statistics are concerned although remaining important in terms of food production.

Conclusion

This chapter has reviewed Maasai livestock development programmes and their ecological implications, and the demography, changing living standards and food system of NCA Maasai. Together these all present a picture of a growing human population with a declining per capita resource base, sustaining itself by traditional means, with little input from goverment or administrative sources, and increasingly vulnerable to damaging exploitation by tourists and organized poachers. Traditional Maasai social structures retain their adaptive functions, but considerable constraints have been imposed over the last 20 years in terms of cultivation ban and stock management restrictions in NCA. An attempt at livestock development resulted in a trebling of the human population 1970–1977 but has otherwise left few lasting benefits. Since the 1960s the standard of living has steadily declined, as evidenced by the livestock per person ratios, by outside observers and by the perceptions of the Maasai themselves. The NCA Maasai are now near the minimum numbers of stock per capita, in terms of both day to day subsistence and of the long-term survival of their herds. Preliminary data from recent surveys suggest the anthropometric status of NCA Maasai is as good as or better than that of the Kajiado Maasai. However, NCA Maasai appear to exist on a diet averaging around 70% recommended dietary energy intake, as do Kajiado Maasai and Ngisonyoka Turkana among other pastoralist groups. A considerable proportion of the NCA Maasai fall below the international anthropometric cut-off points that are taken to indicate malnutrition, as is also the case for Kajiado Maasai and other sub-Saharan pastoralists. Without data on associated morbidity and mortality risks it is not possible to be sure that the standards used are appropriate to these populations. It must therefore be taken as unproven but nonetheless likely that the NCA Maasai represent a nutritionally very vulnerable population, as may several other pastoralist groups. There is no doubt that livestock *per capita* is lower for the NCA than the Kajiado Maasai, as is the contribution of milk to their diet. There is therefore a much greater reliance on grain to supplement the diet in NCA, though grain is considerably harder to come by and NCA Maasai get poorer terms of trade.

The Maasai are not an isolated and primitive group incapable of surviving a technological and cultural impact (cf, other 'tribals' Goodland 1985). They have a sophisticated awareness of, and ability to manoeuvre within, the opportunities and constraints of the wider national and international context. They have for centuries been integrated into a network of neighbouring cultures and land use systems that provides their survival strategies in times of individual or regional disaster. Their strong cultural, social and ecological identity has not only survived but been reinforced by far greater past crises of disease, stock losses, military defeat, social disruption and geographical relocation (Waller 1988). The subsequent alienation of vast areas of critical wet and dry season pastures, the ever-greater imposed interventions, and their progressive marginalisation from national political and economic processes, have altered the range of strategies open to them for coping with disasters and left them less affluent and more vulnerable than they were in the mid-nineteenth century. What has not changed is 'their self-reliant mode of husbandry, their pronounced sense of identity and their inviolate presumption of their own traditions' (Spencer 1988:22). Their reponse to different imposed range development programmes in Tanzania and Kenya has demonstrated their enormous adaptability in the twentieth century context as well, but their land use, stock management and traditional values are extraordinarily resilient to change and disaster. This has inescapable implications for the course of pastoralist development and its interaction with wildlife conservation in NCA.

11

Wildlife conservation and pastoralist development

Merep enkaboboki o icani likai cani The bark of one tree will not stick to another tree

(Tribes cannot assimilate one another's customs – Maasai saying: Mol 1978)

This chapter deals with the problems of integrating conservation and development in NCA. It looks at three main aspects. Firstly we consider general points of conflict and complementarity between wildlife conservation and pastoralist development. Secondly the development of different types of tourism, and the interaction of each with conservation and pastoralism, are examined. Thirdly the possibilities for development of alternative forms of wildlife exploitation (other than tourist viewing) and their compatibility with conservation are discussed.

Development is hard to define or quantify. The history of development interventions in Maasai pastoralism makes it clear that pastoralist development must be in terms of making existing production systems more secure rather than of replacing these systems with imported alternatives. Development is also a vast and complex interdisciplinary topic. In keeping with the overall aim of this book we concentrate here on aspects which concern the ecology of the Maasai and their livestock, and of NCA and its wildlife. For example, the development of Maasai health, education, administration, communications and trade infrastructures are discussed only insofar as they relate to the biology of resource use and environmental impact.

Pastoralist development and wildlife conservation

(a) Conflict and complementarity

The founding principle of NCA contains an apparent paradox, inherent in many third world developmental issues, but nowhere more evident than in joint wildlife conservation/human land use areas. Conservation and development seem to represent diametrically opposed aims, so that management will be at best a compromise and at worst a destructive conflict between the two. The damaging ways in which such conflict can erupt are illustrated by the calculated destruction of wildlife by Maasai in NCA and Amboseli as a form of political protest (chapter 7; Lindsay 1987), by the heavy-handed eradication of cultivation plots by the Conservation Authority in NCA (Makacha and Ole Sayalel 1987) or the excessive concern over and regimentation of Maasai use of Crater salt licks (Machange 1988).

Wildlife conservation and pastoral development are not as mutually incompatible as such confrontations might suggest. The last few years have seen a radical rethinking of the development process in Africa as a whole (Sandford 1983, Cross 1985, Timberlake 1988). Despite a rearguard action by some members of the old school (e.g. Adams 1988), there is now real concern for the environmental, economic and social dangers of many western-initiated developments in the Third World, a concern shared by the most hard-headed aid and financial institutions in the field (Pearce 1988, World Bank 1984, Office of Technology Assessment 1984). There is a growing understanding of the appropriate nature of many traditional forms of wildlife use (McNeely and Pitt 1985, Bell 1987), cultivation and stock rearing (e.g. Ellis and Swift 1988, Cross 1985, Mackenzie D. 1987, Timberlake 1988) as well as of indigenous traditions of land use experiment and willingness to adopt genuinely beneficial innovations (Richards 1985). At the same time there is a fundamental reappraisal of conservation aims and methods, particularly as regards the role of local communities (McNeely and Pitt 1985, Bell 1987, Western 1984). Naturalness and wildness, and the conservation values associated with them, are no longer seen as attributes restricted to ecosystems from which humans have been excluded. In particular, the conservation-oriented exclusion of human populations from ecosystems of high conservation value, with which those populations have a long-standing and close integration, is now recognised as artificial and inappropriate in biological terms. Conservation values are no longer attached primarily to particular species or to a single specific state of a system. Natural processes of change, and the transient states that accompany them, are better understood and more readily accepted. Apart from these biological issues, the rights of local communities, both economic and in terms of quality of life and cultural values, are now taken more seriously (Bodley 1988, Goodland 1985). This enlightened attitude is reinforced by the fact that the long-term political viability of conservation schemes is strongly dependent on enlisting and reinforcing rather than denying the cultural values local communities attach to those natural resources. It is also an attitude that has begun to surface in a number of internal NCAA documents (e.g. Kayera 1985).

In this context many of the apparent conflicts of NCA management disappear. Current land use patterns, and the state of NCA vegetation and wildlife populations show that pastoralism and wildlife conservation work well together. Both rely on the maintenance of similar rangeland conditions, and particularly on the exclusion of large-scale cultivation. This section aims to establish that pastoral development is no threat to conservation. Later sections will argue that the long-term survival of wildlife conservation in NCA is actually dependent on coexistence with Maasai pastoralism. In conservation terms there is little to fear, and much to gain, from the natural course of Maasai pastoralism in the NCA and from maintaining its traditional values. Their respect for wildlife, which can sour where conservation interests are pursued to their serious detriment, survives a transition to a more nationally integrated, market oriented economy (Western 1984). There is plenty of margin for experiment and even error: there is a long way to go before NCA wildlife:livestock biomass ratios decrease to those found in other acclaimed wildlife viewing areas.

The main perceived threat from pastoral development in NCA arises not from current land use but from the possibility that pastoralist livestock densities will grow, pastoralist sedentarisation increase and pastoralist livestock production intensify (Ole Kuwai 1980, Ole Saibull 1978, Pearsall 1957). For example, Field, Moll and Ole Sonkoi (1988) infer a trend to more permanent settlement in NCA, apparently misreading the long-established Crater Highlands practice of transhumance from a fixed dry season boma (where some elders, women and children may live most or all of the year) to a more flexible and mobile wet season grazing camp. The feared outcome of increased sedentarisation is that there will be resulting exclusion of wildlife from vital resources of water, minerals and grazing. Less commonly expressed, but keenly felt by conservationists of the old school, is the fear that an increased Maasai presence will detract from the spectacular 'natural' beauty of the NCA. The very positive reaction of consultant conservation planners, who see limited Maasai presence in the Crater as greatly enhancing tourist experience and understanding, is in strong contrast to this (Taylor 1988). Again the argument centres on whether human and livestock numbers are likely to grow, and to what extent. The trajectory of pastoralist development can be to some extent inferred from the history of interventions in Maasailand and elsewhere (set out in detail in chapter 10): here we discuss the implications of that history for the likely course of pastoral development in NCA.

(b) Development for commercial offtake or for secure subsistence?

Development initiatives in sub-Saharan Africa during the period 1960–1980 concentrated on encouraging large-scale capital intensive commercial enterprises, generally with the urban consumer or export markets in view. These attempts have mostly failed (chapter 10; see also Sandford 1983,

Wyckoff 1985). Special exceptions operate where wealthy individuals are allowed to gain sole access to the title, particularly of high-potential land (Behnke 1984). In general, development policies created adverse patterns of subsidy, marketing, quarantine imposition, etc. which effectively handicapped smallholder subsistence production (Raikes 1981, Sandford 1983).

The unequivocal failure of the overwhelming majority of livestock development projects, together with the severe droughts and famines of the late 1970s and early 1980s, has forced a reappraisal of pastoral development policies. Where the emphasis in Third World livestock development was originally on exotic high yield breeds requiring capital-intensive care, and on high-technology, high-finance methods of production aimed at commercial profits, it has become clear that such enterprises created as many problems as they solved. The emphasis of international aid and national investment in the 1980s shifted to development initiatives concentrating on low cost, low technology, labour intensive 'grass roots' initiatives using local breeds and indigenous techniques and aiming to make subsistence less vulnerable and more environmentally sustainable (e.g. Cross 1985, D. MacKenzie 1987, Timberlake 1988, Ellis and Swift 1988). Accordingly, within the NCAA brief to develop human and livestock resources alongside conservation values, the first priority should be to concentrate on making local subsistence more secure. Development geared to ensuring welfare security and environmental sustainability is not only of lasting worth to both local and national interests. Such development is by definition compatible with conservation, while development geared to commercial exploitation is likely to be the opposite.

(c) The future of Maasai development in NCA

Ecological factors, together with social and cultural structures and values that have evolved to deal with those ecological constraints, have foiled repeated development attempts to intensify livestock production in NCA as for elsewhere in sub-Saharan Africa. In NCA there has been no net increase in livestock numbers or biomass for the last 30 years (chapter 8) despite two decades of attempts at promoting development (chapter 10). These interventions have had two main noticeable effects in NCA. Firstly, there was a trebling of the human population 1970–1977 as Maasai and cultivators responded to the lure of technical developments and the possibility of occupancy rights. Secondly, there has been a paradoxical shift towards smaller management units, brought about by the pressure on an already egalitarian and cooperative society to demonstrate a response to the villagisation campaign (Jacobs 1978). The fact that so much development effort in Maasailand has sunk without trace does not imply that Maasai are incapable of adapting. On the contrary they show themselves to be responsive to incentive and able to incorporate their own social traditions into working compromises with a multiplicity of different imposed administrative structures (chapter 10; see also Waller 1988, Galaty 1980).

Of all the aspects of development which may take place, it is unnecessary to fear a massive increase of livestock densities and an intensification of commercial production to the extent of jeopardising wildlife values in NCA. Such commercialisation could happen only if one or a few powerful individuals were given title to land and allowed to exclude the rest. Livestock population increase and overgrazing, the most commonly invoked threats, have taken place neither in NCA nor in most of Maasailand (chapters 6,8,9). The course of Maasai development within NCA is more likely to be one of small-scale, opportunistic incorporation of those technical, marketing, administrative and political inputs that are perceived by the Maasai as genuinely practical and beneficial. Some examples and guidelines are discussed in chapter 12. Modest development along these lines could make subsistence more secure, but is unlikely either to generate dramatic commercial wealth or to threaten environment and wildlife.

A much more real hazard to conservation may be increasing human population numbers, rather than increasing livestock numbers. In NCA, the installation of technical improvements as a token of future benefits, without the accompanying definition of legitimate user groups, seems to have triggered a trebling of the population over a few years. Since the demise of the MLRDP, NCA population growth has slowed (chapter 10, Table 10.1, Fig. 10.2). The current Maasai population density in NCA is around 2.5 per km^2, considerably lower than that for most joint pastoralist/wildlife systems (see chapter 9) particularly when the comparatively high potential of the NCA land units is taken into account. Maasai population growth is slower than the 2.8% national Tanzanian population growth rate or the 4% quoted for cultivators (e.g. Sindiga 1987).

Pastoralist development does not threaten conservation in NCA. However, immigration and its control may become an important issue. The future definition of user groups and occupancy rights is discussed in chapter 12.

Tourism and conservation

Wildlife, scenery and archaeological values have made NCA a major tourist attraction to both foreign and local people. Tourism, however, is a capricious source of revenue, often dependent on politico-economic factors far from the tourist venue. At the time of our field work tourism was at its lowest ebb in NCA. Today it is a major source of revenue for the area and for Tanzania as a whole. This will naturally sway planners and administrators towards more tourist lodges and perhaps towards excluding Maasai. The development of tourism in NCA must be planned on the basis of a reappraisal of the economics of tourism, of the respective roles of foreign and local tourism, and of their implications for the long-term future of conservation.

(a) Foreign tourism

Foreign tourism has long been seen by the international community, both conservationists and entrepreneurs, as a potentially profitable economic venture which is compatible with conservation in a way that other forms of human land use are not (Eltringham 1984). Tourism is seen as a non-consumptive resource use, and despite mounting evidence as to the potentially serious negative impacts on environment and wildlife in parks like Amboseli (Henry 1977) this is still the assumption for NCA (e.g. Struhsaker *et al.* 1989). From the early 1960s the monetary aspects of tourism have been emphasised by international conservationists advising emerging East African nations.

The economic aspects of tourism in Tanzania and NCA to date are covered in detail below. Briefly, contrary to many advisors' expectations, foreign tourism made a comparatively very small contribution to the Tanzanian economy up to 1987–88. It has also been seen as a mixed blessing by economists, politicians, by some ecologists and perhaps more importantly, by local people. Today it is a boom industry which among other effects is attracting many interested and articulate middle and senior staff from conservation agencies to the tour operators.

Foreign tourism has shown major fluctuations in all East African countries over the last three decades. This is because of changing demands for wildlife viewing versus coastal resort holidays, fluctuations in the value of local and major tourist currencies, and also international perceptions of political instability, together with a small number of widely publicised incidents of armed robbery and murder of tourists (Eltringham 1984, Mascarenhas 1983). In Tanzania, tourism has suffered in others ways as well. The national press carried heated debates suggesting that tourism may be of dubious benefit both economically and socially, and that satisfying tourist requirements uses national funds badly needed elsewhere (Shivji 1975). Maintenance of the tourist infrastructure – roads, vehicle fleets, lodges, fuel, spares, luxury food and drink, wildlife training courses, etc. – is certainly expensive, much of it in foreign exchange (Curry 1980). Only some of the benefits percolate through to non-tourism sectors. For example, the crater rim road, vital for transport of grain to the pastoral settlements in the northern Crater Highlands, has been repaired for use by light tourist vehicles but is frequently closed to supply trucks.

The government has remained committed to developing tourism, but during the 1970s and 1980s many visitors were discouraged by the relative expense and poor availability of commodities considered essential by western tourists. The post-Independence disintegration of the East African Community led among other things to a dispute with Kenya over management of (and revenue from) foreign visitors wishing to tour both countries. The resulting closure of the Kenya/Tanzania border in 1977 caused the numbers of foreign visitors to drop overnight to under one quarter of the previous total (Yeager 1982). Table 11.1 shows the figures are only just recovering. The recent reopening of the border and importation procedures rapidly brought more (mostly short stay) foreign visitors.

It is likely that the current year (1989–90) will see total tourist numbers well above 120 000, with foreigners approaching the 70 000 level reached in 1975 before border closure. Tourist lodges are running near capacity at peak times of year. There is considerable interest in further lodge development: three are planned for the Serengeti, two for Manyara and two for NCA. Both of these are being built on the Crater rim, one at Kimba overlooking the down road, another at Lemala on the opposite side to the present lodge. All lodge developments are being financed through private investment. Touring companies and vehicles have mushroomed along with tourist lodges. There is no regional or local NCA master plan to control growth. Indeed, the siting of current developments is totally against the decision of the NCAA and of the Board of Directors.

As well as the economic considerations involved, foreign tourism can cause ecological problems (Edington and Edington 1986), such as the impact of heavy tourist traffic on fragile soils in semi-arid areas, and the adverse effects of disturbance on rare predators' hunting success and survival (Henry 1977). Foreign tourism in NCA depends on the possibility of seeing scenery, wildlife, indigenous cultures and archaeological remains in an unspoilt setting and this should to some extent guarantee negative feedback constraints on overexploitation. Nonetheless disturbance, erosion, garbage disposal, fuel supply, staff numbers, the associated families and logistic requirements and their resultant impact all create problems. The Crater, as a comparatively small area which is the focus of attention, is particularly vulnerable to tourist impacts. Off-road driving (with extensive resultant scarring and erosion) as well as direct disturbance, chasing and harassment of rhino, lion, cheetah and other species (to get better views or to thrill visitors) are all in theory controlled by the supervision of trained guides. All remain problems in practice.

The indulgent attitude of the NCAA to tourist disturbance – and in the case of some of the conservation consultants the apparent lack of awareness of any

Table 11.1 *Tourist numbers, finance and revenue in NCA*
(a) Tourist numbers, revenue and expenditure 1960–1984

	NCA fees (× 1000)		Tourist numbers		Expenditure (× 1000 Tsh)[b]	
Year[a]	TShs	$[c]		[c]	Development	Recurrent
1960					815	202
1961			6 044		564	361
1962			7 394		1 185	354
1963			11 132		2 285	574
1964			12 137		1 447	708
1965			16 131		540	578
1966			23 571		70	550
1967			24 967		300	650
1968	907			45 000	1 214	650
1969	1 254	60		55 000	1 000	164
1970	1 596	60	70 000	63 000	1 200	90
1971	1 833	90		72 000	1 594	2 866
1972	3 264	120		76 000	800	1 767
1973	3 275	135		76 000	430	3 485
1974	3 608	130		68 000	432	3 647
1975	2 377	115		70 000	800	4 974
1976	2 602	120		95 000	1 404	4 061
1977	3 000	160		23 000	490	4 244
1978	1 568	50		20 000	0	4 244
1979		40		33 000		4 563
1980		60		35 000		
1981		120	[32 000]	33 000		
1982		150	[35 000]	35 000		
1983		240	[31 000]			
1984	[3 500]		[37 000]			

(b) NCAA tourist numbers from internal figures 1984–1988

	Local	Foreign	Total
1984	31 945	22 990	54 945
1985	30 252	26 989	57 241
1986	31 450	31 916	63 366
1987	36 417	41 164	77 581
1988	49 108	54 892	104 000

Notes:
[a] Date at end of financial year.
[b] Figures include operational costs other than for tourism.
[c] Data from bar diagram in Mshanga and Ndunguru, quoted by Chausi 1985.
[ ] Data from NCAA (internal document WP-16b; Jamhuri ya Muungano 1985). Other data from Curry 1980.
Tsh = Tanzanian shillings.

potential problem – contrasts sharply with the conviction expressed by these same authorities that Maasai use of salt licks in the Crater must be banned or severely restricted because of supposed adverse impacts on conservation values. The contradiction here is a reflection of longheld western attitudes to wildlife, in which traditional African inputs are somehow wrong, but western vehicles and usage are right (see also chapter 7 and J. MacKenzie 1987). These beliefs are so deepseated and strongly perpetuated by Wildlife College teaching programmes (as well as erudite consultants) that argument is counterproductive. We mention it only to point out the contradictory attitudes in NCA. Fifty zebra-striped tourist buses a day in the Crater are acceptable: one Maasai grain truck a week, bound for Nainokanoka, is not.

(b) Revenue from foreign tourism and conservation costs

Pennington (1983) found over 90% of Tanzanian secondary school pupils thought conservation areas existed because of their foreign exchange earning potential:

> The main purpose of the National Park is to attract tourists so we will get their foreign exchange

> There will be a decrease in the animals unless action is taken to educate Tanzanian citizens, especially those living near National Parks, that there is profit from having Parks

It is becoming apparent to conservationists that this view is actually dangerous for the long-term future of conservation. Conservation policies cannot be justified on the basis of foreign tourism earnings for most African wildlife areas (Bell 1987). This includes NCA.

Foreign exchange earnings from tourism in Tanzania published in the mid 1980s by the government (Jamhuri ya Muungano wa Tanzania, 1985) are of the order of 150 million Tanzania shillings per year for the whole country (at the 1985 exchange rate of 22/- T.Sh to the $ this would make $7 million approximately). Yeager (1982) gave figures of around $10 million per year for 1977–79. Green (1979) put tourism as the seventh largest source of foreign exchange earnings in Tanzania during the 1970s, but it still contributed only a fraction of a percent (around 0.02% according to the figures in Yeager and Green for total exports). By contrast, tourism contributed about 10% of foreign exchange earnings in Kenya and was consistently outdone only by revenue from coffee exports. In the primarily agricultural economy of Tanzania, foreign tourism has so far contributed relatively very little.

Tanzanian National Parks are mostly far from the coast and thus from the beach holidays currently attracting most holiday makers visiting East Africa

(Ecosystems Ltd. 1980). The northern tourist circuit of Tanzania, of which NCA forms an important part, involves a single route bottleneck pattern which is not wholly satisfactory. However, Kilimanjaro, Manyara, Ngorongoro and Serengeti are famous throughout the Western world and there is great tourist demand to visit these areas. Government currently allows tourists to cross the Kenya–Tanzania border but they must then travel in Tanzania-registered vehicles, mostly arranged by Kenyan tour operators with subsidiary companies in Tanzania. Roads are still poor and facilities few compared with the Kenya circuits.

Ngorongoro attracts by far the largest single number of visitors (over 25% of total park visits). Ngorongoro is clearly the top earner among the Tanzanian parks, and with the increased visitor numbers and the increased park entry fees (payable in foreign exchange) the absolute sums involved are considerable. Mshanga and Ndunguru (1983, quoted in Chausi 1985) gave breakdowns for the number of local and foreign tourists visiting Ngorongoro Conservation Area and the revenue collected from them (Table 11.1). In 1976, before border closure, there were around 95 000 visitors. Of these about 80 000 were foreign and 15 000 local, bringing in the equivalent of about \$150 000 and \$10 000 respectively. In 1977 after border closure foreign visitors dropped to about 12 000, eventually climbing back to around 55 000 in 1988, while local visitors after an initial fall-off having slowly grown to nearly 50 000. As with Kenya (Eltringham 1984) earnings have risen somewhat since the 1970s because of inflation and devaluation. Today non-resident tourists have to pay in foreign exchange. NCA retains some 40% of this foreign revenue as well as all local revenue. In 1987–88 total revenue to the NCAA was 163 million shillings, of which 61 million shillings was foreign exchange (\$500 000). In 1988–89 the total was 176 million shillings of which 74 million shillings was foreign exchange, and the figures are expected to rise further. These represent only the gate fees and bednight levies to the NCAA: total receipts within Tanzania would be of the order of five to ten times this value due to the attraction of Ngorongoro alone, possibly thirty times or more for the whole Tanzanian tour.

Ngorongoro, together with the Serengeti, is widely known throughout Europe and America from books, films and television. However, publicity within Tanzania is much less. The parks hotels account for only some 3% of all Tanzanian nationals' hotel bed-nights overall. The reasons for this are explored in the section on local tourism below.

Although NCA is the biggest attraction among Tanzanian wildlife areas, NCA has only recently begun to pay for its own conservation costs from tourist revenue. Bell (1987) estimates that current expenditure on National Parks in sub-Saharan Africa (excluding South Africa) averages \$50/km^2 while \$200/

km^2 would be necessary for effective conservation. NCA expenditure is probably considerably higher than average, given the size of the NCA Authority (386 staff in 1985) and the degree of international interest. The figures presented in recent internal reports do not quite tally. However, the NCAA budget 1981–84 averaged 12.5 million TShs (Kayera 1985). Expenditure 1983–84 was around 12 million TShs of which government subsidised around 40% and tourist revenue provided around 32% (Kayera 1985, Serengeti workshop 1985 document WP-161b NGO-REP 2). External donations contributed about 1% and there was a deficit of over 20%. Of the total expenditure over 40% went on salaries. In a number of cases money set aside for development projects such as upgrading roads went to NCAA salaries. Most recently Cobb (1989) quotes a 1987 budget of 40 million TShs of which 'the great majority . . . originates from tourism'. This would imply a considerable increase over the 1980–85 budget, though much is due to a greatly devalued shilling. Today, all NCA revenue comes from tourism and for the first time government paid no subsidy in 1988–89. Half the foreign exchange receipts were paid directly to government and the increased tourism in the whole country must have led to an influx of foreign exchange. However, we do not know the cost of maintaining tourist facilities, lodge management, vehicles, fuel, nor indeed of any study of the macro-economics of luxury tourism in East Africa. There is little understanding of the opportunity costs that conservation entails, nor of the social impacts. The conservation policies of NCA impose serious constraints on local people. Most expenditure in African conservation areas is devoted to neutralising opposing interests by law enforcement and public relations: 'under existing wildlife legislation in much of Africa normal rural existence is impossible without breaking the law' (Bell 1987). This is borne out by a recent detailed analysis (Leader-Williams and Albon 1988). The revenue allocation scheme that operates in NCA does not filter back to the people most affected. Few Maasai are employed by NCAA, hotels or service industries, or involved in cultural activities or handicraft sales for tourists (some 75 out of 18 000 – Arhem 1981; see also chapter 9 and later section). Tourism to date has brought the Maasai few benefits. The fall in tourism led to the deterioration of the rim road to Nainokanoka and of the continuing road to Empakaai Crater. This had a major impact on Maasai subsistence by affecting transport of grain and other supplies, and stock marketing. The recent repair of this road allows tourist vehicles to pass, but not grain trucks.

Tourism developments are as likely to damage as to improve matters for Maasai society in Ngorongoro. With the current downward spiral in their living standards, some NCA Maasai, especially younger individuals, are learning that tourists will pay them comparatively large amounts of cash for impromptu displays of their dress, their dancing and other attractive aspects of their culture. This could rapidly grow to a level of roadside soliciting as damaging to Maasai culture as it may ultimately prove unwelcome to tourists. Other Maasai may see involvement in organized poaching as a ready source of much-needed cash. Traditional Maasai social control, largely vested in the elders, is eroded by socioeconomic changes that give younger or less responsible individuals an independent cash income. The progressive replacement of traditional forms of social control by newly evolving networks as new money finds its way into the system is a worldwide phenomenon. It is not of more specific concern to the NCA Maasai than it is elsewhere, and it does not necessarily herald a problematic loss of values nor the emergence of a damagingly exploitative approach. What is of concern is the emerging pattern of exploitation of the Maasai as a tourist attraction, while denying their rights both of land use and of the financial benefits derived from the tourists they help to attract, and the general disregard for any damaging effects on their culture and subsistence. At the root of this problem is a Conservation Area Authority which has no mechanism of dialogue with the Maasai, nor any extension staff capable of counselling and influencing directions of change in the light of local and national goals, and of experience with comparable phenomena elsewhere.

(c) Local tourism

Bell (1987) suggests that local tourism and local appreciation of wildlife should be given much more weight than they are usually accorded. As well as fulfilling often-overlooked local demands for the aesthetic experience offered by conservation areas, tourist developments can double as excellent educational facilities. They illustrate for example the need to conserve not only wildlife, but also water and soil resources, in an area such as NCA which supplies clean and reliable water to adjacent areas year round from its forested catchments.

However, visitors to the Parks must go by vehicle, and comparatively few Tanzanians possess these. There is an all-weather road with bus services for most of the year, but commercial buses are rare and unreliable. While entrance fees for nationals are modest, lodge rates are very high. A dinner for one at Ngorongoro Lodge would usually cost around 30% of one month's salary for any professional Tanzanian. A full board night for one couple would cost over two months' salary. There is no accommodation for students, hitch hikers or the great mass of ordinary Tanzanians. Wildlife and wilderness viewing is not a common pastime. In Africa 'because of the extended family system, spare time and extra money are usually spent with relatives: recreation is a low priority, even with those who can afford it' (Lusigi 1980).

Twenty-eight years after Independence, all tourism promotion and policy is still directed at foreigners. The casual local Tanzanian visitor is neither planned for nor catered for in NCA. Fosbrooke developed a youth hostel in NCA in the 1960s. This burnt down in the late 1970s and has not been reconstructed. There are plans for new luxury lodges, but no thought has been given to low cost hostels (though see Taylor 1988). This failure to encourage educational tourism for Tanzanian schoolchildren and adults has long-term implications for conservation. Future governments less committed to conservation, or in a time of waning foreign tourism, might see conservation in general and the NCA in particular as an expensive luxury that Tanzania can ill afford.

NCA is without doubt the single greatest attraction among the Tanzanian wildlife areas for both foreign and local paying visitors, even if this does not mean much in terms of revenue. It is clear that NCA has an important part to play in Tanzanians' appreciation of their national heritage, as well as its recognised role as a World Heritage Site. The obvious benefits of watershed conservation, and the intangible values to both local and international communities of wildlife and wilderness conservation, are justification for the continued existence of NCA. However, conservation will only work in the long term if the management system gains acceptance from the local and national community (Bell 1987). Where these communities see no obvious economic benefit, acceptance is more likely to require compromise with forms of land use other than the foreign tourism role of conservation areas (Bell, 1987; Abel and Blaikie 1986). This has two main implications for the future of conservation in the NCA. Firstly, local tourism must be encouraged, both for recreational and for environmental education purposes. Secondly, conservation management must seek compromises with alternative and compatible forms of land use. While these may exclude extensive cultivation or commercial wildlife harvesting for meat, they obviously include pastoralism. They may even allow subsistence or trophy hunting. The next section investigates the type and extents of developments in pastoralism and wildlife exploitation that have been suggested for this or other joint conservation/human land use areas.

Wildlife exploitation

East African government and conservation agencies have in the past tended to see wildlife conservation and wildlife harvesting as mutually exclusive, and viewing as the only possible use of wildlife compatible with total conservation. However, the history of conservation in England and Africa, as well as current conservation practice in South Africa, Zimbabwe and Malawi, show that wildlife conservation and hunting are interdependent in many ways.

The history of wildlife conservation in Tanzania's Maasailand includes several attempts at what has often been called rational wildlife development – using the wildlife, as well as looking at it. These utilisation projects vary from controlled trophy hunting to large- and small-scale cropping for meat and skins. Such developments are now advocated as part of an integrated and long-term conservation approach (East 1988). This section looks at the possible integration of conservation and hunting schemes in NCA, whether at the level of commercial harvesting for meat, subsistence hunting for meat or hunting for trophies.

(a) Commercial harvesting

The idea of a ten percent cull from over a million animals a year seems tempting, with the added attraction that it apparently offers something for nothing. Wildlife cropping operations have been discussed in some detail, for example by Parker (1981, 1985) and Eltringham (1984). Constraints on and potential for wildlife cropping in the NCA/Serengeti region has been specifically analysed in background papers presented to the Serengeti workshop by Parker (1985) and Ndolanga (1985), summarised in Malpas and Perkin (1986). The fact that the NCA/Serengeti herds migrate over some 25000 km^2 of wilderness makes them economically impossible to exploit with a processing unit that could provide the standards of hygiene (water, cooling, fly- and dust-free conditions) and transport necessary for a commercial operation. Alternatively, a fixed unit could only operate a few weeks a year. The social organisation of the wildebeest would mean massive disruption for even a low level of cropping. Game meat operations face strong resistance from the established butcher trade and demand is typically restricted to a small luxury tourist market. From experience in this and other East African areas it is generally agreed that commercial cropping can only succeed where small mobile cropping operations produce a small supply of game meat. There are further preconditions: different (lower) public health standards must apply to the game meat and it must be targeted at a tourist or luxury export market. The latter is dwindling with new EEC game meat regulations, and any project must concentrate on trophy values, such as zebra skins, that bring much higher returns than meat (e.g. Wildlife Services Ltd. experimental cropping in Loliondo GCA: Parker 1981, 1985, Bindernagel 1977). Although no details are available on the current Tanzania Game Division TAWICO cropping operation, its early phase from 1970 to 1973 produced similar conclusions. It now crops largely for zebra skins, which subsidises the limited harvesting of wildebeest.

There is a strong argument for leaving the vast herbivore populations of the NCA/Serengeti subject to natural biological processes (Sinclair 1983b). There are no ecological reasons to cull. With the possible exception of trampling in the Angata Salei corridor, the animals are not causing habitat degradation, and this area could be partly protected by the Maasai resuming their former practice of fencing the Angata Kiti and curbing the wildebeest influx, thereby protecting the soils as well as reserving some high quality grazing for cattle. There is no protein-hungry population in the immediate vicinity that could be efficiently and satisfactorily fed from the culled meat. Within NCA a limited cropping operation should be confined to supplying the local tourist lodges with meat and skins for immediate consumption and sale.

(b) Subsistence hunting

Subsistence hunting is very much a part of most African traditional cultures. The pastoral Maasai are an exception, having a cultural distaste for game meat that amounts to a general prohibition except in times of famine. However, even the pastoral Maasai have cultural, kinship and subsistence links with Dorobo hunters (chapter 3). With the growth of wildlife conservation in this century subsistence hunting became illegal in much of East Africa, or subject to licenses barely attainable by the common man. However, subsistence hunting using traditional techniques in general has a negligible impact on wildlife populations in conservation terms. Its prohibition cannot be justified on the basis of conservation needs. MacKenzie, J.M. (1987) following on Graham (1974) makes a fascinating case for the original emergence of wildlife conservation in Africa as a direct extension of the ideology of the European tradition of the Hunt, with hunting becoming the privilege and the symbol of power and wealth, and the exclusion of the lower classes or subject races being an integral part of a ritual process. Hunting by local people was regarded as neither sporting nor sustainable, though hunting by the colonial elite was seen as noble, manly and even environmentally beneficial. In practice access (or right to hunt) was denied firstly to natives and secondly to lower-ranking colonials. Such rights were eventually restricted to top-ranking officials, visiting foreign government elite and foreign royalty, the new elite of scientific research workers, and those wealthy enough to buy access (MacKenzie 1987).

This situation has changed. Conservation in Africa may be viable in the long term only if conservation resources can be used and appreciated by the local community for their aesthetic, cultural and/or economic values. The local community commonly forfeits land and traditional uses, and sometimes risks crop loss or even physical injury because of the conservation area (Bell 1987). Many of the costs of conservation, but few of the benefits, accrue to the local community. Local people continue to place a high value on access, gathering and hunting as part of their traditional cultures and everyday subsistence (chapters 6 and 7, Chamshama, Kirkhof and Singunda 1989). This attitude should be acknowledged and enlisted in support of conserving wilderness areas, not denied so as to generate anti-conservation feeling and action.

Current illegal subsistence hunting in NCA carries no threat to conservation values (chapter 7). In theory it should be possible to legalise, license and thus control subsistence hunting in a number of peripheral zones in NCA. Bell (1987) and Martin (1986) discuss projects of this sort designed to bring both conservation and wildlife utilisation back to community level in Malawi and Zimbabwe, and the Tanzanian Government is preparing detailed plans for similar schemes in village lands around wildlife areas such as the Selous Game Reserve. While some zones of NCA should remain sacrosanct (such as the Crater and plains) other parts such as the Endulen–Kakesio zone and the Forest Reserve might benefit from the higher level of patrols, and from Maasai cooperation in control of licensed hunting. This would only follow if the NCA Maasai perceive genuine gains from the issue or use of hunting licences, but it could mark a return to the kind of anti-poaching cooperation described by Fosbrooke as operating between Maasai and NCAA in earlier times.

(c) Trophy hunting

Bell (1987:93) feels that not only subsistence hunting, but also ivory hunting should become the legal preserve of the local community. He calculates that by making the trade legal both the hunters and the conservation agency benefit economically, and the local community has a strong incentive to police the area. With their tradition of roving *murran* groups the Maasai could patrol very effectively, while the rangers (with current low pay, low prestige and low morale) do not. However, given the conservation status and symbolic value of rhino and elephant it is perhaps unlikely that such a system could or should be operated in NCA. New Convention on International Trade in Endangered Species (CITES) agreements governing exploitation and trade preclude organised ivory hunting. Trophy hunting for other species should nonetheless be given consideration, and the Malawi parallels described by Bell (1987) should be monitored and evaluated carefully. The future of conservation areas throughout Africa generally will come to depend more and more on the extent to which the local communities can enjoy both tangible and intangible values associated with the area (Bell 1987), and NCA is no exception.

Integrating land use in NCA

Several forms of land use in NCA have now been considered. The demands made on resources by pastoralism, wildlife viewing by foreign and local people, commercial and subsistence harvesting of wildlife for meat, and ivory hunting have all beeen described, as have their potential economic and ecological results. Some forms of land use are clearly incompatible with successful conservation in NCA, most notably large-scale cultivation (see chapters 9, 10) and current levels of ivory poaching. However, of the range of land uses that NCA can support, what is the best combination and compromise? Which are likely to be mutually compatible, or even to reinforce one another's success?

Wildlife viewing is generally accepted as fully compatible with conservation. However, foreign tourism may not be sufficiently reliable in NCA to justify exclusion of other forms of land use from areas other than the craters. Local tourism is vital for conservation awareness and must be developed, but has been severely limited by constraints of finance, transport, accommodation and traditions of leisure pastimes. Money is now available in NCAA to establish a base for Tanzanian tourism and conservation education. Wildlife harvesting for meat is not practicable or desirable on a commercial scale. However, subsistence hunting for meat if regulated could be not only compatible with but even a potential incentive for wilderness conservation and anti-poaching vigilance by the Maasai. Trophy hunting could be feasible in some areas, for example Endulen. Pastoralism is highly compatible with wildlife and wilderness conservation. Small scale cultivation of boma scars by pastoralists if regulated could be both compatible and desirable in terms of easing dry season subsistence. Large-scale cultivation by contrast excludes wildlife completely as well as bringing about quasi-permanent loss of natural vegetation, especially woodlands, and must be prohibited.

The conclusion that emerges is that wildlife conservation and pastoralism must continue to coexist as the central forms of land use in NCA. Despite areas of conflict they complement and reinforce one another's claim to NCA resources. Conservation has long-term and global-scale worth; wildlife tourism has short-term financial profit; pastoralism has both immediate and lasting local returns. All three dictate – and justify – the exclusion of large-scale cultivation. Hunting and cultivation on subsistence scales could and perhaps should be planned for buffer zones in consultation with the pastoralist community. Given that pastoralism must remain a central form of land use in NCA, and that NCA Maasai do currently face considerable subsistence problems, the next chapter goes on to look at possible technical interventions that might be both compatible with conservation and desirable in terms of raising productivity and easing subsistence.

Conclusion

With the synthesis of NCA conservation aims, Maasai land use and more general development processes that this book sets out, it is unthinkable that the Maasai should be excluded from NCA. Pastoralist land use presents no threat to wildlife populations or the environment in NCA. Pastoralist development in NCA will concentrate on making subsistence more secure rather than on pursuing commercial offtake. This general aim is entirely compatible with conservation. On the political front, if the conservation values of NCA are to survive the Maasai must be involved at all levels. It is their occupation of the area together with the wildlife that ultimately justifies its retention as a grazing system rather than its conversion to either Disneyland or cultivation in a relatively poor and highly pragmatic developing nation. The Maasai respect for wildlife and the strong aesthetic as well as practical sense of their environment are such a natural basis for local conservation support that it is counterproductive as well as hypocritical and unethical to exclude them.

Foreign tourism has very recently become a major source of foreign exchange. However, revenue from foreign tourism has only just begun to cover the costs of conservation administration, and this does not take into account the opportunity costs of other potential forms of land use in NCA, particularly the conservation-compatible and environmentally sustainable livelihood of 25 000 people. In the context of a developing country with limited resources, compatible forms of land use, notably pastoralism, must be encouraged rather than barely tolerated.

Local tourism has been neglected, but its development will be necessary to raise national awareness of the wider benefits of conservation. Utilisation of wildlife through licensed subsistence hunting should be investigated for specific zones of NCA, particularly those currently most heavily exposed to subsistence poaching. The legalised exploitation of ivory is altogether a less likely development as is the commercial harvesting of wildlife for meat. Any development projects should be designed jointly with the Maasai and should bring benefits to them as much as to the Conservation Authority.

12

Development interventions

Murua rraga mimanya Stay uninhabited
(Maasai wish or blessing, spoken to the landscape: Kipury 1983)

Given that any lasting development in the pastoralist system is likely to be towards a more secure subsistence rather than towards intensive commercial production, could technical inputs make a useful contribution that is also compatible with conservation? Our study of livestock performance suggested that NCA livestock mortality and fertility rates as well as milk production are average for subsistence pastoralism. Declining exchange rates of stock for grain, deteriorating conditions of transport and grain supply, together with the general ecological good health of the NCA, suggest that subsistence problems arise more as a result of the lack of access to grain, the growing population and the adverse market conditions than from any inherent inefficiency in NCA Maasai livestock production. However, it might be possible to solve some of the problems of pastoral subsistence in NCA by improving husbandry and introducing technical interventions. Even if livestock development for intensive commercial offtake is unlikely, development interventions might make Maasai pastoralist subsistence more viable within the constraints of conservation requirements and policies. Are there aspects of management of water resources, range, grazing, breeding control, breed improvement, disease management, or marketing of milk, animals and other produce that could be improved while remaining compatible with conservation? In this chapter potential technical and administrative interventions are examined insofar as they affect Maasai subsistence ecology, and the wildlife and other conservation values of NCA.

Water development

Miarakinoyu enkare oloipang'i Water cannot be forced up a ridge
(Maasai proverb: Kipury 1983)

Page-Jones (1948) gave an early account of the natural water resources in Tanzanian Maasailand and reviewed works carried out during the period 1920–44. Most recently, Cobb (1989) has updated a previous comprehensive report by Kametz (1962), cataloguing springs, lakes, waterholes, pans, and water developments as well as the main areas of traditional Maasai-dug and maintained wells and troughs in NCA. Water sources in different land units show major differences of quality, quantity and permanence. In particular, comparatively high water quality and quantity is found in the Crater Highlands compared to the plains.

The water supply systems constructed by the Serengeti Compensation Scheme, and later by the NCAA, may be classified as those supplying NCAA staff and tourist facilities, those for livestock and wildlife, and those for domestic use by the Maasai. Cobb (1989) inspected 22 of the 29 supply systems constructed by or for NCAA. Of the ten found to be functioning, three supplied NCAA and tourist facilities. The remaining three domestic and four livestock water supply systems currently supplying Maasai pastoralists and their herds are clearly inadequate for a population of some 22 600 people and 275 000 livestock. However, water development was not seen as a top priority by most NCA Maasai, except where pipelines already existed (Cobb 1989) and where there had been considerable population increases since the 1960s (Arhem 1981a). Cobb (1989) found that most NCA Maasai saw water development as secondary to their main problems of WaSukuma raiding (Osinoni–Kakesio area), MCF (Ndutu–Olduvai–Angata Kiti–OlDoinyo Ogol area) and *Eleusine jaegeri* spread (Highlands). Arhem (1981a) found that over half the NCA villages considered current water supplies to be adequate. However, Oloirobi and Olbalbal, with considerable population increases, and many settlements 10–15 km from permanent water, reported a widely felt need for new or rehabilitated water supplies. In Kakesio the current dry season water supply depends on wells hand-dug in dry river beds, and the Kakesio Maasai were reported by Arhem (1981a) to be seeking construction of a new dam. However, more recently Cobb (1989) reported strong reservations on the part of the Maasai about even repairing the three currently nonfunctional major hafirs (excavated water tanks) near Kakesio, as this zone is chronically vulnerable to intertribal raiding, and 'nobody would dare take his cattle there'. In Alaililai where around 80 scattered bomas depend on one waterpoint, the Maasai seek rehabilitation of the six additional pipes originally installed (Arhem 1981a,b, Cobb 1989).

Water development to date has been piecemeal, coming under several authorities (NCAA, MLRDP, the District Council, etc.), sometimes with specific funding for particular developments, more often competing with other projects for funds from a mixture of sources (government, tourist revenue, foreign government aid organisations, missions and charities). Any future water programme needs coherent planning, funding and implemention. Data on hydrogeology (aquifers and water tables) and hydrology (precipitation and stream flow) are not adequate to plan a coordinated programme of water development. However, some of the necessary data collection is being established under the NEMP, and Cobb (1989) discusses guiding principles for any such future programme.

Sandford (1983) stresses the importance of taking advice from the local pastoralists on such matters as flood levels in stream beds, dry season drawdown levels in shallow wells, dry season stock distribution, physical facilities needed for watering stock, etc. Their wealth of experience and management information on the topic compensates for the lack of long-term survey and monitoring data and improves the chance of designing appropriate inputs based on short term surveys in a highly variable environment. The Serengeti Compensation Scheme water developments in NCA perhaps suffered less from lack of consultation with local pastoralists over such issues than have projects elsewhere, thanks to the role of Henry Fosbrooke as both anthropologist and administrator in NCA. Cobb (1989) stresses the need for consultation with the Maasai over water development in general. We go further to recommend a study of their considerable environmental knowledge of NCA to fill in for the absence of long-term data on hydrology.

Water development must ensure separation of domestic water sources from those for livestock and wildlife. Any water delivery points must be designed in the light of past experience to avoid damage or pollution by livestock, wildlife or people. Domestic water points must deliver water of acceptable quality, both bacteriological and chemical, particularly with respect to the salinity and fluoride content. These are dangerously high in any water from permanent springs or deep groundwaters in the plains to the west and northwest of the Crater Highlands. Development must take into account that demand increases with supply, and that domestic demand increases with the changing lifestyles brought about by development, particularly with sedentarisation. Water will be a key issue for the new lodges currently planned for the Crater rim (see for example Fosbrooke 1975).

Water development for livestock (as separate from human use) is a controversial issue. Cobb (1989) suggests a phased programme concentrating initially on rehabilitating failed past developments, and specifically recommends that no separate provision be made for wildlife as apart from livestock watering points.

Livestock-oriented water development projects have been unsuccessful more often than not, and the same mistakes have been repeated in the same areas time and again (Sandford 1983:63–4). Typically such projects either fail to deliver water of adequate quality or quantity; or create problems of local sedentarisation, overgrazing and environmental degradation; or entail problems with loan repayment (e.g. Page-Jones 1948 and Jacobs 1978 for Maasailand). Once established, pumps and boreholes rapidly fail due to lack of fuel, spares and/or maintenance skills. Cobb (1989) lists water supply systems which were non-functional by the time of his survey.

However, our study suggested that the energy loss incurred by long treks to water (and by the consequent restrictions on time spent feeding) was the single most important constraint on milk production, particularly in Gol but also in Ilmesigio and to a lesser extent in Sendui. Reducing this constraint would release resources for useful production (milk, meat and calves) or at least lessen the energetic costs to animals barely surviving on submaintenance diets. Even where livestock water supply systems deliver water too saline or alkaline for long-term use, or of limited quantity, they nonetheless may extend the use of valuable grazing for a few weeks, critical to dry season stock survival. This was the case for example during our study near Lemuta borehole in Gol during the August 1981 dry season. This borehole has since been blocked up (Cobb 1989) but at the time of our study it provided livestock with water for some weeks while water suitable for human consumption was fetched from surface pools 10 km distant.

Over the last 25 years since the Serengeti Compensation Scheme water points were installed, intermediate technology has produced new options. Getting fuel and spare parts is likely to be a recurrent problem in Tanzania for the foreseeable future. However, low-maintenance wind pumps recently installed (Rodgers 1982b) and functioning satisfactorily (Cobb 1989) can only circumvent to a limited extent the problems surrounding boreholes such as Ndureta. In late 1988 the Ndureta windmill failed to deliver water. It was not until seven months later that technicians discovered the simple leather seals had worn out. Maintenance is a human problem of responsibility, interest, supervision and training through appropriate extension services.

Opening water points causes overgrazing and trampling in the immediate vicinity, and may bring about range degradation over a wider area as a result of unprecedented growth in livestock numbers and extension of the period of use. Some vegetation changes inevitably take place but these are by no means necessarily disastrous (Sandford 1983:76; Homewood and Rodgers 1987). A 'sacrifice area' or circle of over-use appears over a limited radius around the waterpoint. This is usually a few hundred metres, but may stretch to a few kilometres during drought and dieoff periods in arid areas. The effect is minimised by the Maasai custom of siting bomas several kilometres from waterholes and moving their stock to water every second day, so maintaining wider access to adequate grazing (cf. some Indian pastoralists relying on forest forage resources – Rodgers 1988). Cobb (1989) reports a brief survey of ground cover which suggests that such impacts are not a serious problem around NCA waterholes, as might be expected from the more general research on resilience and stability of these grasslands (Chapter 6).

In general the Maasai users of any one NCA watering facility currently form a sufficiently small and well integrated group to resolve most problems of shared use internally with a maximum of cooperation. Sabotage and vandalism of water development inputs have been rare in NCA (though in the 1960s *murran* punctured the gravity feed pipe supplying rival groups in the Embulbul Depression). NCA conservation policies generally preclude the establishment of potentially problematic outside commercial concerns, cultivation or settlement, which often monopolise water developments elsewhere. The major exception to this so far has been the siting of administrative and tourist facilities, which may compete not only with Maasai but also with wildlife for scarce and localised water resources (e.g. the tapping of Lerataat to supply the tourist lodges, NCAA HQ and bomas on the Crater rim – Fosbrooke 1975; current plans for future Crater rim lodges).

Range and grazing management

Range management is a further area for possible technical intervention. The tremendous diversity of rangeland types is perhaps NCA's most important asset as a grazing resource, allowing utilisation across different seasons and conditions. Natural grassland productivity is also outstanding (chapter 6). Sandford (1983:88–168) gives a review of possible methods for improvement of semi-arid rangeland. Most involve quite drastic measures of bush clearance, fertiliser or selective herbicide work. It is hard to see how any one of NCA rangeland types could be much improved by such methods without affecting conservation values, to say nothing of the economics and practical problems involved. The exception to this is the use of fire. Burning is a traditionalist pastoralist range management tool which was forbidden by the NCAA during the 1970s. The benefits of controlled burning in limiting disease-bearing tick populations, releasing nutrients from old unpalatable forage, stimulating new pasture growth, and forestalling destructive wildfires are widely accepted (chapter 6). There are also potential problems in the use of fire, with forest edge encroachment effects as well as potentially undesirable changes in runoff and soil erosion. However, much of the present NCA (and Serengeti) landscape is the result of centuries of pastoralist burning as well as livestock and wildlife grazing. Fire-associated changes in species composition and productivity are beneficial to livestock and large ungulate populations. Both livestock and wildlife populations would undoubtedly benefit from renewed careful use of burning within NCA (chapter 6). It could be a significant factor in controlling disease-bearing ticks in Ilmesigio and in generating palatable young growth in *Eleusine* tussock grassland in Sendui.

Other than burning, traditional management of range resources involves moving stock. There is no evidence that any system of paddocking and rotation could improve on traditional patterns of migration and transhumance (Sandford 1983:102–4) despite the enthusiasm of administrative staff for such systems (Ole Kuwai 1981, Chausi and Robberecht 1985). The USAID-funded Maasai Range Development Programme and the Kenya group ranches were experiments in the application of western ideas of range and livestock management to Maasailand. They involved subdivision and definition of user groups, and of wet and dry season grazing areas, together with provision of new water development projects, imposition of agreed stocking levels and in the Tanzanian case, establishing group commercial herds. Their failure to bring about the planners' aims has been discussed in chapter 10.

Recently a number of NCAA staff have suggested specific grazing plans or policies. Ole Kuwai (1980) in a background document for the management plan study draws the parallel between current stock movements and wildlife migrations, and points out that the traditional transhumance system (as in Ilmesigio and Sendui) is equivalent to a two pasture grazing plan. He proposes the formal separation of stock and wildlife zones with a reduction from 13 to seven of Dirschl's land zones open to livestock. He goes on to suggest an alternative grazing plan based on the 'South African switchback plan'. This involves two pastures, each of which is grazed and rested for specified and alternate three and six month periods, but he offers no rationale for this. He also outlines a flexible rotation scheme which would place all animals in one unit until they have used up 50% of available forage, moving them to the next until again they have used 50% forage and so on. This would be combined with a rotation of stock species. Once again there is no clear rationale for this suggestion.

Chausi (1985) suggests a centripetal rotation grazing programme whereby NCA would be divided into short, medium and tall grasslands each of which would be grazed in specified seasons. The scheme does not seem to allow for the wildebeest and MCF problem, but in other ways closely resembles what happens currently on an informal basis at least for Ilmesigio and Sendui bomas. Specification of particular transition dates and boundaries would seem merely to remove much-needed flexibility without conferring any obvious advantage. Chausi sees implementation of some such programme going hand in hand with shifts from milk to meat production, from 'year-long livestock grazing to holistic range resources management', to breed improvements, sanctions against immigrants, and a number of other development policies investigated in detail below.

No advantage has been demonstrated for formal grazing or paddock rotation schemes over the traditional flexible transhumance and migration patterns (Sandford 1983; Jahnke 1982). In practice no such schemes are likely to be implemented in NCA in the foreseeable future. It is very likely that some currently restricted areas will remain so. Thus livestock are likely to remain banned from all of Olduvai other than for access to dry season wells, and from Empaakaai and Ngorongoro Craters (chapter 7). Despite strong resistance to the idea by traditional conservationists, controlled dry season grazing is now seen as feasible in the Forest Reserve, but would require clarification of the permit system and careful monitoring (e.g. Chamshama, Kerkhof and Singunda 1989). Olmoti Crater is valuable dry season grazing and should be opened officially to pastoralist use. Effects should be monitored at least for a trial period.

Sandford (1983) concludes that grazing management is most effective where those who own and herd the livestock are also the decision makers. Where this is not the case there are likely to be major conflicts and problems of implementation. This is so in the case of NCA, where the Authority has at times considered taking over such decisions, as well as currently restricting or banning use of specific areas. Social and legal considerations form a major component of grazing management. At present, other than restricted access conservation areas such as the Crater, NCA is treated as a unit. There is no official recognition that movements of both people and livestock (and indeed wildlife populations) necessarily cut across both 'village' boundaries within NCA and more importantly across the boundaries delimiting NCA from adjacent parts of Arusha Region. These issues become important if and when management policies begin to try to limit the numbers of and utilisation by people and stock in specified areas. This brings us back to the question as to who is entitled to use grazing resources within NCA, and with what degrees of differential rights, and who should regulate this use. Since its inception NCAA has had a paternalist approach; there is a strong need for the development of self-regulation on a village basis. The problems surrounding this question have been discussed in earlier chapters. They involve issues of social, political and administrative organisation, and of equity, that are touched on in the final section of this chapter.

Livestock development

(a) Breeds and breeding management

The relative merits of local and 'improved' breeds in an environment such as NCA have already been discusssed in chapter 8. Breed improvement has already been attempted in NCA with provision of 'improved' breeding bulls at subsidised prices. Wealthier herdowners are willing to experiment but so far there are problems with reduced disease and drought resistance among 'improved' breeds (cf. Trail and Gregory 1984, Western and Finch 1986). Some owners of crossbreeds report no difference in yield under normal management (Field, Moll and Ole Sonkoi 1988). The Highlands could offer limited scope for breed improvement linked to intensive beef or dairy production. This would be at the expense both of conservation values and of the broader pastoralist system, and for a number of reasons is unlikely to be profitable (see sections on ranching below). Breed improvement is unlikely to provide any immediate solutions in NCA.

The Maasai control which male stock are castrated, which bulls in the herd have access to oestrus cows, and also potentially determine when cows are allowed to breed. Further restrictions on breeding season might possibly enhance calf survival and milk yields, but any advantage would be outweighed by the loss of some spread of calf births with its benefits of risk aversion and year-round milk yields.

(b) Livestock disease

This is a major problem in the middle altitude areas such as Oloirobi, Ilmesigio and to a lesser extent throughout NCA. Currently there is an unsatisfactory or even dangerous transition situation whereby traditional methods (e.g. burning to reduce tick populations) have been discouraged while veterinary medicine and particularly dipping are not adequately established and maintained (chapter 8; Sutherst 1987; Machange 1988). The deaths of half the Oloirobi boma herd over a two-year period were largely attributable to this loss of control over disease transmission (chapter 8; Rodgers and Homewood 1986) as was the complete loss of the NCAA dairy herd (Field and Moll 1987, Field, Moll and Ole Sonkoi 1988). There is certainly room for technical intervention here, but the economy of Tanzania and of NCA in particular is not yet able to maintain such services effectively in the long term. A return to traditional methods, especially burning, and where possible resumption of more complete transhumance, is likely to provide the best and most immediate solution to tickborne diseases among other problems (chapters 8,9).

(c) Ranches and dairy farms

A state-run beef ranch was set up in the 1970s just inside the southwest border of NCA. Chausi (1985) and Field and Moll (1987) among others see meat production as the desirable course of development in NCA. Could the future of NCA Maasai lie in such development, zoned to dovetail with conservation needs? Raikes (1981) gives an overview of the history, current status and future prospects of such enterprises in East Africa. Well over 95% Tanzanians depend for their supply of meat on the offtake from the traditional herd kept by agropastoralists and pastoralists. However, the majority of expenditure on livestock development since independence has gone on large state and parastatal ranches and dairy farms. A comparatively small proportion goes on technical services and marketing for the 99% of the national herd not kept on ranches. Despite this funding, Tanzanian ranches run at a loss (Raikes 1981:168). Jahnke (1982) shows that newly-established ranches such as those in Tanzania do not currently show levels of productivity any higher than pastoralist systems, and are unlikely to do so in the foreseeable future due to lack of personnel experienced in ranch management, lack of incentive, frequent personnel transfers, and lack of established infrastructure compared to long-established ranches. De Leeuw, Bekure and Grandin (1988) document running costs and production indices for commercial ranches compared with pastoralist concerns in Kenya Maasailand, and show that while productivity is very similar the commercial ranches which have to pay for labour and capital goods end up with a low net income by comparison (chapter 9; Table 9.2). Raikes (1981) acknowledges that there is a place for a small ranch sector in Tanzania (perhaps a quarter the current number of ranches) supplying the tourist and upper income urban butcher trade. NCA is remote from all markets other than its own hotels with their seasonal tourist influx. There is no case for a beef ranch in NCA on commercial or demonstration grounds. The NCAA demonstration beef ranch at Endulen lost 900 out of 1000 animals due to disease. It runs at a loss and has no educational value.

The NCAA has run a dairy herd with milk yields almost ten times greater than those of the pastoralist cattle. Yet it has not been a success, whether on veterinary or on economic grounds. Could NCA support a thriving dairy industry compatible with conservation? For geographical, ecological and historical reasons examined by Raikes (1981) the dairy industry has never established itself on any national scale in Tanzania, and there are few grade cattle in the country. The siting of Tanzanian areas of dairying potential, and of NCA in particular, precludes transport of most dairy products to the capital and to other administrative and industrial centres, by contrast with the situation in Kenya. Even with intensive management of the vegetation, only a small part of NCA could support grade or even partly improved dairy cattle without major capital intensive investment, and such cattle are not readily available. Commercial dairying is unlikely to be a viable development in NCA in the foreseeable future for milk or for stored products such as cheese and butter.

However, subsistence might be made more secure for milk based pastoralists by developing forms of milk processing and storage to even out seasonal availability of subsistence dairy products. There is rather little milk surplus from Maasai cattle even during wet periods, which may account for the virtual absence of traditional use of stored products. Hand churns for local butter and cheese production are a possible development, although problems of restricted scale, and virtual absence of storage and transport facilities, would limit any such production to local use only. Olmakutian (Ilmesigio) Maasai enquired about hand churns with interest during our study. Jacobs (1978) mentions the extensive use of such equipment in the 1950s, as does Hoben (1976) and Field, Moll and Ole Sonkoi (1988) recommend their reintroduction.

Marketing of livestock and livestock products

Tanzania, unlike Kenya, is hampered by an unfavourable geographical distribution of areas of cattle-rearing potential relative to markets and points of export (Raikes 1981). In Tanzania the colonial government attempted to increase yields from African herds by setting up or expanding local markets, initiating grazing controls and destocking campaigns, providing water development programmes and introducing compulsory dipping. Prices never provided much incentive to sell in Tanzania. In Kenya, the colonial government tried hard to exclude Africans from protected European economic activities such as dairying and quality beef production. Raikes (1981) contrasts the different outcomes in the two countries. African smallholder livestock production and net exports of livestock products have expanded rapidly in Kenya since Independence, but Tanzanian livestock production both for meat and milk has stagnated or declined while imports have risen. This bears out the idea that:

> Exclusion from a visibly profitable activity is a much more effective stimulus to production than encouragement (or enforcement) to produce what is not profitable. Raikes 1981

Raikes (1981) suggests that a more appropriate marketing policy could have generated a better national meat supply from the traditional sector and points out that the established practice of effectively taxing local sales to subsidise the parastatal ranch sector is the worst possible policy.

From the government's point of view, marketing in the traditional sector (such as NCA Maasai livestock) is difficult because of the type and quality of animal provided; the incidence of disease; the enormous variations from season to season and year to year in supply of animals; and the costs of collecting animals from, and disseminating information to a dispersed population. The political weakness of the Maasai is also a factor: they have limited representation and little influence (chapter 3) and as such their livestock marketing has received no positive incentives. NCA is particularly inaccessible compared to other livestock areas. Currently it contains only perhaps 1% or less of the Tanzanian cattle population (Jahnke 1982, Raikes 1981) in contrast to more heavily stocked areas with better communications such as Sukumaland. Auctions organised by the NCAA are meant to be held in regular rotation in three NCA villages. In practice buyers often fail to attend, when they do only a proportion of the stock brought for auction are actually sold, and not surprisingly herdowners have lost faith in the system. Prices are low and fees high. Sandford (1983) quotes government fees of 2–8% (averaging 4–5%) of the cost of the animal for official auctions in Tanzania in the 1970s, rates double those charged by private traders. Government fees tend to be high because they may impose higher standards, pay higher or more regular wages, and be less able to control corruption or inefficiency among their employees than are private traders (Sandford 1983). Hess (quoted in Hoben 1976) and Raikes (1981:107) portray a secondary role of official cattle markets as an opportunity for District Council and Party tax collection and suggest this has discouraged attendance, particularly in Maasai areas where taxes and demands were higher and more regularly associated with auctions. Hoben (1976) quotes Hess as recording Tanzanian District Council livestock market prices ca. 1966–76 as held down to TSh 250–300/- per animal of which TSh 50/-, or 20% of the sale price, would have to be paid back in fees. There are clear indications that the majority of NCA stock sales bypass the official system and have done so for 20 years or more. There is a thriving cross border trade smuggling cattle to Kenya in exchange for 25–50% higher prices (Raikes 1981:209; Hoben 1976 quotes Hess as estimating 60–80% higher prices) paid in harder currency which can buy a wider range of goods at lower prices. Field and Moll (1987) quote NCAA officials as estimating 70% NCA stock sales to pass through this channel (Fig. 8.2).

While regional and national statistics suggest that sales offtake from NCA is particularly low (chapter 8) our study established rather consistent offtakes of around 8% when both official and unofficial transactions including internal use are taken into account. Our study showed no surplus of unwanted male stock waiting to be sold. Steers which are successfully reared are sold to buy grain, or exchanged for heifers. We found that although herd performance in terms of fertility and mortality was average for pastoral cattle, all study herds showed a net decline during the study because of offtake to buy grain.

At first sight the conditions of exchange have improved since the 1960s (six bags of maize/head of cattle during 1978–80 cf. two bags/head in 1960–65 –

Arhem 1981). However, this impression may be misleading. The 1960 figures may have been affected by an enormous influx of poor condition stock during widespread temporary drought conditions (chapter 8). This is supported by figures over a wider time period for Mbulu and Hanang Districts showing a comparable low exchange rate of two to four bags/head cattle in the drought years 1974–76, while pre-drought 1957 figures were 15 bags/head cattle and 1977–78 exchange values stood at eight to ten bags/head cattle. The Mbulu and Hanang values also show that NCA livestock/grain exchange rates are relatively unfavourable to the Maasai, largely because of transport problems and the lack of nearby grain producing areas. Lastly, the number of cattle per head of population has decreased from around 15 to around six. Better transport and communications could make a major difference to the poor terms of trade.

Limiting immigration into NCA

Pastoralism poses no threat to conservation in NCA, but the future joint land use in NCA depends on Maasai, government and Conservation Authority achieving a satisfactory compromise as to who has rights of access, residence and grazing in NCA. If wrongly handled, definition of user group and occupancy rights could result either in a monopoly or in further massive immigration, either outcome being damaging to Maasai and wildlife interests alike.

Just as for Kenya's group ranches, any attempt to define or register individuals with rights to use NCA will be complicated by traditional social mechanisms allowing access through ties of age-set, clan and section (Grandin and Lembuya 1987) as well as by fundamental problems of equity (Sandford 1983, Graham 1988). Sandford (1983) discusses the whole range of options, from the inclusive (which admits any who so wish to legal membership, or at most applies a criterion of current residence at the time of adjudication) to the exclusive (which applies selection criteria such as early volunteering, area of origin, scale of prior investment in the area, current livestock and other holdings, livestock management competence, etc.). In the latter case only a tiny proportion of those earning their living from the land acquire rights to it, and the rest are dispossessed.

Traditional systems lie between the two extremes, with access and user rights dependent on kinship, geographical residence and social networks allowing flexibility in time of need. Sandford (1983) suggests that such systems allow relatively equal opportunities within society and also allow for admission of new members. Their very flexibility makes formal definition for purposes of regulation extremely difficult. Kinship and evidence as to past use are in practice complex and difficult criteria to apply (see Grant 1954 for an early survey on occupancy of Serengeti/Ngorongoro). Sandford (1983) points out that official administration is often reluctant to use kinship as the basis for land tenure. The section system has in the past formed the basis of organisation in warfare and raiding – often against the groups from which government officials come – and also represents a network that makes considerable demands on any Maasai officials.

It is beyond the scope of this book to recommend guidelines along which user groups and rights of access to NCA should be defined. However, any attempt to do so should bear in mind that NCA represents a point of convergence from which different sections fan out into more arid wet season grazing areas. It contains representatives of several different sections, all of which must be considered, and each of which will retain close ties with other section members beyond the boundaries of NCA. The NCA Maasai will have quite clear ideas on rights of access and on carrying capacities of the area under different conditions. They must be consulted both on membership and on control of access. Traditional localities, to the extent that they represent geographical divisions (chapter 3), might well provide an initial basis for defining access rights. It will be essential to bear in mind that:

1. There are extensive movements of human, livestock and wildlife populations across NCA boundaries on both seasonal and year-to-year bases.
2. Such movements are a necessary part of the grazing system affecting an area that probably extends across national boundaries into Kenya.
3. So far the cross-boundary movements of the Maasai and their stock do not seem to have had any unfavourable impact on either conservation values or pastoralist subsistence – if anything quite the reverse.
4. Those veterinary, water and range developments that have taken place have not been accompanied by any overall increase in stock numbers, but together with social and administrative changes were associated with a rapid and massive increase in the human population of NCA.

Until a great deal more is understood about Maasai decision making and regulation of grazing resources it would be wise to leave grazing management including access rights largely in the hands of the pastoralists. Past experience suggests current grazing practices are in no way detrimental to the NCA environment, but the pitfalls of attempting to define and limit the human and livestock populations using NCA, and to formalise internal boundaries further, could entail serious ecological impacts.

It is also necessary to bear in mind that there is a considerable non-Maasai population of NCAA and tourist-associated staff and their dependents, traders, and others amounting to over 3000 in 1987 (NCDP 1987, Cobb 1989). This group shows if anything a more rapid rate of increase than the Maasai and to date has taken the lion's share of basic resources (see for example chapter 6) and development inputs. It is as important to control the growth of this group, perhaps even to reduce it, as to control Maasai immigration.

It is necessary to involve the Maasai not only in planning and managing immigration and grazing, but also in planning any developments in research, wildlife use, and tourism. Participatory management is an ideal hard to achieve (Sandford 1983; Aronson 1985). While traditional forms of decision-making organisation may be by far the most effective in, for example, grazing management, new structures and roles may be essential for making unprecedented technical and administrative decisions (Sandford 1983). The speed with which new decision-making organisations can establish themselves within pastoralist groups varies enormously and it takes time to establish working teams and to accumulate practical experience. Sandford (1983) sees equity, rather than efficiency, as the criterion in choice of using existing organisations for development purposes instead of establishing new ones. Traditional structures among the Maasai are highly egalitarian within any one age and sex class, but not between age and sex classes (Talle 1988). For example, women have no formal voice in decisions outside the household. Aronson (1985) points out that under such circumstances problems arise when outside agencies foster the development of decision-making bodies that are meant to be egalitarian with respect to all sections of society. Channels of communication between government and pastoralists (see e.g. Oba 1985) may contribute to these problems, notably when they are monopolised by one or a few powerful groups who secure their own interests to the detriment of the rest (Sandford 1983, Aronson 1985, Wyckoff 1985). This has become evident in parts of Tanzanian Maasailand, an example being the Lolkisale bean agribusiness. It could happen in NCA, if a small group were able to control rights of access, or rights to lease parts of NCA to various tourist or research enterprises. In theory the elders' councils and the NCA village councils (their modern compromise with Tanzanian administrative structure) are perhaps best constituted to formulate equitable and practicable procedures in the first instance. However, their power and efficacy are currently limited by the way NCA operates (chapters 3,4).

The grievances of the Maasai – lack of channels of communication, lack of input to the management decision-making process, an authoritarian and inflexible Conservation Authority – can be resolved both by changes in the administrative structure and by careful selection of administrative staff. If more recognition is given to the role of Maasai in NCA, to the place of NCA in Tanzanian ecology, and to the importance of conservation-oriented educational tourism in a national curriculum that until now has strongly emphasised political or economic (rather than environmental) development issues, it will be correspondingly easier to recruit and motivate the high calibre of staff that the Conservation Authority need.

Summary and Conclusion

1. There are a number of relatively simple interventions that could improve conditions of livestock production without jeopardising conservation values. The concept of zones suggested in all past management plans needs to be retained, albeit in modified form.

2. There should be wildlife priority zones such as the Crater, catchment zones such as the Forest Reserve, pastoralist zones such as Oloirobi, but this zonation need not imply expropriation by one or other form of land use. Just as pastoralist zones should continue to act as buffer zones and migration corridors for wildlife, so some wildlife zones will need to provide for fuelwood, timber and plant collection, possibly hunting.

3. Careful water development, including repair to existing systems, could ease both domestic and livestock management problems in the dry season, and improve livestock survival and milk production without causing environmental problems.

4. Range and grazing management can be efficient and sustainable as they stand, and are best left to the Maasai. Relaxation of burning and specified grazing access restrictions (particularly of forest and Olmoti Crater dry season access) would improve grazing conditions and check disease transmission.

5. There are few improvements that can realistically be made to livestock breed management. Disease management is however a priority, perhaps best tackled through a resumption of traditional disease avoidance and control techniques.

6. There is no place for commercial beef or dairy enterprises in NCA under present conditions of production and marketing. However, there is scope for improvement of marketing from the traditional herds. If the funds that currently go to bolstering Tanzania's generally unsuccessful state or parastatal ranching concerns were to go to subsidising marketing arrangements for pastoralist cattle, this could improve meat supply to surrounding areas as well as easing subsistence in NCA.

7. Transport and communications are a priority that will facilitate development of conservation education, tourism and of the livestock/grain exchange trade vital for Maasai subsistence in NCA.

8. The desire of planners to define that population with access rights to NCA and to define stocking rates and stock quotas will be an inevitable source of future conflict. Adverse ecological impacts can only be avoided by involving the Maasai at all stages of planning and management.

9. Maasai involvement in planning within NCA raises issues of equity, and of channels of communication within Maasai society and between Maasai and government. Given that Maasai society remains egalitarian within age groups, traditional councils of resident elders may be the best qualified to control access of both people and stock to NCA.

10. Past attempts to stop movements of Maasai and stock between adjudicated group ranches in Kenya and across defined zones in the Tanzanian Maasai Rangeland Project have failed. The best compromise may be to attempt to ensure that this control remains a matter of decision for councils representing elders throughout NCA rather than allowing small groups to manipulate the situation for their own interests.

13

Viewpoint

Migil enkaputi te nkupes Do not break a relationship without good cause
(Maasai saying: Kipury 1983)

This book is not merely an abstract ecological debate about land use policies and development objectives. At the time of writing Tanzania's politicians and administrators are moving towards a decision that will affect the future of 23 000 Maasai pastoralists within the Ngorongoro Conservation Area. Ngorongoro itself is not just another piece of African real estate. It is a World Heritage site and has been described as the eighth wonder of the world. It is probably the most important wildlife tourist destination in East Africa.

The Maasai themselves are the heirs to several thousands of years of pastoralist involvement in Ngorongoro. They have helped shape the present environment and have long helped protect the wild animal populations. They are pure pastoralists with mixed species herds practising a spectrum of nomadic and transhumant husbandry. They are among the best known of Africa's peoples and epitomise the ecological and cultural value of pastoralism.

Academic questions aside, the critical issue is whether the Maasai should stay in NCA, or whether their impacts on environment and wildlife justify their resettlement elsewhere. These questions hinge on a definition of the values of NCA and on an evaluation of the impacts of pastoralism. We argue that the conservation values of NCA are inseparable from the pastoralist presence, especially in that Ngorongoro illustrates man and nature coexisting in harmony. Secondly we argue from a comprehensive review of the biology of Ngorongoro that there are no negative impacts of Maasai land use on wildlife values.

We strongly maintain there is no justification on conservation or other grounds for expelling the Maasai. There should be a strong political and administrative decision which guarantees the future of the Maasai as pastoralists in NCA. Any move to expel the Maasai will be counterproductive to long-term conservation interests, quite apart from being a major abuse of human rights.

Traditional conservationists talk of the major problems of Ngorongoro Conservation Area. By contrast we see thirty years of relatively successful multipurpose use following on millennia of coexistence in these and surrounding rangelands. We see no reason to expect anything other than further successes.

Of course there will be problems, as there are with any human endeavour. This is a system of great ecological variability, within a social, economic and political system that is also of considerable variability. Change in any one input will suggest problems to managers geared to stable systems. Management will neeed to maintain flexible approaches to problem-solving within a system which monitors patterns of change.

Thirty years of relative success have been achieved with most inputs going to conservation and self-administration by the NCAA. There has been little input into the pastoralist economy, and if anything support to the pastoralist production system has dwindled. There are fewer roads, markets, waterholes, dips, etc. than there were 30 years ago.

We conclude that if management redresses the balance, and restores pastoralist infrastructure, then pastoralist perceptions of problems will also decrease. NCA will continue not just with its major wildlife values, but also with a viable pastoralist community. The future remains unclear. It is initially dependent on a political decision, and ultimately on the ability of the NCA to implement a new policy direction encompassing and supporting Maasai pastoralism.

The issues of Ngorongoro are complex. Policy decisions can only be made with an understanding of law, sociology, politics, economics, environmental sciences, conservation biology as well as a sense of aesthetics, compassion and common sense. We do not pretend to be experts in all these disciplines. We approached the problem as resource ecologists, and have researched it and discussed it for ten years with a huge variety of interested expertise. Our conclusion is that NCA works as a joint land use area, and that management must continue to adapt to the changes in this complex system.

REFERENCES

Abel, N. and Blaikie, P. (1986). Elephants, people, parks and development: the case of the Luangwa Valley, Zambia. *Environmental Management*, **10** (6), 735–51.

Adams, N. (1988). A new plan for Africa. *New Scientist*, **118** (1619), 89.

Aikman, D.I. and Cobb, S.M. (1989), *Water Development*. Technical Report No 8 Ngorongoro Conservation and Development Project, IUCN Regional Office, Nairobi.

Allan, W. (1968). Soil resources and land use in Tropical Africa. In *Conservation of vegetation south of the Sahara. Acta Phytogeographica Suecica*, Uppsala. **54**, ed. I. and O. Hedberg, pp. 9–13.

Altmann J. (1974). Observational study of behaviour: sampling methods. *Behaviour*, **49**, 227–67.

Anacleti, A.O. (1975). Pastoralism and development: Economic changes in pastoral industry in Serengeti 1750–1961, MA Thesis, University of Dar es Salaam.

Anacleti A.O. (1978), Serengeti: its people and their environment. *Tanzania Notes and Records*, **81/82**, 23–34.

Anderson, D. (1984). Depression, dust bowl, demography and drought: the colonial state and soil conservation in East Africa during the 1930s. *African Affairs*, **83** (332), 321–43.

Anderson, D. (1988), Cultivating pastoralists: ecology and economy among the Il Chamus of Baringo 1840–1980. In *The ecology of survival: case studies from north east African history*. pp. 241–260. ed. D. Johnson and D. Anderson, Lester Crook Academic Publishing/Westview Press.

Anderson, D. and Grove, R. (eds) (1987). *Conservation in Africa: peoples, policies and practice*. Cambridge University Press.

Anderson, G.D. and Herlocker, D.J. (1973). Soil factors affecting the distribution of vegetation types and their utilisation by wild animals in Ngorongoro Crater. *Journal of Ecology*, **61**, 627–51.

Anderson, G.D. and Talbot, L.M. (1965). Soil factors affecting the distribution of the grassland types and their utilisation by wild animals on the Serengeti Plains, Tanzania. *Journal of Ecology*, **53**, 33–56.

Andrews, P.J. (1989). Palaeoecology of Laetoli. *Journal of Human Evolution*, **18**, 173–81.

Arhem, K. (1981a). *The ecology of pastoral land use in the Ngorongoro Conservation Area. A background paper for a new management plan for the Ngorongoro Conservation Area Authority*. BRALUP, University of Dar es Salaam.

Arhem, K. (1981b). *Maasai pastoralism in Ngorongoro Conservation Area: sociological and ecological issues*. BRALUP Research Paper **69**: University of Dar es Salaam.

Arhem, K. (1985a). *The Maasai and the state: the impact of rural development policies on a pastoralist people in Tanzania*. IWGIA Document **52**.

Arhem, K. (1985b). *Pastoral man in the garden of Eden. The Maasai of the Ngorongoro Conservation Area Tanzania.* Uppsala Research Reports in Cultural Anthropology.

Arhem, K., Homewood, K. and Rodgers, W.A. (1981). *A pastoral food system: the Ngorongoro Maasai in Tanzania.* BRALUP Research Paper **70**. University of Dar es Salaam.

Arman, P., Hopcraft, D., and McDonald, I. (1975). Nutritional studies on East African herbivores. II. Losses of nitrogen in the faeces. *British Journal of Nutrition*, **33**, 265–76.

Aronson, D. (1985). Implementing local participation: the Niger range and livestock project. *Nomadic peoples*, **18**, 67–76.

Banyikwa, F. (1976). A quantitative study of the ecology of the Serengeti short grasslands. PhD thesis, University of Dar es Salaam.

Barnes, D.L. (1979). Cattle ranching in the semi-arid savannas of East and South Africa. In *Management of semi-arid ecosystems*, ed. B. Walker, pp. 9–54. Amsterdam: Elsevier.

Beckwith, C. and Ole Saitoti, T. (1980). *Maasai.* London: Elm Tree Books.

Behnke, R.H. (1984). Fenced and open-range ranching: the commercialization of pastoral land and livestock in Africa. In *Livestock development in subSaharan Africa.* ed. J. Simpson and P. Evangelou, pp. 261–84. Boulder: Westview Press.

Behnke, R.H. (1985). Measuring the benefits of subsistence versus commercial livestock production in Africa. *Agricultural Systems*, **16**, 109–35.

Bell R.M. (1970). The use of the herb layer by grazing ungulates in the Serengeti. In *Animal populations in relation to their food resources*, pp. 111–23. ed. A. Watson, Oxford: Blackwell.

Bell, R.M. (1982). The effect of soil nutrient availability on community structure in African ecosystems. In *Ecology of tropical savannas*, ed. B. Huntley and B. Walker, pp. 193–216. Springer-Verlag.

Bell, R. (1987). Conservation with a human face: conflict and reconciliation in African land use planning. In *Conservation in Africa: people, policies and practice*, ed. D. Anderson and R. Grove, pp. 79–102, Cambridge University Press.

Belsky, A.J. (1984). Role of small browsing mammals in preventing woodland regeneration in the Serengeti National Park. *African Journal of Ecology*, **22**, 271–79.

Belsky, A.J. (1985). Long term vegetation monitoring in the Serengeti National Park Tanzania. *Journal of Applied Ecology,* **22**, 449–60.

Belsky, A.J. (1986a). Revegetation of artificial disturbances in grasslands of the Serengeti National Park, Tanzania. I. Colonisation of ungrazed plots. *Journal of Ecology*, **74**, 419–37.

Belsky, A.J. (1986b). Revegetation of artificial disturbances in grasslands of the Serengeti National Park, Tanzania. II. Five years of successional change. *Journal of Ecology*, **74** (4), 937–52.

Belsky, A.J. (1986c). Population and community processes in a mosaic grassland in the Serengeti Plain, Tanzania. *Journal of Ecology*, **74** (3), 841–56.

Belsky, A.J. (1987). Revegetation of natural and human-caused disturbances in the Serengeti National Park, Tanzania. *Vegetatio*, **70**, 51–9.

Belsky, A.J. and Amundson, R.G. (1986). Sixty years of successional history behind a moving sand dune near Olduvai Gorge, Tanzania. *Biotropica*, **18** (3), 231–5.

Benefice, E., Chevassu-Agnes, S. and Barral, H. (1984). Nutritional situation and seasonal variations for pastoralist populations of the Sahel (Senegalese Ferlo). *Ecology of Food and Nutrition*, **14**, 229–47.

Berntsen, J. (1976). The Maasai and their neighbours: variables of interaction. *African Economic History*, **2**, 1–11.

Berntsen, J. (1979). Economic variations among Maa-speaking peoples. In *Ecology and History in East Africa, Hadith* vol. 7, ed. B.A. Ogot, pp. 108–28. Nairobi: EDNP.

Bertram, B. (1979). Serengeti predators and their social systems. In *Serengeti: Dynamics of an ecosystem*, ed. A. Sinclair and M. Norton-Griffiths, pp. 221–248. Chicago University Press.

Bindernagel, J.A. (1977). *Wildlife Utilization in Tanzania.* Rome: Food and Agriculture Organisation.

Birley, M. (1982). Resource management in Sukumaland, Tanzania. *Africa*, **52** (2), 1–30.
Bodley, J.H. (1988). *Tribal peoples and development issues*. California: Mayfield Publishing.
Borgerhoff-Mulder, M., Sieff, D. and Merus, M. (1989). Disturbed ancestors: Datoga history in the Ngorongoro Crater. *Swara*, **12** (2), 32–5.
Borner, M. (1981). Black rhino disaster in Tanzania. *Oryx*, **XVI** (1), 59–66.
Borner, M. (1985). The increasing isolation of Tarangire National Park. *Oryx*, **19** (2), 91–6.
Boshe, J.I. (1988). *Wildlife ecology*. Technical Report No 3, Ngorongoro Conservation and Development Project, IUCN Regional Office, Nairobi.
Boutton, T., Tieszen, L. and Ibamba, S. (1988a). Biomass dynamics of grassland vegetation in Kenya. *African Journal of Ecology*, **26**, 89–102.
Boutton, T., Tieszen, L. and Ibamba, S. (1988b). Seasonal changes in the nutrient content of East African grassland vegetation. *African Journal of Ecology*, **26**, 103–16.
Bower, J. (1973). Seronera: excavations at a stone bowl site in Serengeti National Park, Tanzania. *Azania*, **8**, 71–104.
Bower, J.R.F., Nelson, C.M., Waibel, A.F. and Wandibba, S. (1977). The University of Massachusetts Later Stone Age/Pastoral Neolithic comparative study in central Kenya: an overview. *Azania*, **12**, 119–46.
Bower, J. and Nelson, C. (1978). Early pottery and pastoral cultures of the Central Rift Valley, Kenya. *Man*, **13**, 554–66.
Bradley-Martin, E. (1979). *The international trade in rhinoceros products*. A report for the World Wildlife Fund. December 1979.
Branagan, D. (1974). Conflict between tourist interests and pastoralism in the Ngorongoro Highlands of Tanzania. In *Tourism in Africa and the management of related resources*, Proceedings of a seminar at the centre for African Studies. University of Edinburgh, ed. D.N. McMaster, pp. 67–74. Centre for African Studies, University of Edinburgh.
Braun, H.M. (1971). Primary production of grasslands and their utilization by game in the Serengeti National Park, Tanzania. *Netherlands Foundation for the Advancement of Tropical Research. Report for 1970*. pp. 35–37.
Braun, H.M. (1973). Primary production in the Serengeti. Purpose, methods and some results of research. *Annales de l'Université d'Abidjan Série E (Ecologie)*, **IV** (2), 171–188.
Bredon, R. and Marshall, B. (1962). Selective consumption by stallfed cattle and its influence on the results of a stallfed digestibility trial. *East African Agriculture and Forestry Journal*, **27**, 168–72.
Breman, H. and de Wit, C. (1983). Rangeland productivity and exploitation in the Sahel. *Science*, **221**, 1341–7.
Brown, L.H. (1971). The biology of pastoral man as a factor in conservation. *Biological Conservation*, **3**, 93–100.
Casebeer, R.L. and Koss, G.G. (1970). Food habits of wildebeest, zebra, hartebeest and cattle in Kenya Maasailand. *East African Wildlife Journal*, **39**, 26–36.
Caughley, G. (1976). Plant–herbivore systems. In *Theoretical ecology: principles and applications*, ed. R.M. May, pp. 94–113. Oxford: Blackwell Scientific Publications.
Caughley, G. (1977). *Analysis of vertebrate populations*. Chichester, New York, Brisbane, Toronto: Wiley Interscience.
Caughley, G. (1983). Dynamics of large animals and their relevance to culling. In *Management of large mammals in African conservation areas*, ed. R. N. Owen-Smith pp. 115–26. Pretoria: HAUM Educational Press.
Caughley, G., Shepherd, N. and Short, J. (eds) (1987). *Kangaroos: their ecology and management in the sheep rangelands of Australia*. Cambridge University Press.
Chamshama, S.A.O., Kerkhof, P. and Singunda, W.T. (1989). *Community needs for forest and tree products*. Technical Report No 6 Ngorongoro Conservation and Development Project, IUCN Regional Office, Nairobi.
Chausi, E.B. (1985). Range management and ecology in Ngorongoro Conservation Area,

Tanzania. An integrated range resources management prospective (sic). Unpublished MSc in Range Resources, University of Idaho.
Chausi, E. and Robberecht, R. (1985). Seasonal grazing: a practical management system for the Ngorongoro Conservation Area rangeland, Tanzania. *Society for Range Management Annual Meeting*, **38**, 56 (abstract).
Child, G.S. (1965). Some notes on mammals of Kilimanjaro. *Tanganyika Notes and Records*, **64**, 77–90 in revised edition issued 1974.
Chuwa, S., Mwasumbi L.B. and Rodgers, W.A. (1985). *A vegetation checklist for Ngorongoro Conservation Area*. 86pp mimeo. NCAA. Box 1, Ngorongoro, Tanzania.
Clebowski, L. (1979). *Range management report: Oloirobi Village*. Masai Range Project, USAID. Arusha, Tanzania.
Cobb, S.J. (1976). Distribution and abundance of the large mammal community of Tsavo National Park, Kenya. D. Phil Thesis, University of Oxford.
Cobb, S.M. (1989). *Ngorongoro Conservation and Development Project: water development impact assessment in the Ngorongoro Conservation Area Tanzania*. Babtie Shaw and Morton Consulting Engineers/International Union for the Conservation of Nature and Natural Resources, East Africa Regional Office, Nairobi.
Coe, M.J. (1967). *The ecology of the alpine zone of Mt. Kenya*. The Hague: W. Junk.
Coe, M.J. Cumming, D. and Phillipson, J. (1976). Biomass and production of large African herbivores in relation to rainfall and primary production. *Oecologia*, **22**, 341–54.
Collett, D. (1987). Pastoralists and wildlife: image and reality in Kenya Maasailand. In *Conservation in Africa: people, policies and practice*, ed. D. Anderson and R. Grove, pp. 129–48. Cambridge University Press.
Coppock, D., Ellis, J. and Swift, D. (1986). Livestock feeding ecology and resource utilization in a nomadic pastoral ecosystem. *Journal of Applied Ecology*, **23**, 573–83.
Coppock, D., Swift, D. and Ellis, J. (1986). Seasonal nutritional characteristics of livestock diets in a nomadic pastoral ecosystem. *Journal of Applied Ecology*, **23**, 585–95.
Cossins, N.J. (1985). The productivity and potential of pastoral systems. *ILCA Bulletin*, **21**, 10–15.
Coughenour, M., Ellis, J., Swift, D., Coppock, D., Galvin, K., McCabe, J. and Hart, T. (1985). Energy extraction and use in a nomadic pastoral ecosystem. *Science*, **230**, 619–25.
Coughenour, M., McNaughton, S. and Wallace, L. (1985). Responses of an African graminoid (*Themeda triandra* Forsk) to frequent defoliation, nitrogen and water: a limit of adaptation to herbivory. *Oecologia*, **68**, 105–10.
Crosby, A. (1986). *Ecological imperialism: The biological expansion of Europe 900–1900*. Cambridge University Press.
Cross, M. (1985). Waiting for a green revolution. *New Scientist*, **105**, 37–40.
Croze, H. (1974a). The Seronera bull problem. I. The trees. *East African Wildlife Journal*, **12**, 1–27.
Croze, H. (1974b). The Seronera bull problem. II. The elephants. *East African Wildlife Journal*, **12**, 28–47.
Curry, S. (1980). *The state sector and the expansion of international tourism in Tanzania*. Economic Research Bureau Paper **80**.6 University of Dar es Salaam.
D'Hoore, J.L. (1964). *Soil map of Africa 1:5 000 000*. Joint Project No. 11. Commission for Technical cooperation in Africa South of the Sahara. Publication **93**, Lagos.
Dahl, G. (1979). *Suffering grass: subsistence and society in Waso Borana*. Department of Social Anthropology, University of Stockholm.
Dahl, G. and Hjort, A. (1976). *Having herds: pastoral herd growth and household economy*. Stockholm Studies in Social Anthropology.
Davidson, S., Passmore, R., Brock, J. and Truswell, A. (1979). *Human nutrition and dietetics*. 7th edn. Edinburgh, London and New York: Churchill Livingstone.

De Leeuw, P.N., Bekure, S. and Grandin, B. (1988). Some aspects of livestock productivity in Maasai group ranches in Kenya. In *Ecology and management of the world's savannas*, ed. J.C. Tothill and J.J. Mott, pp. 247–51. Farnham UK: Australian Academy of Sciences, Canberra and Commonwealth Agricultural Bureau.

De Leeuw, P.N. and Wilson, R.T. (1987). Comparative productivity of indigenous cattle under traditional management in subSaharan Africa. *Quarterly Journal of International Agriculture*, **25**, 377–90.

De Wit, H.A. (1978). Soils and grassland types of the Serengeti Plains (Tanzania). Dissertation. University of Wageningen. Mededelingen Landbouwhogeschool Wageningen, Netherlands.

Deshmukh, I. (1984). A common relationship between precipitation and grassland peak biomass for East and Southern Africa. *African Journal of Ecology*, **22**, 181–6.

Dirschl, H.J. (1966). *Management and development plan for the Ngorongoro Conservation Area.* Ministry of Agriculture, Forestry and Wildlife, Dar es Salaam.

Douglas-Hamilton, I. (1972). On the ecology and behaviour of the African elephant: the elephants of Lake Manyara. D.Phil. Thesis, Oxford University.

Douglas-Hamilton, I. and Hillman, K. (1981). Elephant carcasses and skeletons as indicators of population trends. In *Report of workshop on low level aerial survey techniques*, pp. 113–30. ILCA Monograph **4**. Addis Ababa.

Drijver, C. (1990). People's participation in environmental projects in developing countries. *Dryland Networks Programme Paper* **17**, 1–17. International Institute for Environment and Development (IIED) London.

Dublin, H. and Douglas-Hamilton, I. (1987). Status and trends of elephants in the Serengeti-Mara ecosystem. *African Journal of Ecology*, **25**, 19–34.

Dyson-Hudson, N. (1980). Strategies of resource exploitation among East African pastoralists. In *Human ecology in savanna environments*, ed. D. Harris, pp. 171–84. London: Academic Press.

East, R. (1984). Rainfall, soil nutrient status and biomass of large African mammals. *African Journal of Ecology*, **22**, 245–70.

East, R. (1988). *Antelopes: Global survey and regional action plans. Part 1. East and north east Africa.* Gland, Switzerland: IUCN.

Ecosystems Ltd (1980). *The status and utilisation of wildlife in Arusha region, Tanzania.* Consultancy report to the Tanzanian Government. Nairobi: Ecosystems Ltd.

Edington, J.M. and Edington, M.A. (1986). *Ecology, recreation and tourism.* Cambridge University Press.

Ehret, C. (1974). Cushites and the highland and plains nilotes to AD 1800.In *Zamani: a survey of East African history*. ed. B.A. Ogot, pp. 150–69. EAPH/Longman.

Elliott, H.F. (1948). Some hints on climbing Maasailand mountains. *Tanganyika Notes and Records*, **26**, 68–76.

Ellis, J.E. and Swift, D.M. (1988). Stability of African pastoralist ecosystems: alternate paradigms and implications for development. *Journal of Range Management*, **41**, 450–9.

Eltringham S.K. (1980). A quantitative assessment of the range usage by large African mammals with particular reference to the effects of elephants upon trees. *African Journal of Ecology*, **18**, 53–72.

Eltringham, S.K. (1984). *Wildlife resources and economic development.* Chichester: Wiley.

Estes, R.D. (1966). Behaviour and life history of the wildebeest (*Connochaetes taurinus* Burchell). *Nature*, **212**, 999–1000.

Estes, R.D. (1967). The comparative behaviour of Grant's and Thomson's gazelles. *Journal of Mammalogy*, **48**, 189–209.

Estes, R.D. (1969). Territorial behaviour of the wildebeest (*Connochaetes taurinus* Burchell 1823). *Zeitschrift Tierpsychologie*, **26**, 284–370.

Estes, R.D. and Goddard, J. (1967). Prey selection and hunting behaviour of the African wild dog. *Journal of Wildlife Management*, **31**, 52–70.

Estes, R.D. and Small, R. (1981). The large herbivore populations of Ngorongoro Crater. *East African Wildlife Journal*, **19**, 175–86.

Evangelou, P. (1984). *Livestock development in Kenya's Maasailand. Pastoralist transition to a market economy*. Boulder, Colorado: Westview Press.

Field, C.R. (1971). Elephant ecology in the Queen Elizabeth National Park, Uganda. *East African Wildlife Journal*, **9**, 99–123.

Field, C.R. (1975). Climate and food habits of ungulates on Galana Ranch. *East African Wildlife Journal*, **13**, 203–20.

Field, C.R. and Moll, G. (1987). *Preliminary report on livestock development in the Ngorongoro Conservation Area, Tanzania*. Ngorongoro Conservation and Development Project, NCAA typescript, 23pp.

Field, C.R., Moll, G., and Ole Sonkoi, C. (1988). *Livestock Development*. Technical Report No 1 Ngorongoro Conservation and Development Project, IUCN Regional Office, Nairobi.

Fimbo, M. (1981). *Institutional context of the Ngorongoro Conservation Area*. Background paper prepared for the 1981 management plan. Institute of Resource Assessment, University of Dar es Salaam.

Finch, V.A. and King, J.N. (1982). Energy-sparing mechanisms as an adaptation to undernutrition and water deprivation in the African zebu. In *Use of tritiated water in studies of production and adaptation in ruminants*. Vienna: International Atomic Energy Agency.

Finch V.A. and Western, D. (1977). Cattle colour in pastoral herds: natural selection or social preference? *Ecology*, **58**, 1384–92.

Flora of Tropical East Africa (1974). *Gramineae, part 2*. W.D. Clayton, S.M. Phillips and S.A. Renvoize. Crown Agents.

Ford, J. (1971). *The role of trypanosomiases in African ecology: a study of the tsetse fly problem*. Oxford University Press.

Forster, M. and Malecela, E.M. (1988). *Legislation*. Technical Report No 5 Ngorongoro Conservation and Development Project, IUCN Regional Office, Nairobi.

Fosbrooke, H. (1948). An administrative survey of the Maasai social system. *Tanganyika Notes and Records*, **26**, 1–51.

Fosbrooke, H. (1962). *Ngorongoro Conservation Area Management Plan*. Ministry of lands, forests and wildlife. Dar es Salaam. mimeo, 46pp plus appendices A–N and 5 maps.

Fosbrooke, H. (1972). *Ngorongoro: the eighth wonder*. Deutsch.

Fosbrooke, H. (1975). Ngorongoro Conservation Area: 1961 to 1971 development. *Tanzania Notes and Records*, **76**, 85–8.

Fosbrooke, H. (1980). *Maasai motivation and its application: a study of Maasai organizational ability as applied to the cattle marketing system 1938–1959*. mimeo: 13pp. Arusha, Tanzania: Development Alternatives.

Frame, G. (1976). An ecological survey and development plan for Empakaai Crater (Ngorongoro Conservation Area, Tanzania). MSc Thesis, Utah State University, Logan, Utah.

Frame, G. (1982). Wild mammal survey of Empakaai Crater Area. *Tanzania Notes and Records*, **88/89**, 41–56.

Frame, G. (1986). Carnivore competition and resource use in the Serengeti ecosystem of Tanzania. PhD Thesis, Utah State University, Logan, Utah.

Frame, G., Frame, L. and Spillett, J. (1975). *An ecological survey and development plan for the Empakaai Crater Ecosystem, Ngorongoro Conservation Area, Tanzania*. Serengeti Research Institute contribution, **212**.

Frost, P., Menaut, J.C., Walker, B., Medina, E., Solbrig, O. and Swift, M. (eds) (1986). Responses of savannas to stress and disturbance. A proposal for a collaborative programme of research. Report of IUBS Working Group 1985. *Biology International*, **10** (special issue).

Fryxell, J. and Sinclair, A. (1988a). Seasonal migration by white-eared kob in relation to

resources. *African Journal of Ecology*, **26**, 17–31.
Fryxell, J. and Sinclair, A. (1988b). Causes and consequences of migration by large herbivores. *Trends in Ecology and Evolution*, **3**, 237–41.
Galaty, J. (1980). The Maasai group ranch: politics and development in an African pastoral society. In *When nomads settle: processes of sedentarisation as adaptation and response*, ed. P. Salzman, pp. 157–72. New York: Praeger.
Galaty, J. (1982). Being 'Maasai'; being 'people-of-cattle': ethnic shifters in East Africa. *American Ethnologist*, **9**, 1–22.
Geertsema, A. (1985). Aspects of the ecology of the serval *Leptailurus serval*. *Netherlands Journal of Zoology*, **35**, 527–610.
Gilchrist, F.M.C. (1962). *Northern Highlands Forest Reserve Management Plan July 1962–June 1966*. Forest Department, Government of Tanzania (included as appendix to NCA management plan of 1966).
Gilchrist, F.M.C. and Mackie, R.I. (1984). *Herbivore nutrition in the subtropics and tropics*. Craighall, South Africa: The Science Press.
Glantz, M. (1987). Drought in Africa. *Scientific American*, **256**, 34–40.
Glover, P.E. (1961). Report on the Ngorongoro Pasture Research Project (typescript) Ngorongoro Conservation Unit.
Goddard, J. (1967). Home range, behaviour and recruitment rates of two black rhinoceros populations. *East African Wildlife Journal*, **5**, 133–50.
Goddard, J. (1968). Food preferences of two black rhinoceros populations. *East African Wildlife Journal*, **6**, 1–18.
Goodland, R. (1985). Tribal peoples and economic development: the human ecological dimension. In *Culture and conservation: the human dimension in environmental planning*. ed. J. McNeely and D. Pitt. Croom Helm.
Graham, A. (1974). Gardeners of Eden.
Graham, O. (1988). Enclosure of the East African Rangelands: recent trends and their impact. *Pastoral Development Network Paper*, No. **25**a, Overseas Development Institute.
Grandin, B.E. (1988). Wealth and pastoral dairy production. A case study from Maasailand. *Human Ecology*, **16**, 1–21.
Grandin, B.E. and Lembuya, P. (1987). The impact of the 1984 drought at Olkarkar Group Ranch, Kajiado, Kenya. *Pastoral Development Network Paper*, No. **23**e. Overseas Development Institute.
Grant, H. St.J. (1954). Report on human habitation of the Serengeti National Park. 20pp. Unpublished report to Trustees of Tanganyika National Parks. Published 1957 by Government Printer, Dar es Salaam.
Green, R.H. (1979). Toward planning tourism in African countries. In *Tourism: passport to development? Perspectives on the social and cultural effects of tourism in developing countries*, ed. E. de Kadt, pp. 79–100. Oxford University Press.
Greenberg, J. H. (1963). The Languages of Africa. *International Journal of American Linguistics*, vol. **29**, publication 25. Bloomington, Indiana and the Hague: Indiana University Research Center in Anthropology, Folkore and Linguistics.
Griffiths, J.F. (1962). The climate of East Africa. In *The natural resources of East Africa*. ed. E.W. Russell. Nairobi: D.D. Hawkins Ltd/East Africa Literature Bureau.
Grimsdell, J.J.R. (1978). *Ecological Monitoring*. Handbook No. 4. Nairobi: African Wildlife Leadership Foundation.
Grimsdell, J.J.R. (1979). Changes in populations of resident ungulates. In *Serengeti: Dynamics of an ecosystem*, ed. A. Sinclair and M. Norton-Griffiths, pp. 353–9. Chicago University Press.
Gulliver, P. (1955). *The family herds*. London: Routledge and Kegan Paul.
Hacker, J.B. and Ternouth, J.H. (1987). *The nutrition of herbivores*. Sydney: Academic Press.
Hamilton, A. (1984). *Deforestation in Uganda*. Nairobi: Oxford University Press.

Hanby, J. and Bygott, D. (1979). Population changes in lions and other predators. *Serengeti: Dynamics of an ecosystem*, ed. A. Sinclair and M. Norton-Griffiths, pp. 249–62. Chicago University Press.

Hanby, J. and Bygott, D. (1989). *An illustrated guidebook to Ngorongoro*. Arusha.

Harker, K.W. Taylor, J.I. and Rollinson, D.H.L. (1954). Studies on the habits of zebu cattle. I. Preliminary observations of grazing habits. *Journal of Agricultural Science (Cambridge)*, **44**, 193–8.

Harris, J.M. (1985). Age and palaeoecology of the Upper Laetoli Beds, Laetoli, Tanzania. In *Ancestors: the hard evidence*. ed. E. Delson, pp. 76–81. New York: Alan R. Liss.

Hay, R.L. (1976). Geology of the Olduvai Gorge. Berkeley: University of California Press.

Hedberg, O. (1951). Vegetation belts of the East African Mountains. *Svensk. Botanisk Tidsskrift*, **45**, 140–202.

Heine, B., Heine, I. and Konig, C. (1988). *Plant concepts and plant use. An ethnobotanical survey of the semi-arid and arid lands of East Africa. Part V. Plants of the Samburu (Kenya)*. Saarbrucken. Fort Lauderdale: Verlag Breitenbach Publishers.

Henin, R. and Egero, B. (1972). *The 1967 Population census of Tanzania. A demographic analysis*. 53pp BRALUP Research Paper No. **19** University of Dar es Salaam.

Henry, W. (1977). Tourist impact on Amboseli National Park. *African Wildlife Leadership Foundation News*, **12** (2), 4–8.

Herlocker, D.J. and Dirschl, H.J. (1972). *Vegetation of the Ngorongoro Conservation Area, Tanzania*. Canadian Wildlife Service Report Series **19**, Information. Canada, Ottawa. 39pp + map.

Hill, A.G. (1985). *Population, Health and Nutrition in the Sahel: Issues in the welfare of selected West African communities*. KPI Routledge and Kegan Paul.

Hillman, K. (1981). *IUCN African rhino survey*. Report to IUCN.

Hoben, A. (1976). *Social soundness of the Maasai livestock and range management project*. USAID Tanzania.

Hogg, R.S. (1985). The politics of drought: the pauperization of Isiolo Boran. *Disasters*, **9**, 39–43.

Hogg, R. (1987). Settlement, pastoralism and the commons: the ideology and practice of irrigation development in Northern Kenya. In *Conservation in Africa: people, policies and practice*, ed. D. Anderson and R. Grove, pp. 293–306. Cambridge: Cambridge University Press.

Hollis, A.C. (1905). *The Maasai: their language and folklore*. Oxford: Clarendon Press.

Homewood, K. (1990). Review of Heine, B., Heine, I. and Konig, C. (1988). *Plant concepts and plant use. An ethnobotanical survey of the semi-arid and arid lands of East Africa. Part V. Plants of the Samburu. Africa*, **60**, (3), 451–2.

Homewood, K. and Hurst, A. (1986). Comparative ecology of pastoralist livestock in Baringo, Kenya. *Pastoral Development Network*, **21b**, 1–41.

Homewood, K. and Lewis, J. (1987). Impact of drought on pastoral livestock in Baringo, Kenya 1983–1985. *Jounal of Applied Ecology*, **24**, 615–31.

Homewood, K. and Rodgers, W.A. (1984). Pastoralism and conservation. *Human Ecology*, **12**, 431–441.

Homewood, K. and Rodgers, W.A. (1987). Pastoralism, conservation and the overgrazing controversy. In *Conservation in Africa: peoples, policies and practice*, ed. D. Anderson and R. Grove, pp111–28. Cambridge University Press.

Homewood, K., Rodgers, W.A. and Arhem, K. (1987). Ecology of pastoralism in Ngorongoro Conservation Area, Tanzania. *Journal of Agricultural Science (Cambridge)*, **108**, 47–72.

Howell, P., Locke, M. and Cobb, S.M. (eds) (1988) *The Jonglei Canal. Impact and opportunity*. Cambridge University Press.

Iliffe, J. (1979). *A modern history of Tanganyika*. Cambridge University Press.

Inglis, J.M. (1976). Wet season movements of individual wildebeest of the Serengeti migratory herd. *East African Wildlife Journal*, **14**, 17–34.
Institute of Resource Assessment (1982). *Ngorongoro Management Plan*. University of Dar es Salaam.
IUCN (1987). *IUCN Directory of Afrotropical Protected Areas*. Gland, Switzerland and Cambridge, UK: IUCN.
Jacobs, A. (1965). The traditional political organisation of the pastoral Maasai. D.Phil. Thesis, Oxford.
Jacobs, A. (1975). Maasai pastoralism in historical perspective. In *Pastoralism in tropical Africa*. ed. T. Monod, pp. 406–22. Oxford University Press.
Jacobs, A. (1978). *A final report of development in Tanzania Maasailand: the perspective over 20 years 1957–1977*. Tanzania: USAID.
Jahnke, H.E. (1982) *Livestock production systems and livestock development in tropical Africa*. Kieler Wissenschaftsverlag vauk.
Jamhuri ya Muungano wa Tanzania (1985). *Hali ya Uchumi wa Taifa Katika Mwaka 1984*. Wizara ya Mipango na Uchumi Dar es Salaam Tanzania. Mpigachapa wa Serikali (Government Printer).
Jarman, P.J. and Sinclair, A.R.E. (1979). Feeding strategy and the pattern of resource partitioning in ungulates. In *Dynamics of an ecosystem*. ed. A. Sinclair and M. Norton-Griffiths, pp. 130–63. Chicago University Press.
Jewell, P.A. (1980). Ecology and management of game animals and domestic livestock in African savannas. In *Human ecology in savanna environments*, ed. D. Harris, pp. 353–82. Academic Press.
Jewell, P.A. and Nicholson, M.J. (1989). Strategies for water economy amongst cattle pastoralists and in wild ruminants. *Symposium of the Zoological Society, London*, **61**, 73–87.
Kabigumila, J.D.L. (1988). The ecology and behaviour of elephants in Ngorongoro Crater. MSc Thesis, University of Dar es Salam.
Kahurananga, J. (1981). Population estimates, densities and biomass of large herbivores in Simanjiro Plains, Northern Tanzania. *East African Wildlife Journal* **19**, 225–38.
Kaihula, S. (1983). Vegetation change in Lerai Forest, Ngorongoro, and its probable causes. MSc Thesis, University of Dar es Salaam.
Kaiza-Boshe, T. (1988). Editorial. *Kakakuona. Magazine of the Wildlife Protection Fund*. Maliasili. Dar es Salaam.
Kametz, H. (1962). *Maasailand comprehensive report, Ngorongoro Conservation Area II*. Water Development and Irrigation Department, Arusha Tanzania.
Kayera, J.A. (1985). *An overview of the Ngorongoro Conservation Area situation*. Background paper for Serengeti Wildlife Research Centre workshop 'Toward a regional conservation strategy for the Serengeti' IUCN Regional Office for Eastern Africa, Nairobi.
Kikula, I.S. (1981). *Woodland and forest a real change in the Ngorongoro Conservation Area, Tanzania*. Background paper for the 1982 management plan. BRALUP, University of Dar es Salaam.
King, R.B. (1980). *Landform and erosion in the Ngorongoro Conservation Area. Background paper for the 1982 management plan*. BRALUP, University of Dar es Salaam.
King, R.B. (1982). Rapid rural appraisal with LANDSAT imagery: the Tanzanian experiment. *Zeitschrift Geomorphologie supplement* BAND, **44**, 5–20.
Kipury, N. (1983). *Oral literature of the Maasai*. Nairobi and London: Heinemann Educational.
Kitomari, N.N. (1986). Preface and opening address (Principal Secretary, Ministry of Natural Resources and Tourism). In *Toward a regional conservation strategy for the Serengeti*, ed. R. Malpas and S. Perkin, pp. 8–9. Report of a workshop held at Serengeti Wildlife Research Centre, Seronera Tanzania. 2–4 December 1985. IUCN Report/Ministry of Natural Resources and Tourism, Government of Tanzania. Dar es Salaam and IUCN regional office

for East Africa, Nairobi, Kenya.

Kiwia, H.Y.D. (1983). The behaviour and ecology of the black rhinoceros *Diceros bicornis* L. in the Ngorongoro Crater: MSc Thesis. University of Dar es Salaam.

Kjaerby, F. (1979). *The development of agropastoralism among the Barabaig in Hanang District.* BRALUP Research Paper No. **56**, University of Dar es Salaam.

Kjekshus, H. (1977). *Ecology, control and development in East African history.* Heinemann.

Kreulen, D. (1975). Wildebeest habitat selection on the Serengeti Plain, Tanzania, in relation to calcium and lactation: a preliminary report. *East African Wildlife Journal*, **13**, 297–304.

Kruuk, H. (1972). *The spotted hyena.* University of Chicago Press.

Kruuk, H. (1975). *Hyaenas.* Oxford University Press.

Kurji, F. (1981). *Human population trends within and around the Ngorongoro Conservation Area: the demographic settings.* BRALUP Research Report No. **44**, New Series, University of Dar es Salaam.

Lamphear, J. (1988). The people of the grey bull: the origin and expansion of the Turkana. *Journal of African History*, **29**, 27–39.

Lamprey, H. (1964). Estimation of the large mammal densities, biomass and energy exchange in the Tarangire Game Reserve and the Maasai Steppe in Tanzania. *East African Wildlife Journal*, **2**, 1–46.

Lamprey, H. (1983). Pastoralism yesterday and today: the overgrazing problem. In *Tropical savannas*, ed. F. Bourliere, pp. 643–66. Amsterdam: Elsevier.

Laws, R.M. (1970). Elephants as agents of habitat and landscape change in East Africa. *Oikos*, **21**, 1–15.

Le Houérou, H.N. and Hoste, C. (1977). Rangeland production and annual rainfall relations in the Mediterranean Basin and in the African Sahelo–Sudanian zone. *Journal of Range Management*, **30**, 181–9.

Le Houérou, H.N. (1980). (ed.) *Browse in Africa: the current state of knowledge.* Addis Abbaba: ILCA.

Leader-Williams, N. and Albon, S. (1988). Allocation of resources for conservation. *Nature*, **336**, 533–5.

Leakey, L.S.B. (1965). *Olduvai Gorge 1951–1961*, vol. 1. Cambridge University Press.

Leakey, M.D. (1966). Excavation of burial mounds in Ngorongoro Crater. *Tanzania Notes and Records*, **66**, 123–35.

Leakey, M.D. (1971). *Olduvai Gorge*, vol. 3. Cambridge University Press.

Leakey, M.D. and Harris, J.M. (eds.) (1987). *Laetoli: a Pliocene site in Northern Tanzania.* Oxford: Clarendon Press.

Ledger, H. (1977). The comparative energy requirements of penned and exercised steers for long term maintenance at constant liveweight. *Journal of Agricultural Science*, **88**, 27–33.

Legislative Council of Tanganyika (1956). *The Serengeti National Park.* Sessional paper No. **1**. Government Printer, Dar es Salaam.

Lewis, J.G. (1977). Game domestication for animal production in Kenya: activity patterns of eland, oryx, buffalo and cattle. *Journal of Agricultural Science (Cambridge)*, **89**, 551–63.

Lewis, J.G. (1978). Game domestication for animal production in Kenya: shade behaviour and factors affecting the herding of eland, oryx, buffalo and zebu cattle. *Journal of Agricultural Science (Cambridge)*, **90**, 587–95.

Lind, E. and Morrison, M. (1974). *East African Vegetation.* London: Longman.

Lindsay, K. (1987). Integrating parks and pastoralists: some lessons from Amboseli. In *Conservation in Africa: people, policies and practice* ed. D. Anderson and R. Grove, pp. 149–67. Cambridge University Press.

Little, M.A. (1980). Designs for human biological research among savanna pastoralists. In *Human Ecology in Savanna Environments*, ed. D. Harris. Academic Press.

Little, M., Galvin, K. and Leslie P. (1987). Health and energy requirements of nomadic Turkana

pastoralists. In *Coping with uncertainty in food supply*, eds. I. de Garine and G.A. Harrison, pp. 290–317. Oxford University Press.
Little, P. (1983). The livestock–grain connection in northern Kenya: an analysis of pastoral economics and semi-arid land development. *Rural Africana*, **15–16**, 91–108.
Lundgren, B. (1978). *Soil conditions and nutrient cycling under natural and plantation forests in Tanzanian Highland.* Reports in forest ecology and forest soils No. **31**. Department of forest soils, Swedish University of Agricultural Sciences, Uppsala.
Lundgren, B. and Lundgren, L. (1972). Comparison of some soil properties in one forest and two grassland ecosystems on Mt Meru, Tanzania. *Geografiska Annaler A.*, **54**, 3–4.
Lusigi, W. (1980). New approach to wildlife conservation in Kenya. *Ambio*, **10**, 87–92.
Lusigi, W. (1981). *Combatting desertification and rehabilitating degraded production systems in Northern Kenya.* IPAL Technical Report A-4, Man and Biosphere programme, UNESCO.
Mace, R. (1988). A model of herd composition that maximises household viability, and its potential application in the support of pastoralists under stress. *Pastoral Development Network*, **26b**, Overseas Development Institute London.
Machange, J.H. (1987). *A preliminary consultancy report on dry season livestock/wildlife interactions.* Ngorongoro Conservation and Development Project. IUCN Regional Representative Office, Nairobi, Kenya.
Machange, J.H. (1988). *Livestock/wildlife interactions.* Technical Report No. 4, Ngorongoro Conservation and Development Project, IUCN Regional Office, Nairobi, Kenya.
MacKenzie, D. (1987). A hoof up for Africa's livestock. *New Scientist*, **114**, 38–9.
MacKenzie, J.M. (1987). Chivalry, social Darwinism and ritualised killing: the hunting ethos in Central Africa up to 1914. In *Conservation in Africa: people, policies and practice*, ed. D. Anderson and R. Grove, pp. 41–62. Cambridge University Press.
MacKenzie, W. (1973). *The livestock economy of Tanzania.* Economics Research Bureau. University of Dar es Salaam.
Maddock, L. (1979). The 'migration' and grazing succession. In *Serengeti: Dynamics of an ecosystem*, ed. A. Sinclair and M. Norton-Griffiths, pp. 104–29. University of Chicago Press.
Makacha, S. and Frame, G. (1977a). *Kitete corridor observation recommendations (Mbulu District, Arusha Region, Tanzania).* Serengeti Research Institute, Seronera Tanzania.
Makacha, S. and Frame, G. (1977b). *Lositete conservation recommendations (Mbulu District Arusha Region Tanzania).* Serengeti Research Institute Seronera Tanzania.
Makacha, S. and Frame, G. (1986). *Population trends and ecology of Maasai pastoralists and livestock in Ngorongoro Conservation Area Tanzania.* Serengeti Wildlife Research Institute Contribution No. **338**, NCAA publication No. **84**. Arusha, Tanzania: NCAA.
Makacha, S. Mollel, C.L. and Rwezaura, J. (1979).The conservation status of the black rhinoceros in Ngorongoro Crater Tanzania. *African Journal of Ecology*, **17**, 97–103.
Makacha, S., Msingwa, M. and Frame, G. (1982). Threats to the Serengeti herds, *Oryx*, **XVI**. 437–44.
Makacha, S. and Ole Sayalel, P. (1987). *The problem of agriculture at Ngorongoro.* NCAA internal document (Law Enforcement) mimeo, 13pp. NCAA, Arusha, Tanzania.
Malpas, R. and Perkin, S. (eds) (1986). *Towards a regional conservation strategy for the Serengeti.* IUCN Report/Ministry of Natural Resources and Tourism, Government of Tanzania Dar es Salaam and IUCN Regional Office for East Africa, Nairobi, Kenya.
Martin, R.B. (1986). Communal area management plan for indigenous resources (Project CAMPFIRE) in *Conservation and wildlife management in Africa*, ed. R. Bell and E. McShane-Caluzi, pp. 279–96. Washington: US Peace Corps.
Mascarenhas, A. (1983). Ngorongoro: a challenge to conservation and development. *Ambio*, **12**, 146–52.
McCabe, J., Schofield, E.G. and Pederson, G. (1989). *Food security and nutritional status.* Technical Report No 10, Ngorongoro Conservation and Development Project, IUCN

Regional Office. Nairobi, Kenya.
McCown, R., Haaland, G. and de Haan, C. (1979). The interaction between cultivation and livestock production in semi-arid Africa. *Ecological Studies*, **34**, 297–332.
McCracken, J. (1987). Colonialism, capitalism and ecological crisis in Malawi: a reassessment. In *Conservation in Africa: people, policies and practice*, ed. D. Anderson and R. Grove, pp. 63–78. Cambridge University Press.
McDonald, I. and Frame, G. (1988). The invasion of introduced species into nature reserves in tropical savannas and dry woodlands. *Biological Conservation*, **44**, 67–94.
McKay, A.D. (1971). Seasonal and management effects on the composition and availability of herbage, steer diet and liveweight gains in a *Themeda triandra* grassland in Kenya. *Journal of Agricultural Science (Cambridge)*, **76**, 1–26.
McLaughlin, R.T. (1970). Nairobi National Park Census, 1968. *East African Wildlife Journal*, **8**, 203.
McNaughton, S.J. (1979). Grassland–herbivore dynamics. In *Serengeti: Dynamics of an ecosystem*, ed A. Sinclair and M. Norton-Griffiths, pp. 46–81. University of Chicago Press.
McNaughton, S.J. (1983). Serengeti grassland ecology: the role of composite environmental factors and contingency in community organisation. *Ecological Monographs*, **54**, 291–320
McNaughton, S.J. (1985). Ecology of a grazing ecosystem: the Serengeti. *Ecological Monographs*, **55**, 259–94.
McNaughton, S.J. (1988). Mineral nutrition and spatial concentrations of African ungulates. *Nature*, **334**, 343–45.
McNaughton, S.J. (1990). Mineral nutrition and seasonal movements of African migratory ungulates. *Nature*, **345**, 613–15.
McNeely, J.A. and Pitt, D. (1985). *Culture and conservation: the human dimension in environmental planning*. Croom Helm.
Meadows, S. and White, J. (1979). Structure of the herd and determinants of offtake rates in Kajiado District in Kenya 1962–1977. *Pastoral Development Network*, vol. 7d. London: Overseas Development Institute.
Mefit Babtie (1983). *Development studies in the Jonglei canal area. Range ecology survey, livestock investigation and water supply*. A report to the Jonglei Executive Organ, Government of Sudan. 10 volumes. Consultancy report edited by S.M. Cobb. Glasgow: Mefit Babtie.
Mehlmann, M. (1977). Excavations at Nasera Rock, Tanzania. *Azania*, **12**, 111–18.
Mehlmann, M. (1979). Mumba-Hohle revisited: the relevance of a forgotten excavation to some current issues in East African prehistory. *World Archaeology*, **11**, 80–94.
Merker, M. (1904). *Die Masai*. D. Reimer: Berlin.
Mol, F. (1978). *Maa: a dictionary of the Maasai language and folklore. English–Maasai*. Marketing and Publishing, Nairobi.
Moris, J. (1981). A case in rural development: the Maasai Range development project. In *Managing Induced Rural Development*, ed. J. Moris, pp. 99–113. Bloomington Indiana: International Development Institute.
Mshanga, P. and Ndunguru, J. (1983). *Ten year report of the Ngorongoro Conservation Area* (unpublished report). NCAA, Tanzania.
Mturi, A.A. (1976). New hominid from Lake Ndutu, Tanzania. *Nature*, **262**, 484–85.
Mturi, A.A. (1981). *The archaeological and palaeontological resources of the Ngorongoro Conservation Area*. Ministry of National Culture and Youth, Dar es Salaam. Background paper for the 1982 management plan.
Mwalyosi, R.B. (1977). Vegetation changes in Lake Manyara National Park. MSc Thesis, University of Dar es Salaam.
Mwalyosi, R.B. (1981). Ecological changes in Lake Manyara National Park. *African Journal of Ecology*, **19**, 201–4.
Mwalyosi, R.B. (1987). Decline of *Acacia tortilis* in Lake Manyara National Park Tanzania.

African Journal of Ecology, 25, 51–4.
NCAA Board of Directors (1980). Minutes of 8th-meeting of NCAA Board, Arusha 31.12.1980: Minute 1.8: Mapendekezo ya muda mrefu – Lazima binadamu waondoke.
NCAA (1980). *Hesabu ya mifugo eneo la hifadhi ya Ngorongoro*. October 1980.
NCAA (1987). Livestock and wildlife census April 1987 (NEMP Total Crater Count).
NCDP (1987) 1987 Wet season ground census: preliminary report. Ngorongoro Conservation and Development Project, NCAA, P.O. Box 1 Arusha or IUCN Regional Office, Nairobi.
NCDP (1989). *Annual report of activities February 1988–January 1989*. Ngorongoro Conservation and Development Project, IUCN Regional Office, Nairobi.
Ndagala, D. (1982). Operation *Imparnati*: the sedentarisation of the pastoral Maasai in Tanzania. *Nomadic Peoples*, **10**, 28–39.
Ndagala, D. (1990a). Pastoral territoriality and land degradation in Tanzania. In *From water to world-making*, ed. G. Palsson, pp. 175–88. Scandinavian Institute of African Studies.
Ndagala D. (1990b). Pastoralists and the State in Tanzania. *Nomadic Peoples*, 25–27, 51–64.
Ndolanga, M.A. (1985). *The role of wildlife utilization in a conservation and development strategy for the Serengeti Region*. Background paper for the workshop 'Towards a regional conservation strategy for the Serengeti'. IUCN Regional Office, Nairobi.
NEMP (1987). *Ngorongoro Ecological Monitoring Program semi-annual report*. NCAA, Tanzania.
NEMP (1989). *Semi-annual report: April 1989*. NCAA, PO Box 1, Ngorongoro Crater.
Nestel, P.S. (1985). Nutritional status of Maasai women and childen in relation to subsistence food production. PhD Thesis, London University.
Nestel, P.S. (1986). A society in transition: developmental and seasonal influences on the nutrition of Masai women and children. *Food and Nutrition Bulletin*, **8**, 2–18.
Nestel P. (1989) Food intake and growth in the Maasai. *Ecology of Food and Nutrition*, **23**, 17–30
Newbould, J. (1961). Ngorongoro pasture research and range management scheme: outline of present position and future plans. In *Report on the Ngorongoro Pasture research project*, ed. P. Glover, Appendix B, Ngorongoro Conservation Unit.
Niamir, M. (1990). *Herders' decision-making in natural resources management in arid and semi-arid Africa*. Community Forestry Note 4. Rome: Food and Agriculture Organization.
Nicholson, S. and Entekhabi, D. (1986). The quasi-periodic behaviour of rainfall variability in Africa and its relationship to the Southern Oscillation. *Archiv. für Meteorologie, Geophysik und Bioclimatologie: Series A, Meteorology and Atmospheric Physics*, **34**, 311–48.
Norton-Griffiths, M. (1978). *Counting animals*, 2nd edn. Nairobi, Kenya: African Wildlife Foundation.
Norton-Griffiths, M. (1979). The influence of grazing browsing and fire on the vegetation dynamics of the Serengeti In *Serengeti: Dynamics of an ecosystem* ed. A. Sinclair and M. Norton-Griffiths, pp. 310–52. University of Chicago Press.
Norton-Griffiths, M, Herlocker, D. and Pennycuick, D. (1975). The patterns of rainfall in the Serengeti ecosystem, Tanzania. *East African Wildlife Journal*, **13**, 347–74.
Noy-Meir, I. (1975). Stability of grazing systems: an application of predator prey graphs. *Journal of Ecology*, **63**, 459–81.
Noy-Meir, I. (1978). Grazing and production in seasonal pastures: analysis of a simple model. *Journal of Applied Ecology*, **15**, 809–35.
Noy-Meir, I. (1982). Stability of plant–herbivore models and possible applications to savanna. *Ecological Studies*, **42**, 591–609.
Oba, G. (1985). Local participation in guiding extension programs: a practical proposal. *Nomadic Peoples*, **18**, 27–46.
Odner, K. (1972). Excavations at Narosura, a stone bowl site in the southern Kenya highlands. *Azania*, **7**, 25–92.
Office of Technology Assessment (1984). *Africa tomorrow: issues in technology, agriculture and*

US foreign aid. A technical memorandum. Office of Technology Assessment, US Congress Washington DC.

Ogallo, L. and Nassib, I. (1984). Drought pattern and famines in East Africa during 1922–1983. Extended abstracts. *Second WMO Symposium on Meteorology*. Geneva.

Ole Kuwai, J.L. (1980). *A preliminary survey of Ngorongoro Conservation Area.* 31pp mimeo, NCAA.

Ole Kuwai, J.L. (1981). *Livestock development in the Ngorongoro Conservation Area.* Background paper for BRALUP management plan. University of Dar es Salaam.

Ole Parkipuny, L. (1981). *On behalf of the people of Ngorongoro: a discussion of the question, does the future of Ngorongoro lie in livestock vs wildlife or livestock and wildlife.* Background paper for a new management plan in Ngorongoro. BRALUP, University of Dar er Salaam.

Ole Parkipuny, L. (1983). Maasai struggle for home rights in Ngorongoro. Anthropology of Human Rights Symposium. *XI International Congress of Anthropological and Ethological Sciences*, Quebec City and Vancouver.

Ole Saibull, S.A., (1978). The policy process: the case of conservation in the Ngorongoro Crater Highlands. *Tanzania Notes and Records*, **83**, 101–15.

Ole Saitoti, T. (1986). *The worlds of a Maasai warrior: an autobiography.* London: André Deutsch.

Oliver, R. (1982). The Nilotic contribution to Bantu Africa. *Journal of African History*, **23**, 433–42.

O'Rourke, J., Frame G. and Terry P. (1975). Progress on control of *Eleusine jaegeri* Pilg. in East Africa. *Proceedings of the Academy of Natural Sciences*, **21**, 67–72.

O'Rourke, J., Terry P. and Frame G. (1976). Experimental results of *Eleusine jaegeri* Pilg. control in East African highlands. *East African Agriculture and Forestry Journal*, **41**, 253–65.

Orr, J.B. and Gilks, J.L. (1931). Studies of nutrition: the physique and health of two African tribes (Maasai and Kikuyu). *Medical Research Council (GB) Special Report Series*, **155**, 1–93.

Owen-Smith, R. (ed.) (1983). *Management of large mammals in African Conservation Areas.* HAUM Educational Publishers: Pretoria.

Pacey, A. and Payne, P. (eds) (1985). *Agricultural development and nutrition.* Hutchinson with FAO/UNICEF.

Page-Jones, F.H. (1948). Water in Masailand. *Tanganyika Notes and Records*, **26**, 51–9.

Parker, I.S.C. (1981). Perspectives on Game Cropping. In *The status and utilisation of wildlife in Arusha region, Tanzania.* Ecosystems LTD 1980, Annex 1. Consultancy report to the Tanzanian Government. Nairobi: Ecosystems Ltd.

Parker, I.S.C. (1984). Perspectives on wildlife cropping or culling. In *Conservation and wildlife management in Africa*, ed. R. Bell and E. McShane-Caluzi, pp. 233–54. Washington US Peace Corps Office of Training and Program Support, Forestry and Natural Resources Sector.

Parker, I.S.C. (1985). *Three points relevant to game cropping in the Serengeti Region.* Background paper for the Serengeti workshop: Towards a regional conservation plan for the Serengeti. IUCN regional office for East Africa, Nairobi.

Payne, W. and MacFarlane, J. (1963). A brief study of cattle browsing behaviour in a semi arid area of Tanganyika. *East African Agriculture and Forestry Journal*, **29**, 131–3.

Pearce, D. (1988). Economists befriend the Earth. *New Scientist*, **120** (1639), 34–9.

Pearsall, W.H. (1957). *Report on an ecological survey of the Serengeti National Park, Tanganyika.* London: Fauna Preservation Society. Later reprinted in *Oryx*, **4**, 71–136.

Peden, D.G. (1987). Livestock and wildlife population distributions in relation to aridity and human population in Kenya. *Journal of Range Management*, **40**, 67–71.

Pellew, R. (1983). The impacts of elephant, giraffe and fire upon the *Acacia tortilis* woodlands of the Serengeti. *African Journal of Ecology*, **21**, 41–74.

Pennington, H. (1983). *A living trust: Tanzanian attitudes towards wildlife conservation.* College of African Wildlife Management, Mweka.

Pennycuick, C. (1979). Energy costs of locomotion and the concept of 'foraging radius' In *Serengeti: Dynamics of an ecosystem*, ed. A. Sinclair and M. Norton-Griffiths, pp. 164–84

Chicago University Press.
Pennycuick, L. (1975). Movements of the migratory wildebeest population in the Serengeti area between 1960 and 1973. *East African Wildlife Journal*, **13**, 65–87.
Pennycuick, L. and Norton-Griffiths, M. (1976). Fluctuations in the rainfall of the Serengeti ecosystem. *Journal of Biogeography*, **3**, 125–140.
Perkin, S. (1987). Wet season ground census 1987: preliminary report. Ngorongoro Conservation and Development Project.
Peterson, D. (1978). Seasonal distribution and interactions of cattle and wild ungulates in Masailand, Tanzania. MSc Thesis, Virginia Polytechnic Institute.
Peterson, D. and McGinnes, B.S. (1979). Vegetation of south Masailand, Tanzania. A range habitat classification. *East African Agriculture and Forestry Journal*, **44**, 252–71.
Phillipson, D.W. (1977). *The later prehistory of Eastern and Southern Africa.* London: Heinemann.
Phillipson, D.W. (1985). *African Archaeology*. Cambridge University Press.
Pickering, R. (1960). *The topography and geology of the Ngorongoro Conservation Area.* Report to NCA.
Pickering, R. (1968). *Ngorongoro's geological history.* Ngorongoro Conservation Area Booklet No 2. East African Literature Bureau, Nairobi.
Pienaar, U. de V. (1983). Management by intervention: the pragmatic/economic option. In *Management of large mammals in African Conservation Areas*, ed. R. Owen-Smith, pp. 23–36. Pretoria: HAUM Educational.
Poirier, F. (1987). *Understanding human evolution.* Prentice Hall.
Potts, R. (1988). *Early human activities at Olduvai.* New York: Aldine de Gruyter.
Pratt, D.J. and Gwynne, M.D. (eds) (1977). *Rangeland management and ecology in East Africa.* London: Hodder and Stoughton.
Pullan, N. (1978). Condition scoring of white Fulani cattle. *Tropical Animal Health and Production*, 10, 118–20.
Pusey, A. and Packer, C. (1987). The evolution of sex-biased dispersal in lions. *Behaviour*, **101**, 275–310.
Raikes, P. (1981). *Livestock development and policy in East Africa.* Centre for Development Research, Copenhagen. Scandinavian Institute of African Studies, Uppsala.
Rasmusson, E. (1987). Global climate change and variability: effects on drought and desertification in Africa. In *Drought and Hunger in Africa*, ed. M. Glantz, pp. 3–22. Cambridge University Press.
Read, D. and Chapman, P. (1982). *Waters of the Sanjan.* Nairobi: General Printers.
Reck, H. (1933). *Oldoway: die Schlucht des Urmenschens.* Leipzig: F.A. Brockhaus.
Rees, W.A. (1974). Preliminary studies into bush utilization by cattle in Zambia. *Journal of Applied Ecology*, **11**, 207–14.
Richards, P. (1985). *Indigenous agricultural revolution.* Hutchinson.
Rigby P. (1985). *Persistent Pastoralists: Nomadic societies in transition.* London: Zed Books.
Riney, T. (1982). *Study and management of large mammals.* Chichester and New York: John Wiley and Sons.
Robertshaw, D. and Katongole, C.B. (1969). Adrenocortical activity and intermediary metabolism of *Bos indicus* and *Bos taurus* in the high altitude tropics. *International Journal of Biometeorology*, **13**, 101.
Robertshaw, P. (ed.) (in press) *Early pastoralists of south western Kenya.*
Robertshaw, P. and Collett, D. (1983a) The identification of pastoral peoples in the archaeological record: an example from East Africa. *World Archaeology*, **15**, 67–78.
Robertshaw, P. and Collett, D. (1983b). A new framework for the study of early pastoral communities in East Africa. *Journal of African History*, **24**, 289–301.
Rodgers, W.A. (1981a). The status of natural resources in Ngorongoro Conservation Area. Background paper to IRA/UNESCO Management plan for Ngorongoro. Institute of Resource Assessment, University of Dar es Salaam, Tanzania.

Rodgers, W.A. (1981b). The distribution and conservation status of colobus monkeys in Tanzania. *Primates*, **22** (1), 33–45.

Rodgers, W.A. (1982a). Decline in large mammal populations on the Rukwa Valley grasslands in South West Tanzania. *African Journal of Ecology*, **20**, 13–22.

Rodgers, W.A. (1982b). *Replacement of two defunct diesel pumps with windmill pumps on water boreholes*. EEC Village Micro Project Proposal.

Rodgers, W.A. (1988). Pastoralism and conservation: can they coexist? *Fourth International Rangeland Congress*. New Delhi.

Rodgers, W.A. and Homewood, K.M. (1982). Species richness and endemism in the Usambara mountain forests, Tanzania. *Biological Journal of the Linnean Society*, **18**, 197–242.

Rodgers, W.A. and Homewood, K.M. (1986). Cattle dynamics in a pastoralist community in Ngorongoro Tanzania during the 1982–1983 drought. *Agricultural Systems*, **22**, 33–51.

Rodgers, W.A. and Nicholson, B.D. (1973). *Guidelines for long term development of selected game areas with suggested management and development plans* Ministry of Natural Resources and Tourism (Game Division Projects), Dar es Salaam. 51pp.

Rodgers, W.A. and O'Rourke, M. (1987). *A survey of livestock disease in pastoralist communities in Ngorongoro Conservation Area, Tanzania*. NCAA.

Rodgers, W.A. Owen, C. and Homewood, K.M. (1983). Biogeography of East African forest mammals. *Journal of Biogeography*, **9**, 41–54.

Rogerson, A. (1970). Food intake and energy utilization by cattle. *East African Agriculture and Forestry Journal*, **36**, 195.

Rogerson, A., Ledger, H. and Freeman, G. (1968). Food intake and liveweight gain comparisons of *Bos indicus* and *Bos taurus* steers on a high plane of nutrition. *Animal Production*, **10**, 373–80.

Rollinson, D., Harker, K., Taylor, J. and Leech, F. (1956). Studies on the habits of zebu cattle. IV Errors associated with recording techniques. *Journal of Agricultural Science*, **47**, 1–5.

Rose, G. (1975). Buffalo increase and seasonal use of Ngorongoro Crater. *East African Wildlife Journal*, **13**, 385–8.

Rossiter, P.B., Jessett, D.M. and Karstad, L. (1983). Role of wildebeest fetal membranes and fluids in the transmission of malignant catarrhal fever virus. *The Veterinary Record*, **113**, 150–2.

Said, A.N. (1971). In vivo digestibility and nutritive value of Kikuyu grass *Pennisetum clandestinum* with a tentative assessment of its yield of nutrients. *East African Agriculture and Forestry Journal*, **37**, 15–21.

Sandford, S. (1982). Pastoral strategies and desertification: opportunism and conservatism in dry lands. In *Desertification and development: dryland ecology in social perspective*. ed. B. Spooner and H. Mann, pp. 61–80. Academic Press: London.

Sandford, S. (1983). *Management of pastoral development in the third world*. London: Overseas Development Institute, Chichester and New York: John Wiley and Sons.

Schaller, G. (1972). *The Serengeti Lion*. University of Chicago Press.

Schmidt, W. (1975). Plant communities on permanent plots of the Serengeti Plains. *Vegetatio*, **30**, 133–45.

Scott, G. (1985). Rinderpest in the 1980s. *Progress in Veterinary Microbiology and Immunology*, **1**, 145–75.

Serengeti Committee of Enquiry (1957). *Report of the Serengeti Committee of Enquiry*. Government Printer, Dar es Salaam.

Serengeti Ecological Monitoring Programme (1987). *Annual Report*. Serengeti Research Institute.

Serengeti Ecological Monitoring Programme (1988). *Six-monthly report*. Serengeti Research Institute.

Shepherd, N. and Caughley, G. (1987). Options for the management of kangaroos. In

Kangaroos: their ecology and management in the sheep rangelands of Australia, ed. G. Caughley, N. Shepherd and J. Short, pp. 188–219. Cambridge University Press.
Shivji, I.G. (ed.) (1975). *Tourism and social development*. Dar es Salaam: Tanzania Publishing House.
Simpson, J. (1984a). Overview. In *Livestock development in subSaharan Africa. Constraints, prospects, policy*, ed. J. Simpson and P. Evangelou, pp. 1–4. Boulder, Colorado: Westview Press.
Simpson, J. (1984b). Problems and constraints, goals and policy: conflict resolution in development of subSaharan Africa's livestock industry. In *Livestock development in subSaharan Africa. Constraints, prospects, policy*, ed. J. Simpson and P. Evangelou, pp. 5–20. Boulder, Colorado: Westview Press.
Simpson, J. and Evangelou, P. (eds) (1984) *Livestock development in subSaharan Africa. Constraints, prospects, policy*. Boulder, Colorado: Westview Press.
Sinclair, A. (1975). The resource limitation of trophic levels in tropical grassland ecosystems. *Journal of Animal Ecology*, **44**, 497–520.
Sinclair, A. (1977). *The African Buffalo*. University of Chicago Press.
Sinclair, A. (1979). Dynamics of the Serengeti ecosystem pp. 1–30 In *Serengeti: Dynamics of an ecosystem*, ed. A. Sinclair and M. Norton-Griffiths, pp. 1–30. Chicago University Press.
Sinclair, A. (1983a). The function of distance movements in vertebrates In *The ecology of animal movement*, I. Swingland and P.J. Greenwood, pp. 240–58. Oxford: Clarendon Press.
Sinclair, A. (1983b). Management of African conservation areas as ecological baseline controls. In *Management of large mammals in African Conservation Areas*. ed. R. Owen-Smith, pp. 13–22. Pretoria: HAUM Educational Publishers.
Sinclair, A. (1985). Does interspecific competition or predation shape the African wildlife community? *Journal of Animal Ecology*, **54**, 899–918.
Sinclair, A. Dublin, H. and Borner, M. (1985). Population regulation of Serengeti wildebeest: a test of the food hypothesis. *Oecologia*, **65**, 266–8.
Sinclair, A. and Fryxell, J. (1985). The Sahel of Africa: ecology of a disaster. *Canadian Journal of Zoology*, **63**, 987–94.
Sinclair, A. and Norton-Griffiths, M. (eds) (1979). *Serengeti: Dynamics of an ecosystem*. Chicago University Press.
Sinclair, A. and Norton-Griffiths, M. (1982). Does competition or facilitation regulate migrant ungulate populations in the Serengeti? A test of hypotheses. *Oecologia*, **53**, 364–9.
Sindiga, I. (1984) Land and population problems in Kajiado and Narok, Kenya. *African Studies Review*, **27**, 23–39.
Sindiga, I. (1987) Fertility control and population growth among the Maasai. *Human Ecology*, **15**, 53–66.
Spencer, P. (1973) *Nomads in alliance: symbiosis and growth among the Rendille and Samburu of Kenya*. Oxford University Press.
Spencer, P. (1988). *The Maasai of Matapato: a study of rituals of rebellion*. International African Library, Manchester University Press.
Spinage, C. (1973). A review of ivory exploitation and elephant population trends in Africa. *East African Wildlife Journal*, **11**, 281–9.
Ssemakula, J. (1983). A comparative study of hoof pressures of wild and domestic ungulates. *African Journal of Ecology*, **21**, 325–8.
Stelfox, J.B. (1986) Effects of livestock enclosures (bomas) on the vegetation of the Athi Plains (Kenya). *African Journal of Ecology*, **24**, 41–5.
Struhsaker, T., Odegaard, A., Ruffo, C. and Steele, R. (1989). Forest Conservation and Management. Technical Report No. 5, Ngorongoro Conservation and Development Project, IUCN Regional Office, Nairobi.
Sutherst R.W. (1987). Ectoparasites and herbivore nutrition In *The Nutrition of Herbivores*. ed.

J.B. Hacker and J.H. Ternouth, pp. 191–210. Sydney: Academic Press.
Sutton, J.E.G. (1974). The settlement of East Africa. In *Zamani. A survey of East African History*, Nairobi and London: 2nd edn, ed. B.A. Ogot, pp. 70–97. East African Publishing House/Longmans.
Sutton, J.E.G. (1978). Engaruka and its waters. *Azania*, **13**, 37–70.
Sutton, J. (1984). Irrigation and soil conservation in African agricultural history: with a reconsideration of the Inyanga terracing (Zimbabwe) and Engaruka irrigation works (Tanzania). *Journal of African History*, **25**, 25–41.
Swift, J. (1983). *The start of the rains*. Research memo, Sussex: Institute for Development Studies.
Swingland, I. and Greenwood P.J. (eds) (1983) *The ecology of animal movement*. Oxford: Clarendon Press.
Synnott, T.J. (1979). *A report on the status, importance and protection of the montane forests*. IPAL Technical Report D-2a. Paris: UNEP-MAB.
Talle, A. (1988). *Women at a loss. Changes in Maasai pastoralism and their effects on gender relations*. Stockholm Studies in Social Anthropology no. 19.
Talle, A. (1990). Ways of milk and meat among the Maasai. In *From water to world making. African models and arid lands*, ed. G. Palsson, pp. 73–93. Scandinavian Institute of African Studies, Uppsala.
Tanzania Notes and Records (1974) *Kilimanjaro*. (Revised edition of the March 1965 issue of Tanganyika Notes and Records no. 64). 162pp. Dar es Salaam. The Tanzania Society.
Taylor, M.E. (1988). *Multiple land-use planning and management*. Technical Report No. 2, Ngorongoro Conservation and Development Project. IUCN Regional Office, Nairobi.
Timberlake, L. (1988). Sustained hope for development. *New Scientist*, **119**, 60–3.
Tolsma, D, Ernst, W., and Verwey, R. (1987). Nutrients in soils and vegetation around two artificial waterpoints in Eastern Botswana. *Journal of Applied Ecology*, **24**, 991–1000.
Tomikawa, M. (1970). The distribution and the migrations of the Datoga tribe. *University of Kyoto African Studies*, **5**, 1–46.
Toulmin, C. (1983). Herders and farmers or farmer-herders and herder-farmers? *Pastoral Development Network*, **15**d, 1–22. Overseas Development Institute, London.
Trail, J. and Gregory, K. (1984). Animal breeding in subSaharan Africa: towards an integrated programme for improving productivity. In *Livestock development in subSaharan Africa. Constraints, prospects, policy*, ed. J. Simpson and P. Evangelou, pp. 107–122. Boulder, Colorado: Westview Press.
van Orsdol K.G. (1981) Lion predation in Rwenzori National Park, Uganda. PhD Thesis, Cambridge University.
Wagenaar-Brouwer, M. (1985). Preliminary findings on the diet and nutritional status of some Tamasheq and Fulani groups in the Niger Delta of central Mali. In *Population, Health and Nutrition in the Sahel*, ed A. Hill, pp. 226–53. Routledge and Kegan Paul.
Walker, B. and Noy-Meir, I. (1982). Aspects of the stability and resilience of savanna ecosystems. *Ecological Studies*, **42**, 556–90.
Waller, R.D. (1976). The Maasai and the British 1895–1905. *Journal of African History*, **17**, 529–53.
Waller, R.D. (1979). *The lords of East Africa: the Maasai in the mid-nineteenth century (c. 1840–1880)*. PhD Thesis, Cambridge University.
Waller, R.D. (1985). Ecology, migration and expansion in East Africa. *African Affairs*, **84**, 347–70.
Waller, R.D. (1988). *Emutai*: crisis and response in Maasailand 1883–1902. In *The Ecology of Survival: case studies from Northeast African History*. ed. D. Johnson and D. Anderson, pp. 73–114. Lester Crook Academic Publishing/Westview Press.
Waller, R.D. (1990). Tsetse fly in western Narok, Kenya. *Journal of African History*, **31**, 81–101.

Warren, A. and Agnew, C. (1988). *An assessment of desertification and land degradation in arid and semi-arid areas.* Paper No. 2, International Institute for Environment and Development (IIED) Drylands Programme.

Watson, R.M., Graham, A. and Parker, I. (1969). A census of the large mammals of Loliondo Controlled Area, Northern Tanzania. *East African Wildlife Journal*, **7**, 43–59.

Watson, R.M. and Kerfoot, O. (1964). A short note on the intensity of grazing of the Serengeti Plain by plains game. *Zeitschrift für Säugetierkunde*, **29**, 317–20.

Western, D. (1971). The human–animal equation. *African Wildlife Leadership Foundation News*, **6**(2), 3–6. Nairobi, Kenya.

Western, D. (1973). *The structure, dynamics and changes of the Amboseli ecosystem.* PhD Thesis, University of Nairobi.

Western, D. (1975). Water availability and its influence on the structure and dynamics of a savanna large mammal community. *East African Wildlife Journal*, **13**, 265–86.

Western, D. (1982). Amboseli National Park: enlisting landowners to conserve migratory wildlife. *Ambio*, **11**, 302–8.

Western, D. (1984). Amboseli National Park: Human values and the conservation of a savanna ecosystem. In *National parks, conservation and development: the role of protected areas in sustaining society.* ed. J. McNeely and K. Miller, pp. 93–100. IUCN/UNEP conference. Smithsonian Institute Press.

Western, D. and Dunne, T. (1979). Environmental aspects of settlement site decisions among pastoral Maasai. *Human Ecology*, **7**, 75–98.

Western, D. and Finch, V. (1986). Cattle and pastoralism: survival and production in arid lands. *Human Ecology*, **14**, 77–94.

Western, D. and Ssemakula, J. (1981). The future of savanna ecosystems. *African Journal of Ecology*, **19**, 7–19.

Western, D. and van Praet, C. (1973). Cyclical changes in the habitat and climate of an East African ecosystem. *Nature*, **241**, 104–6.

Western, D. and Vigne, L. (1984). The status of rhinos in Africa. *Pachyderm*, **4**, 5–6.

White, F. (1983). *The vegetation of Africa.* Paris. UNESCO. Natural Resources Research Memoir 20.

Williamson, G. and Payne, W.J.A. (1978). *An introduction to animal husbandry in the tropics.* 3rd edn. London and New York: Longman.

Wilson, K. (1990) Ecological dynamics and human welfare. PhD Thesis, University of London.

Wilson, R.T. Diallo, A. and Wagenaar, K. (1985). Mixed herding and the demographic parameters of domestic animals in arid and semi-arid zones of tropical Africa. In *Population, Health and Nutrition in the Sahel.* ed. A. Hill, pp. 116–38. Routledge and Kegan Paul.

Wood, P.J. (1974). The forest glades of Kilimanjaro (revised edn). *Tanzania Notes and Records*, **64**, 108–10.

World Bank (1984). *Toward sustained development in sub-Saharan Africa: a joint program of action.* Washington DC: World Bank.

Wyckoff, J. (1985). Planning arid land development projects. *Nomadic Peoples*, **19**, 59–69.

Yeager, R. (1982). *Tanzania: an African experiment.* Boulder, Colorado: Westview Press.

AUTHOR INDEX

SUBJECT INDEX

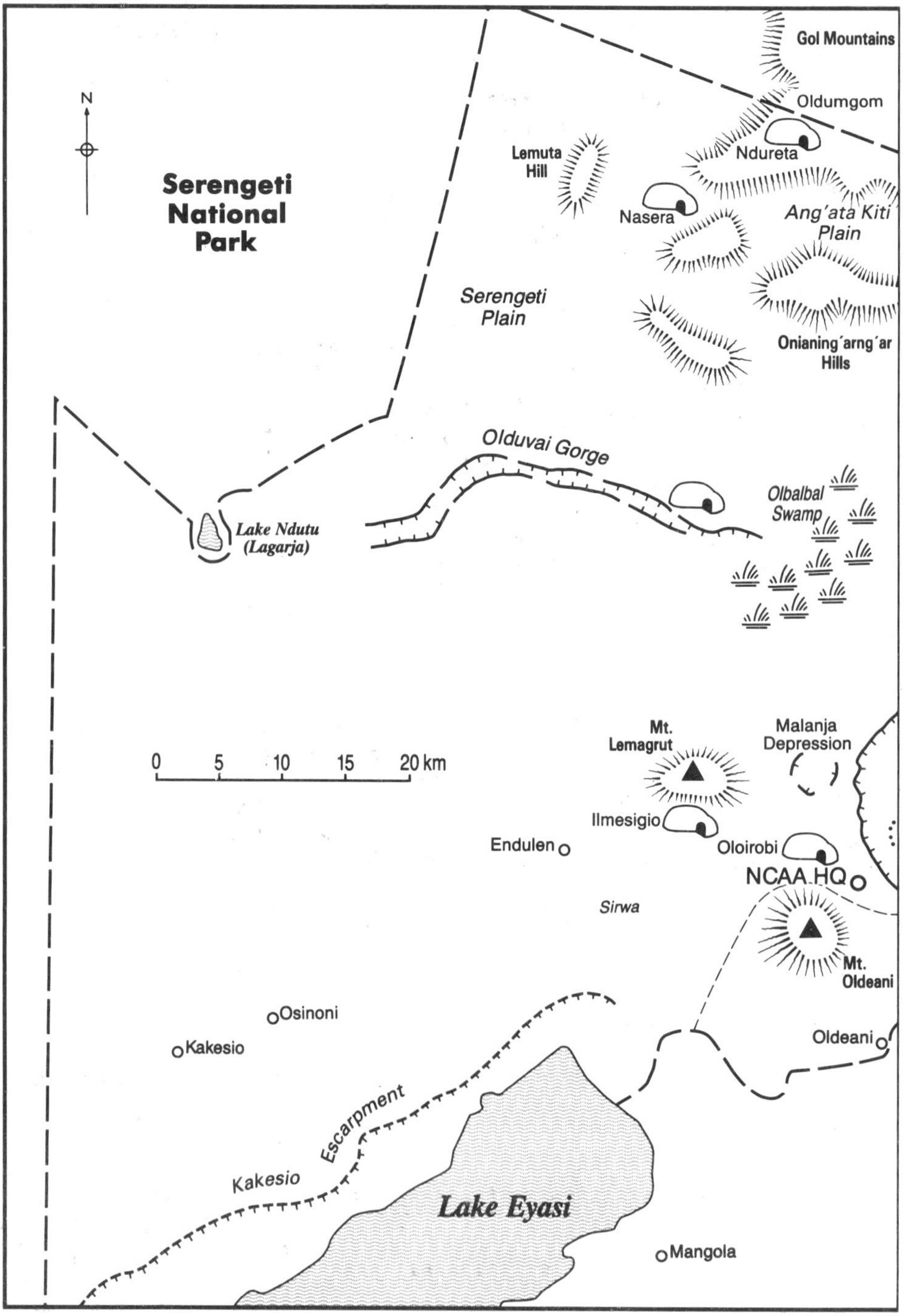

Ngorongoro Conservation Area, showing the main features and place names used in the text